STREAM ECOLOGY
Application and Testing of
General Ecological Theory

[$_\wedge^P$ lenary Symposium on the testing of General Ecological Theory in Lotic Ecosystems]

STREAM ECOLOGY
Application and Testing of General Ecological Theory

Edited by
JAMES R. BARNES
Brigham Young University
Provo, Utah

and
G. WAYNE MINSHALL
Idaho State University
Pocatello, Idaho

PLENUM PRESS • NEW YORK AND LONDON

Library of Congress Cataloging in Publication Data

Plenary Symposium on the Testing of General Ecological Theory in Lotic Ecosystems
(1981: Provo, Utah)
  Stream ecology.

  "Based on the proceedings of the Plenary Symposium on the Testing of General
Ecological Theory in Lotic Ecosystems held in conjunction with the 29th annual
meeting of the North American Benthological Society, April 28, 1981, Provo, Utah."
  Includes index.
  Bibliography: p.
  1. Stream ecology—Congresses. I. Barnes, James R. II. Minshall, G. Wayne. III.
North American Benthological Society. IV. Title.
QH541.5.S7P55 1981   574.5′26323   83-11145
ISBN 0-306-41460-0

Based on the proceedings of the Plenary Symposium on the Testing of General
Ecological Theory in Lotic Ecosystems held in conjunction with the 29th annual
meeting of the North American Benthological Society, April 28, 1981, Provo, Utah

©1983 Plenum Press, New York
A Division of Plenum Publishing Corporation
233 Spring Street, New York, N.Y. 10013

Printed in the United States of America

199767

PREFACE

    Most of the papers included here were part of the Plenary Symposium on The Testing of General Ecological Theory in Lotic Ecosystems held in conjunction with the 29th Annual Meeting of the North American Benthological Society in Provo, Utah, April 28, 1981. Several additional papers were solicited, from recognized leaders in certain areas of specialization, in order to round out the coverage. All of the articles have been critiqued by at least two or three reviewers and an effort was made to rely on authorities in stream and theoretical ecology. In all cases this has helped to insure accuracy and to improve the overall quality of the papers. However, as one of our purposes has been to encourage thought-provoking and even controversial coverage of the topics, material has been retained even though it may upset certain critical readers. It is our hope that these presentations will stimulate further research, encourage the fuller development of a theoretical perspective among lotic ecologists, and lead to the testing of general ecological theories in the stream environment.

James R. Barnes
G. Wayne Minshall

Avondale, Pennsylvania
March 10, 1982

CONTENTS

STREAM ECOLOGY: AN HISTORICAL AND CURRENT PERSPECTIVE

James R. Barnes and G. Wayne Minshall

Department of Zoology, Brigham Young University, Provo
Utah 84602 and Department of Biology, Idaho State
University, Pocatello, Idaho 83209

Stream ecology is coming of age. It has progressed from
descriptive studies to the testing of general theoretical
constructs in ecology and the formulation of new concepts in less
than 25 years. Stream ecology can now make significant
contributions to the development and testing of general
ecological theory. As can be judged from the title of this book,
our purpose is three fold: (1) to call attention to this new
dimension in stream ecology, (2) to indicate the type of work
that is currently being done by stream workers who are addressing
basic ecological questions, and (3) to suggest new directions for
future emphasis.

One need only examine the most popular general ecology and
limnology texts to realize the absence of stream ecology as a
constructive force in the development of general ecological
theory and in the study of aquatic environments (Table 1). It
seems that not until recently have lotic ecologists recognized
the opportunities that exist in streams to test and develop
principles which have broad applicability to ecology. But in
defense of stream ecologists, they have led the way at the
ecosystem level of study in recognizing the watershed as the
basic unit of study, the role of geomorphology in ecosystem
functioning, and the importance of the terrestrial-riparian-
aquatic linkage.

Stream ecology as a discipline has its historical roots in
the fields of aquatic entomology and fishery biology. Early
studies involved mainly the taxonomy, food habits, morphological
adaptations to living in running water, biogeographical
distributions, and habitat requirements of aquatic insects and

fish (e.g., Needham and Lloyd, 1916; Needham, 1938; Berg, 1948).
Although it was early recognized that stream communities are
complex, the insect taxonomic problems alone were a major
stumbling block to definitive studies. By the 1940's lotic
invertebrate communities were being studied but mainly as a food
base for game fish (e.g., the basic question being, is the insect
standing crop sufficient to support a given fish population?
Allen, 1951).

Through the 1960's, stream ecologists were still struggling
with sampling and taxonomic problems but they were beginning to
study factors controlling community structure. The commonly held
idea was that abiotic factors controlled stream community
structure. Animal-substrate relationships in particular received
attention (see Minshall, 1983). Also, during this period the
importance of detritus as a factor in controlling distributions
was being studied and the microdistributions of closely related
species in streams started to receive attention (see Cummins,
1964; Hart, this volume).

In 1970, Hynes published his benchmark volume "The Ecology
of Running Waters", which represented a synthesis of the world's
stream literature through early 1966. This compilation was
dominated by descriptive studies. Hynes' volume marks a turning
point in stream ecology. The work of preceding generations
became readily available to both aquatic and general ecologists.
With this volume, classical limnologists had enough information
about streams to include sections on flowing water in limnology
courses. Starting in the mid-1960's separate stream ecology
courses came into existence and students with interests other
than aquatic insects and fish began to enroll.

Also in the early 1970's, two other important synthesis
volumes were published: (1) "River Ecology and Man" (Oglesby et
al., 1972) and (2) "River Ecology" (Whitton, 1975). Both books
represent collections of articles in which stream workers
presented up-to-date discussions of their own specialities. Two
other important books need mentioning. Macan's (1963)
"Freshwater Biology", examined the ecological work done on
streams within a general freshwater biological context. Elliott
(1971) published "Some Methods for the Statistical Analysis of
Samples of Benthic Invertebrates". This latter work provides a
basic framework of sampling and statistical procedures for
workers in aquatic ecology.

With the beginning in the early 1970's of the environmental
movement and the concern about water pollution, a whole suite of
water pollution literature developed. The development of this
area of aquatic ecology was aided by Hynes (1960) who synthesized
the early water pollution literature in his book "The Biology of

Polluted Waters". The aquatic community response to a
perturbation was usually measured by the effect on species
diversity. Such studies have rarely been utilized in
understanding the general ecology of stream systems.

Major stream ecology advances in recent years have been
primarily at the ecosystem level of organization and oriented
towards: (1) realization of the detritus-based economy of stream
ecosystems and an improved understanding of the relative
importance of autochthonous and allochthonous inputs (e.g.
Kaushik and Hynes, 1971; Minshall, 1978; Anderson and Sedell,
1979), (2) application of the functional feeding group approach
to the examination of benthic macroinvertebrate guild
relationships in lotic communities (Cummins, 1973, 1974; Hawkins
and Sedell, 1981; Minshall et al., 1982), (3) the
conceptualization of various-sized streams of a river system as
an integrated ecological unit or "River Continuum" (Vannote et
al., 1980; Minshall et al., 1983), and (4) the observation that
over the river continuum nutrients cycle in a "spiraling" fashion
(Webster, 1975; Wallace et al., 1977; Newbold et al., 1982;
Elwood et al., 1983). In this book, further refinements and
important implications of these advances are discussed (Bott,
Cummins et al., Fisher, Gregory, this volume).

In recent years we have seen the application of the
experimental approach to stream ecology and the beginning of the
exploration of such topics as competition, predation, and
behavior. But as the papers by Allan, Hart, McAuliffe, and
Peckarsky (this volume) reveal, stream ecologists are just
beginning to scratch the surface in these areas. The prevailing
view still is that abiotic variables over-ride biotic ones in
determining stream community organization. Recent work by these
and other authors (including Bruns and Minshall, this volume)
provide evidence that factors such as competition and predation
may exert a strong influence in the determination of stream
community structure.

In addition, other facets such as succession, density
dependence and independence, primary production, diversity,
community formation and organization, and ecosystem stability are
coming under critical examination. The preliminary results, as
exemplified by the efforts of Fisher, Shiozawa, Bott, Stanford
and Ward, Minshall et al., and Webster et al. (this volume),
indicate that streams are excellent systems for study of basic
ecological questions and can provide additional tests of these
ideas.

In future, stream ecologists will have to be much more aware
of what is going on in other areas of ecology than in the past,
not only so that we may better integrate those findings and ideas

into our own work, but also so that we can make our own
contribution to the development and refinement of the principles
on which the discipline of ecology is founded. On the other
hand, all ecologists need to become aware of the developments in
stream ecology. If these are ignored and if the important
terrestrial-aquatic linkage is not recognized, then future
ecologists will not realize the potential that the study of
flowing water systems has to offer to the development of general
ecological theory. Without this integration, understanding of
ecosystem functioning will never approach completeness.

Let this symposium volume serve notice that we have arrived
at the threshold of an exciting new era in stream ecology and let
the presentations contained herein indicate the dimensions of
that science and some of its future directions.

REFERENCES

Allen, K. R. 1951. The Horokiwi stream; a study of a trout
     population. New Zealand Marine Dept. Fish. Bull. 10.
Anderson, N. H. and J. R. Sedell. 1979. Detritus processing by
     macroinvertebrates in stream ecosystems. Ann. Rev. Entomol.
     24:351-377.
Berg, K. 1948. Biological studies of the River Susaa. Folia
     Limnol. Scand. 4:1-318.
Cummins, K. W. 1964. Factors limiting the microdistribution of
     larvae of the caddisflies Pycnopsyche lepida (Hagen) and
     Pycnopsyche guttifer (Walker) in a Michigan stream. Ecol.
     Monogr. 34:271-295.
Cummins, K. W. 1973. Trophic relations of aquatic insects.
     Ann. Rev. Entomol. 18:183-206.
Cummins, K. W. 1974. Structure and function of stream
     ecosystems. BioScience 24:631-641.
Elliott, J. M. 1971. Some methods for the statistical analysis
     of samples of benthic invertebrates. Freshwat. Biol. Assoc.
     Sci. Publ. No. 25. 148 pp.
Elwood, J. W., J. D. Newbold, R. V. O'Neill, and W. Van Winkle.
     1983. Resource spiraling: an operational paradigm for
     analyzing lotic ecosystems. In: T. D. Fontaine and S. M.
     Bartell (eds.), The dynamics of lotic ecosystems. Ann Arbor
     Science, Ann Arbor, Michigan.
Hawkins, C. P. and J. R. Sedell. 1981. Longitudinal and
     seasonal changes in functional organization of
     macroinvertebrate communities in four Oregon streams.
     Ecology 62:387-397.
Hynes, H. B. N. 1960. The biology of polluted waters.
     Liverpool University Press. 202 pp.
Hynes, H. B. N. 1970. The ecology of running waters. Univ.
     Toronto Press, Toronto, Ontario. 555 pp.

Kaushik, N. K. and H. B. N. Hynes. 1971. The fate of dead leaves that fall into streams. Arch. Hydrobiol. 68:465-515.

Macan, T. T. 1963. Freshwater ecology. J. Wiley and Sons, Inc. New York. 338 pp.

Minshall, G. W. 1978. Autotrophy in stream ecosystems. BioScience 28:767-771.

Minshall, G. W., J. T. Brock, and T. W. LaPoint. 1982. Characterization and dynamics of benthic organic matter and invertebrate functional feeding group relationships in the Upper Salmon River (USA). Internat. Rev. Ges. Hydrobiol. 67:793-820

Minshall, G. W. 1983. Aquatic insect-substrate relationships. In: V. H. Resh and A. M. Rosenberg (eds.), Ecology of aquatic insects: a life history and habitat approach. Praeger Publ. In Press.

Minshall, G. W., R. C. Petersen, K. W. Cummins, T. L. Bott, J. R. Sedell, C. E. Cushing, and R. L. Vannote. 1983. Interbiome comparison of stream ecosystem dynamics. Ecol. Monogr. 53:1-25.

Needham, J. G. and J. T. Lloyd. 1916. The life of inland waters. Comstock Publ. Co., Ithaca, New York. 438 pp.

Needham, P. R. 1938. Trout streams. Comstock Publ. Co., Ithaca, N.Y. 233 pp.

Newbold, J. D., R. V. O'Neill, J. W. Elwood and W. Van Winkle. 1982. Nutrient spiralling in streams: implications for nutrient limitation and invertebrate activity. Amer. Nat. 120:628-652.

Odum, E. P. 1959. Fundamentals of ecology. 2nd ed. Saunders, Philadelphia. 546 pp.

Oglesby, R. T., C. A. Carlson, and J. A. McCann (eds.). 1972. River ecology and man. Academic Press, New York. 465 pp.

Vannote, R. L., G. W. Minshall, K. W. Cummins, J. R. Sedell, and C. E. Cushing. 1980. The river continuum concept. Can. J. Fish. Aquat. Sci. 37:130-137.

Wallace, J. B., J. R. Webster, and W. R. Woodall. 1977. The role of filter feeders in flowing waters. Arch. Hydrobiol. 79:506-532.

Webster, J. R. 1975. Analysis of potassium and calcium dynamics in stream ecosystems on three southern Appalachian watersheds of contrasting vegetation. Ph.D. Dis., Univ. Georgia, Athens. 323 pp.

Welch, P. S. 1952. Limnology, 2nd ed. McGraw-Hill Book Co., New York. 583 pp.

Whitton, B. A. 1975. River ecology. University of California Press, Berkeley, California. 724 pp.

SUCCESSION IN STREAMS

Stuart G. Fisher

Department of Zoology
Arizona State University
Tempe, Arizona 85287

INTRODUCTION

Succession is one of the oldest, most persistent, least resolved concepts in ecology. Throughout its history, the concept has been largely botanical. Early phytosociologists such as Cowles (1901), Clements (1916), and Gleason (1926) viewed succession exclusively as temporal change in terrestrial plant communities. Drury and Nisbet (1973) recently emphasized that successional ideas have been derived from and tests should be restricted to temperate forests. Given the parochial nature of the field, any attempt to apply successional concepts to running waters is, to say the least perilous. On the other hand, cross-fertilization by two disparate scientific fields is often fruitful to both in generating, if not always answering, interesting questions.

In this paper, I shall explore the extent to which stream ecology and ecological succession may provide reciprocal insights and thus mutual benefit. Two aspects of stream succession will be explored: site-specific temporal succession and longitudinal succession. Site-specific change in communities with time following disturbance is classical succession. Many data document this process in a variety of ecosystems, including streams. Attempts to superimpose spatial pattern on the process in terrestrial ecosystems have been diversionary; however, in streams longitudinal spatial pattern is fundamental and provides the context in which all stream processes operate. Finally, the problem of time scales will be discussed. Streams are influenced by a set of periodic (yet stochastic) variables different from those in terrestrial forests. Analysis of differences in these

7

periodic functions may thus provide insight into the general
successional process.

In what follows, I will consider not only how successional
theory might influence how we think about streams, but the
converse – which is the charge of this volume – how streams might
contribute to our understanding of succession as a general
ecological concept.

Succession can be defined most simply as change in community
composition after disturbance at a site. Ecosystems may be
disturbed by such stochastically recurrent agents as fire,
hurricanes, and floods and several human influences such as
dredging, forest cutting, and application of toxic chemicals.
Recovery of communities can and should be reckoned by such
measures as species composition, richness, and dominance. There
is little argument that succession as defined above occurs, but
there is much dissension over the generality of successional
pattern in both like and disparate ecosystem types. Drury and
Nisbet (1973) suggest that generalization may be attainable only
if our purview is restricted to temperate forests. McIntosh
(1980), in a thoughtful review, wonders whether a general pattern
exists to be discovered.

A variety of ecological processes has been subsumed under
the banner of succession. Lake eutrophication has been forced
into the scheme as has seasonal change in phytoplankton
communities in otherwise undisturbed lakes (Lewis, 1978).
Seaward changes in stream communities are often termed
longitudinal succession. While precedent exists in terrestrial
ecology (Pickett, 1976), inclusion of spatial pattern stretches
the classic definition of the process, yet must be considered in
the analysis of flowing waters. To complicate matters and extend
the definition still further, collective properties of
communities such as ecosystem metabolism and nutrient cycling,
have been incorporated as descriptors of the successional process
(Odum, 1969). Given this historic license with both the
definition and appropriate dependent variables, and considering
the wide range of systems studied, it is not surprising that the
definitive conceptual synthesis has been elusive.

Succession involves two basic elements: colonization and
subsequent change. There is some question whether the former is
necessary in naturally disturbed systems where no disturbance may
be sufficiently severe to eliminate any species completely. In
these cases, succession occurs by adjustments of dominance. If
disturbance is severe enough to remove species locally (e.g., the
Mt. St. Helens eruption) colonization is germane to successional
rate and pattern. Colonization by definition is by individuals
not initially present at the site and is thus a stochastic

feature of the process that cannot be predicted solely by
features of the disturbed site. Few ecologists suggest that
colonization patterns alone can account for observed successional
patterns and none consider the terms to be synonomous.

Tansley (1929) divides the causes of succession beyond
colonization into two categories: allogenic and autogenic.
Allogenic forcing functions are largely external to the system
and include seasonal, climatic, and geomorphic cycles. These
driving forces are abiotic in contrast to autogenic changes which
are induced by the biota (e.g., competition, predation). While
change at a site after disturbance may be shaped by allogenic
forces (the disturbance itself is usually allogenic), it is the
complex of autogenic causes that has received the most attention
of biologists. Connell and Slatyer (1977) separate these
autogenic elements of community change into interactive and
non-interactive components. In essence, their null hypothesis is
that succession is a consequence of individual life history
characteristics of the biota. Alternative hypotheses are
interactive - that extant communities either inhibit or
facilitate ascension of their successors. Much of the
controversy in successional theory revolves around forms of these
hypotheses. In reality, allogenic and autogenic causes overlap.
Seasonal change (e.g., in temperature) may alter competition
coefficients. Competition for resources (e.g., a dissolved
nutrient) is mediated through the abiotic environment (described
by the concentration of that nutrient). It is reasonable to
assume that in some ecosystems, biologic interactions are intense
while in others, communities are shaped by allogenic factors.
Connell and Slatyer's hypotheses are thus not mutually exclusive.

Controversy over the relative importance of autogenic
mechanisms in succession is not new and stems at least from the
Gleason-Clements schism of the early part of this century. The
Gleasonian individualistic school of succession is today well
represented by Horn (1974), Drury and Nisbet (1973), Connell and
Slatyer (1977), Picket (1976) and others. These evolutionary
ecologists view succession as a consequence of adaptation and
natural selection at the level of the individual. The community
then, is a coincidence. Successional change is viewed in terms
of species composition and causes of succession include
consequences of life history attributes of individuals.

The Clementsian, holistic school, features of which are best
represented today by E. Odum (1969), H. Odum (1971) and Margalef
(1968), views the community as an organic entity and succession,
described in terms of collective or emergent properties of
ecosystems, as an analogue of organismal development. Mechanisms
of community change are not specified, but implicitly involve
selection above the level of the organism. Clearly the latter

approach is not in current vogue and will remain in disfavor
until plausible mechanisms are offered; however, the approach has
made a useful contribution to succession on two counts. First,
by routinely including collective attributes such as
productivity, energy flow, and nutrient cycling in analyses of
succession, this school has provided empirical data useful in
understanding and managing natural and manipulated ecosystems.
In addition to this pragmatic payoff, ecosystem biology
emphasizes abiotic features more so than does evolutionary
population biology or community ecology. It is these abiotic
characteristics that likely mediate organismal interactions. If
succession involves more than colonization dynamics and
subsequent life history consequences, interactive causes may best
be resolved by close attention to the ecosystem context. This is
particularly true of streams where affector and affected
organisms (and communities) may be spatially separate but linked
through abiotic influences transmitted in a highly oriented
fashion by flowing water.

SITE SPECIFIC TEMPORAL SUCCESSION IN STREAMS

     Streams can provide a unique opportunity to study
site-specific temporal succession. Natural disturbance by
scouring floods occurs over a range of severity at statistical
frequencies. Return intervals for severe washout events usually
extend two years or more and thus provide ample time for multiple
generations of predominant organisms. Colonization by aerial
vectors, drift, or upstream movements can be easily measured. In
spite of these favorable features, stream succession has received
little attention. This is partly because those stream types
historically favored by stream ecologists are least suited to
successional analysis. Large rivers may flood predictably but
never with sufficient intensity to drastically alter the biota.
Small temperate streams flood, but do so too frequently and
provide little opportunity to examine uninterrupted recovery
sequences. Further, frequently flooding streams do so with
return periods shorter than life cycles of predominant taxa.
Finally, seasonal change is marked in many streams and clouds
patterns attributable to succession alone. Seasonal change in
terrestrial systems lends a stochastic element to early
successional sequences (Keever, 1950) but in general, terrestrial
seres encompass many seasonal cycles. The ideal stream for
successional study is one in which 1) disturbance by flooding is
severe, 2) the biota experiences several generations between
disturbance events, 3) no minor disturbances occur between
catastrophic events, and 4) seasonal effects are minimal.
Seasonal change can be ameliorated in regions of even climate
(e.g., the tropics) or in streams where the successional play is
completed within a single season (e.g., hot desert streams).

While desert streams are well suited to successional studies,
they have received little attention until recently (Gray and
Fisher, 1981; Fisher et al., 1982). At this juncture, inferences
about stream succession must be gleaned largely from studies of
temperate, mesic stream ecosystems.

The literature of stream ecology relevant to site-specific
succession is largely anecdotal and of limited insight. Seldom
if ever have streams been used to test a priori hypotheses about
general successional theory. A brief review is still in order to
establish the current boundaries of the field.

Streams are certainly subject to periodic disturbance.
Flooding has been most thoroughly documented but responses to
dredging (Hannan and Dorris, 1970), pesticides (Dimond, 1967),
coal ash (Cherry et al., 1979), drought (Harrison, 1966) and hail
storms (Brock and Brock, 1969) have also been described. In
addition, colonization of recently inundated new substrates, both
natural (Kennedy, 1955) and artificial (e.g., Wiegert and
Fraleigh, 1972; Gale and Gurzynski, 1979; Cattaneo and Ghittori,
1975) has been documented. These agents of disturbance affect
streams in ways which are both qualitatively and quantatively
different and leave disturbed ecosystems in quite different
states vis a vis subsequent recovery.

Perhaps the most significant variable shaping stream
succession is the degree to which the stream and nearby systems
have been reset. Bare substrate recolonization studies have
usually been performed in otherwise intact streams where
potential colonizers are abundant. Similarly, dredging usually
occurs in mid-channel leaving peripheral areas intact. Pesticide
application may remove insects but leave algal communities
untouched. Floods also differentially affect the community
depending upon their intensity. Siegfried and Knight (1977)
describe the rapid recolonization of Sagehen Creek, California
following a typical annual washout. In contrast, Collins et al.
(1981) report deep channel cutting and extensive erosion of bank
deposits during a 100-year flood in an Arizona stream. Recovery
of some taxa in that system may require decades.

I am aware of few documented cases where flood damage is
sufficiently great to remove all biota and initiate primary
succession. Floods of such intensity rarely occur and are seldom
studied by ecologists. An exception is Minshall's current work
on recolonization of the Teton River following the Teton Dam
disaster in Idaho (Minshall et al., this volume). Intensity
alone is probably insufficient to denude a stream completely
unless flood duration is long. For insects at least, the stream
ecosystem includes the riparian corridor which during much of the
year is replete with adult stages which remain untouched by

floodwaters and oviposit immediately upon their recession. This
is true of Sonoran Desert streams where floods remove nearly 99%
of the insect fauna but last only a few hours (Fisher et al.,
1982). Many insects of this desert system reproduce continuously
and adults of many taxa are aerial at any given time during the
year (Gray, 1981). Recolonization by oviposition is consequently
rapid.

Given this variation in disturbance intensity and spatial
coverage, it is not surprising that rates of recovery also vary
widely. Flood damaged streams reattain normal conditions in time
spans ranging from a month to several years. As general
successional theory would have it, this is a function of life
span of the taxa considered, thus diatom communities recover in a
month or two (Wiegert and Fraleigh, 1972; Brock and Brock, 1969)
while fishes may require several years (Hanson and Waters, 1974).
Insects are somewhat intermediate depending upon taxon and
voltinism (Thorup, 1970; Hoopes, 1974; Gray, 1981). Recovery is
accelerated if colonization routes are open. The four to five
year recovery period for fish reported by Hanson and Waters
(1974) is for a brook trout population which experienced no
recruitment by immigration. Larimore et al. (1959) report a two
week recovery time for the fish community of a newly inundated
Vermillion River tributary in Illinois where all of the
population increase was by immigration.

Initial colonization is perhaps the best documented
component of succession in streams. Blum (1956) suggests that
algal colonization and dominance occur synchronously and argues
that therefore succession does not occur. Other studies
demonstrate that initial colonization is itself quite variable
and can affect subsequent dominance depending upon the
colonization regime. Brock and Brock (1969) report recovery of a
Yellowstone spring by growth of survivors of a hail storm, not
colonizers. Jones (1978) and Eichenberger and Wuhrmann (1975)
present data to show that rate of early algal growth depends upon
continued seeding. Similarly, Kaufman (1980) demonstrated faster
algal colonization of glass substrates in intact than
artificially depopulated stream channels. Early community
structure may be determined by abiotic conditions extant during
initial colonization. For example, slow current velocity may
enhance attachment of algae while rapid velocities favor growth
following attachment (Reisen and Spencer, 1970). Further, algal
colonization may not occur during winter in spite of high algal
standing crops at that time (Gale and Gurzynski, 1979).
Invertebrate colonization rate and pattern also depend greatly on
such parameters as season of disturbance (Hynes, 1958). A
similar effect has been demonstrated for fishes. By differential
impact on eggs, autumn floods in Sierra Nevada streams decimate
brook trout while spring floods cause high mortality in rainbows

(Seegrist and Gard, 1972). These multifarious observations
suggest that initial colonization patterns have a highly
stochastic component depending on timing, magnitude and spatial
extent of the disturbance. If alternate pioneer communities
converge on fewer, later community types, then the sere is not
determined at its outset as Blum (1956) hypothesized. In
Sycamore Creek, Arizona, initial post-flood colonizers are always
diatoms yet these assemblages are replaced by bluegreen- or
Cladophora-dominated assemblages depending upon local conditions
(Fisher et al., 1982).

In sum, post-disturbance colonization of streams is far from
deterministic. There may be several routes to one subsequent
community type or several subsequent community types generated
from one pioneer assemblage. Egler's (1954) initial floristics
model, which says that all species of a sere are present at the
outset, may apply, yet this is difficult to test in a system
dominated by small, easily overlooked organisms. Even if this
model applies, initial abiotic conditions during colonization and
the regime of continued seeding add a stochastic element to
community composition early in stream succession.

Beyond initial colonization, direct evidence for interactive
autogenic succession is slim. Dickman (1974) showed that diatom
inhibition greatly altered the community structure of pioneer
non-diatom algae as well as their ultimate standing crops. Both
Moore (1976) and Eichenberger and Wuhrmann (1975) have shown that
spontaneous detachment of algal mats is requisite to
establishment of subsequent algal communities not subject to
detachment. Harrison (1966) reported dominance of "oddball"
species such as Simulium spp. early in succession and complete
disappearance later in newly inundated stream channels. Our work
at Sycamore Creek shows that while all algae are present shortly
after flood recession, dominance shifts dramatically within the
first month of recovery in ways not predictable from life history
data alone (Fisher et al., 1982). Insect colonization on the
other hand follows the initial floristics model for the stream as
a whole, yet isolated backwaters and pools may show changes in
community composition circumstantially attributable to predation
and marked by local extinctions (Gray and Fisher, 1981).

In short, disturbed streams recover in a stochastic manner
influenced by timing and severity of the disturbance and
availability of nearby colonizers. While the initial floristics
model of Egler probably applies, augmentation of early stages by
repeated colonization is important in shaping subsequent
community structure. Beyond colonization, evidence for or
against interactive models is lacking. While one might guess
that Rocky Mountain trout streams exhibit a truncated,
non-interactive succession highly shaped by colonization, less

frequently disturbed warm water streams may not fit this pattern.
If interactive models hold for any stream (or for any ecosystem)
they might best be tested in streams of the latter type.

   Collective properties of stream communities have received
little insightful attention in a successional context. Those
elements most commonly resolved are those which Peters (1976)
rightly labels most tautologic. For example, biomass,
chlorophyll a and primary production increase during stream
succession (Kaufman, 1980,; Fraleigh and Wiegert, 1975) as might
be expected in virtually all ecosystems. It is difficult to
imagine their doing otherwise. Recovery functions are largely
asymptotic. Species diversity is an exception which often
declines to moderate levels after an early increase (Cattaneo and
Ghittori, 1975; Siegfried and Knight, 1977).

   With the exception of artificial substrate studies, changes
in collective properties during succession seldom have been
documented in streams, and the field can provide little test of
Odum's (1969) ecosystem development generalizations. While
primary production, ecosystem respiration, P to R ratios and
nutrient cycling attributes have received wide attention in
streams, these measures have been made infrequently in a
successional context. At Sycamore Creek, where stream succession
can readily be observed, Odum's generalizations hold except for
those attributes most influenced by the open nature of stream
ecosystems (Fisher et al., 1982). Thus for example, P/R does not
approach unity and net production is continually lost as fluvial
export. Further, there is no evidence that succession results in
an ecosystem more resistant to perturbation. Dominant organisms
at all stages remain characteristic of frequently disturbed
ecosystems. Disturbance by flooding is simply too severe to be
biologically ameliorated.

   General successional theory thus has much to offer stream
ecology. The process can be documented in a variety of stream
types (in some more easily than others), and markedly affects
stream ecosystem structure and function. It is incumbent upon
stream ecologists to report results of their work, whatever the
focus, in a successional framework, or to demonstrate the
irrelevance of that framework to the specific questions asked.
At the minimum, time since disturbance should be included among
other descriptors, regardless of the focus of the study at hand.

   But alas, beyond expanding the idea to open ecosystems,
stream ecology has not yet done much to illuminate general ideas
of site-specific temporal succession. Longitudinal succession,
on the other hand, is a process unique to streams and may provide
a more lucrative opportunity.

LONGITUDINAL SUCCESSION

   Longitudinal succession has been loosely defined as the
sequence of communities in streams from headwaters to large
rivers. The concept is independent of time and does not involve
recovery from a disturbance. Longitudinal pattern is well
documented for streams and is most often attributed to abiotic,
physiographic variables. Fish community distribution has been
attributed to slope (Huet, 1959), depth (Sheldon, 1968), stream
order (Kuehne, 1962) and temperature (Burton and Odum, 1945), all
of which change longitudinally. Butcher (1933) maintained that
change in current velocity explains macrophyte longitudinal
distribution in rivers. Holmes and Whitton (1977) include
changing substrate type as an important influence on macrophyte
longitudinal succession. Invertebrates also change seaward in
response to abiotic variables such as substrate, temperature,
discharge, current velocity and food availability (Illies and
Botosaneanu, 1963; Andrews and Minshall, 1979).

   In all of these cases, community structure at a given point
in the longitudinal gradient is interpreted in terms of
site-specific environmental factors. While these factors are
predominantly abiotic, biologic interactions such as competition
and predation at the site might also be included (see chapters by
Hart and Allan in this volume). Allan (1975) invokes competition
between congeners to explain several sharp transitions in the
longitudinal distribution of insects in Cement Creek, Colorado.

   The literature explicity treating longitudinal succession in
streams thus views communities as independent points on a
gradient much as a forest ecologist would describe an
elevational gradient in terrestrial vegetation. The role of
upstream communities as determinants of downstream community
structure and function is largely ignored, much as the
terrestrial ecologist would discount the role of spruce-fir
forests in shaping pinyon-juniper communities. Yet running water
is a unidirectional, highly competent transport vector linking
communities on the longitudinal stream gradient. The potential
for an upstream community to influence those downstream through
modifications of the aqueous medium far exceeds that of the
terrestrial analogue which links contiguous communities through
downslope winds. It is somewhat surprising that stream ecology
has not more fully exploited this most unique property of
streams. The strong unidirectional linkage of sequential
communities through flowing water is a powerful aid to
interpretation of longitudinal pattern. Viewing longitudinal
succession apart from community interaction limits explanation of
community structure along the gradient to life history attributes
and ecoclinal adaptation. This is tantamount to accepting
Connell and Slatyer's null hypothesis - that communities are not

influenced by their predecessors, either temporal or, in this case, farther upstream in the watershed.

It is somewhat unfortunate that the term succession has been so widely applied to biotic zonation in streams, and not restricted to temporal sequences as it is in other ecosystems. Yet its use to describe spatial pattern may be legitimate when one community influences the structure and function of the next in (spatial) sequence. This possibility is strong enough in streams to warrant closer examination.

In the past decade or so, conceptual advances in three areas of stream ecology inadvertantly provide the substance to flesh out the concept of longitudinal succession. These are 1) terrestrial dependence of stream metabolism, 2) the river continuum concept and 3) nutrient spiralling. Each of these advances orients causation, rather than merely community type, on the longitudinal axis. These conceptual advances may be traced to Noel Hynes, who two decades ago proposed the idea that many streams are dependent upon terrestrial detritus for energy (Hynes, 1963). His early work on processing of allochthonous leaf material (Hynes and Kaushik, 1969) stimulated much subsequent research on detrial processing and detrital fluxes, the latter including work which quantified detrital inputs and outputs in the form of organic matter budgets (Hall, 1972; Fisher and Likens, 1973; Sedell et al., 1974; Fisher, 1977). Budget studies reveal that energy in the form of detritus moves in a spatially oriented fashion from forest to small streams in response largely to gravity. Secondly, the kinds and numbers of organisms present in recipient lotic ecosystems depend upon detrital input quantity and quality. Finally, detrital material not utilized in a given stream reach may be exported to the next reach downstream, again in a spatially oriented fashion, presumably affecting organisms there. Hynes' work also stimulated studies of detrital processing (Cummins, 1974; Petersen and Cummins, 1974; Suberkropp and Klug, 1976) which also indicate that upstream communities are shaped by detrital inputs, and in altering this material, affect particle size, modify transport dynamics, and presumably influence the structure and function of downstream communities (Wallace, 1977).

The river continuum concept (Vannote et al., 1980) incorporates elements of classical river zonation, but it does more than this in formalizing the spatially oriented processing sequence in streams in an attempt to explain downstream patterns in stream communities. In so doing, a framework is developed to explain observed longitudinal community transitions in terms of both physiographic and biologic variables. The stream does indeed provide an ecocline in space as early workers have shown. Community change can be interpreted as individual adaptation to this cline, yet interaction between communities through organic matter processing and transport is marked. This interaction is inequitable in favor of upstream effects on downstream communities. Reverse interactions are known, but weak. The analogy with community interactions in time at a single site is good with two qualifications: 1) temporal abiotic changes in site-specific succession are largely biotically induced, whereas the stream gradient is under geomorphic control, and 2) interaction between communities in time is not reversible. Later communities never affect earlier ones, yet downstream communities may affect those upstream, e.g., by migrations (Hall, 1972).

The nutrient spiralling concept (Newbold et al., 1981) provides a mechanism for dealing with complex interactions in time and space, and it broadens the suite of interactions to include inorganic nutrients. This concept has strong grounding in engineering science and incorporates ideas long the stock in trade of wastewater managers (Wuhrmann, 1974), but strong roots also extend to Professor Hynes. Kaushik and Hynes (1971) showed that decomposing leaf litter in streams caused a net decrease in dissolved inorganic nitrogen in the medium. This process, well documented by soil ecologists considerably earlier (e.g. Bocock, 1964), has profound consequences for macroinvertebrates which consume this nitrogen-augmented and thus higher quality food.

The nutrient spiralling group however, focused not on the detrital subsystem, but on the interactions between the detrital component and nutrients in the dissolved state. Recent nutrient spiralling models emphasize the dynamic nature of streams as processors of both organic and inorganic materials and provide a mechanism for describing the "tightness" with which a given element cycles or, because there is a significant spatial element in streams, spirals. Extension of this model to non-steady state systems implies that upstream communities alter nutrient concentrations and ratios in transport coincident with organic matter processing. In autotrophic streams, macrophytes and algae also contribute to nutrient uptake and release (Crisp, 1970; Grimm et al., 1981; Edwards, 1969; Vincent and Downs, 1980). Change in nutrient climate is quickly exported downstream where it can influence the biota there through nutrient depletion or enrichment. Potential for community interaction is thus great. The nutrient spiralling concept provides the conceptual framework for resolving this interaction in space.

These recent advances in stream ecology have established that community structure and function are shaped by spatially oriented inputs, that communities alter the medium through various activities and that this modified medium moves downstream and influences communities there. (In all of these considerations "community" need not connote more than spatially

and temporally coincident species populations.) At any point in
a stream then, community characteristics are influenced by biotic
and abiotic site characters (competition, predation; stream
velocity, slope, depth), and by biologic activities of upstream
communities. The upstream community is thus analogous to
temporally antecedent communities in site-specific terrestrial
succession.

A redefinition of succession is thus in order. To encompass
both temporal and spatial succession, succession may be defined
as a sequence of communities in which the ascendancy of each is
influenced by its predecessors. The sequence can be resolved in
time or space. The influence of predecessors can be weak or
strong, direct or indirect. I suspect that in the rushing
mountain trout stream, the influence of upstream communities is
very weak and indirect; however, in autotrophic desert streams,
upstream communities may strongly influence those in lower
reaches. In Sycamore Creek, for example, dissolved inorganic
nitrogen declines rapidly below spring heads as a result of
assimilation by diatoms and Cladophora. Heterocystous (and
presumably nitrogen-fixing) bluegreen algae are dominant
downstream where dissolved nitrogen is undetectable (Fisher et
al., 1982).

The conservative reader will not welcome this extension of
the succession concept to include spatial change. The extension
is modest in that streams are the only ecosystems that readily
come to mind in which causation is sufficiently oriented in space
to qualify for inclusion. I believe the major benefit of
extending the definition to longitudinal succession is pragmatic.
Several questions which are difficult to resolve in site-specific
succession can be more easily addressed when reframed in the
spatial context. For example, it is not always easy to determine
how one terrestrial community modifies the environment at a site
and thereby influences its successors. More often than not this
involves a host of influences which operate simultaneously (e.g.
competition, shading, soil development, nutrient enhancement) and
are difficult to separate. Interaction between stream
communities occurs through the flowing medium which can be
sampled, analyzed, and ultimately, manipulated. Spatial
disjunction thus facilitates identification and quantification of
interaction vectors.

Secondly, streams can be manipulated to differentiate
between physiographic effects and biologic interactions. If
community development were observed in a single pass, uniform
artificial stream channel, one would predict no longitudinal
heterogeneity of communities if physiography were the sole
determinant of longitudinal community pattern. Significant
longitudinal change would require rejection of this null

hypothesis and suggest several alternate hypotheses which could
be easily tested by monitoring transport fluxes between
communities. Alternatively, a legendary Round River, which
exhibited physiographic variation but recirculated water, could
be constructed to test the hypothesis that biologic community
interactions shape longitudinal pattern. If this system did not
differ from an identical system in which water was not
recirculated, then the hypothesis could be rejected in favor of
alternates implicating physiography.

There is of course an obvious complication to this simple
resolution of longitudinal succession in streams, namely that
temporal, site-specific succession is occurring simultaneously.
If, for example, an entire drainage were scoured by flood,
colonization and subsequent recovery would begin simultaneously
at each point on the stream. Pattern and rate would depend upon
availability of colonizers, local site conditions, resource
availability, and so forth. Pattern and rate would also depend
on position in the stream relative to other communities. In
other words, upstream communities in the rapid growth phase of
early succession may lower dissolved nutrients or increase fine
particulates in transport and thereby influence the rate and
pattern of succession downstream. Artificial channels could be
used here as well to test Connell and Slatyer's facilitation and
inhibition models by examining rate and pattern of development at
a given site with and without an upstream biota. Similar
experiments could be performed in natural ecosystems utilizing
either natural or planned experiments involving partial scouring.
Implications of these results are highly significant to stream
ecology. If the facilitation model holds, streams will tend to
recover from the bottom up; if inhibition applies, from the top
down. But most importantly, investigation of succession in
streams will help separate abiotic and biotic factors and
identify their mode of action. Results cannot fail to be
significant to general successional theory and are best gained
through a study of streams.

SPACE AND TIME IN STREAMS

While it is possible to reframe the idea of longitudinal
succession in a form conducive to testing general ecological
theory, two problems persist: stream patchiness and successional
time scales. Both can be addressed in the context of
site-specific temporal succession.

The patch problem is a recurrent one in ecosystems where
small organisms occur in a heterogeneous environment. The
problem is not conceptual but methodological. In short, how can
we define organismal interactions when spatial relationships are

destroyed upon sampling? In Sycamore Creek, for example, algae
are distributed in several distinct patch types, each
macroscopically apparent, each consisting of the same species but
differing in terms of relative species composition. Cholorophyll
a and biomass are also quite different between patches (Busch,
1979; Busch and Fisher, 1981). When we speak of successional
change at a given site, we usually lump all organisms from bank
to bank in a reach of some reasonable length (on the order of
meters), and consider this to comprise the community of interest.
In reality, this large sample coalesces several distinct,
separable community types, each of which may represent a point in
the successional continuum. At Sycamore Creek, transitions
between patches at a given point on the stream bed are
predictable but probabilistic and occur at variable rates (Fisher
et al., 1982). Thus pioneer diatom communities usually give way
to a mixed Cladophora-diatom assemblage but may go directly to an
Anabaena-dominated community. Transitions are rapid except in
the center of the stream where velocities are high and pioneer
diatoms persist indefinitely. A stochastic element enters when
heavy mats break loose from the substrate, wash downstream and
expose clean substrate which is then colonized by diatoms. In
essence then, the stream is a mosaic of algal patches in an array
of successional states. While we can describe mean conditions at
a given point and time, several interesting processes escape this
level of resolution. One could relegate this to a mere sampling
problem and dismiss it were it not for parallel advances in
terrestrial ecosystem research. Horn (1975) for example has
described forest succession as a matrix of replacement
probabilities for individual tree species. Williams et al.
(1969) have extended this analysis to the level of the small
patch as we have done. It would be delightful to apply Horn's
approach to algal communities, yet methodological problems appear
intractable. This is an area where glass slide colonization
studies might make a significant contribution. Methodologic
difficulties wane somewhat when we consider larger organisms and
disappear with mobile ones, yet even invertebrates and
macrophytes exhibit patchy distributions (Reice, 1980; Bilby,
1977) which thereby compartmentalize the stream into several
relatively discrete communities. No terrestrial ecologist would
attempt to describe the plant community of California by applying
a stratified random sampling program statewide. Until stream
ecologists abandon this approach, questions dealing with control
of community structure will elude us.

The final point I wish to make concerns successional time.
Succession occurs until a disturbance restores initial
conditions. The interval between resetting events defines
successional time. In order for detectable community change to
occur, organismal life spans must be shorter than the period
between disturbances. A grassland burned yearly will remain a

grassland. Longer lived trees will not occur unless the burning
interval is lengthened. This of course assumes that some tree is
adapted to the unburned site. Similarly, high gradient mesic
streams can be viewed as ecosystems which are too frequently and
severely disturbed (by scouring) to permit observable succession.
Succession would then be considered truncated and typical
residents would be called pioneers. Alternatively, we might view
such streams as undisturbed. In the torrential setting, spates
do not reset the system, nor would the absence of spates result
in a change in community composition. The characteristic
community thus could be viewed as a climax which persists in
what, in other contexts, might seem to be a severe environment.
The point however, is moot. Organisms present in that
environment are adapted to it in, for example, morphology,
behavior, life history characteristics, etc. These adaptations
may define "pioneer" in other systems where non-pioneers also
sometimes exist, but "pioneer" loses definition where they do
not. The essential point is that in some ecosystems no ecocline
in time is generated to which residents cannot adjust. This may
result from dampened abiotic fluctuation, limited biologic
modification of the site, or both. In lower gradient streams,
biotic modification of the site may occur, given enough time, and
this modification may generate an ecocline in time so steep as to
foster species replacement. This may be especially pronounced in
autotrophic systems where plants modify a host of physical,
chemical and biological features of the system. More frequent
disturbance would truncate this successional sequence, less
frequent floods would extend it.

A spectrum of lotic succession thus exists. Heterotrophic
New England streams with weekly spates may exhibit little or no
successional change. Autotrophic Arizona streams which may flood
once a year show marked succession. Flooding less frequent than
once per year often results in loss of the stream by drying, thus
lotic successional time is usually resolved in a year or less.

Longer cycles also exist. Detrital turnover times for small
Northwestern U.S.A. streams approach 100 years (Naiman and
Sedell, 1980) and catastrophic floods with a similar return
frequency may scour these streams to bedrock, initiating another
long accumulation period (Swanson and Lienkaemper, 1978). Desert
streams of Arizona may also participate in long term cycles.
Approximately 100 years ago, several drainages in the Southwest
supported large marshes or cienegas. At the turn of the century,
a combination of overgrazing and climatic change led to the
demise of most of these by headward arroyo cutting and eventual
drainage of cienegas (Hastings, 1959). Several streams,
particularly in southeastern Arizona are now incised in several
meters of old cienega deposits. Data are few but in all
probability, cienegas probably resist floods up to a 100 year

magnitude. Rebuilding is even more poorly known but we can
surmise that several years of minimal flooding will permit
luxuriant growth of macrophytes which entrap sediments, reduce
stream gradient, and eventually pond water in marshes.

Ramon Margalef in his provocative 1960 paper suggested that
the climax community of streams is the terrestrial climax
characterizing the region. Stream succession as discussed here
is thus viewed as the early stages of a hydrosere. While this
contention begs the question somewhat, the cienega is indeed the
terrestrial climax for perpetually wet sites in the arid
Southwest. Perhaps more importantly, studies of streams in this
long term perspective reveal that many are not at steady state at
any point in time. In most ecosystems that alone is prima facie
evidence for succession.

REFERENCES

Allan, J. D. 1975. The distributional ecology and diversity of
    benthic insects in Cement Creek, Colorado. Ecology
    56:1040-1053.
Andrews, D. A. and G. W. Minshall. 1979. Longitudinal and
    seasonal distribution of benthic invertebrates in the little
    Lost River, Idaho. Amer. Midl. Nat. 102:225-236.
Bilby, R. 1977. Effects of a spate on the macrophyte vegetation
    of stream pool. Hydrobiologia 56:109-112.
Blum, J. L. 1956. The application of the climax concept to
    algal communities of streams. Ecology 37:603-604.
Bocock, K. L. 1964. Changes in the amounts of dry matter,
    nitrogen, carbon and energy in decomposing woodland leaf
    litter in relation to the activities of the soil fauna. J.
    Ecol. 52:273-284.
Brock, T. D. and M. L. Brock. 1969. Recovery of a hot spring
    community from a catastrophe. J. Phycol. 5:75-77.
Burton, G. W. and E. P. Odum. 1945. The distribution of stream
    fish in the vicinity of Mountain Lake, Virginia. Ecology
    26:182-194.
Busch, D. E. 1979. Patchiness of diatom distribution in a
    desert stream. J. Az. Nev. Acad. Sci. 14:43-46.
Busch, D. E. and S. G. Fisher. 1981. Metabolism of a desert
    stream. Freshwat. Biol. 11:301-308.
Butcher, R. W. 1933. Studies on the ecology of rivers. I. On
    the distribution of macrophytic vegetation in the rivers of
    Britain. J. Ecol. 21:58-91.
Cattaneo, A. and S. Ghittori. 1975. The development of
    benthonic phytocoenosis on artificial substrates in the
    Ticino River. Oecologia 19:315-327.
Cherry, D. S., S. R. Larrick, R. K. Guthrie, E. M. Davis and F.
    F. Sherberger. 1979. Recovery of invertebrate and

vertebrate populations in a coal ash stressed drainage
system. J. Fish. Res. Bd. Canada. 36:1089-1096.

Clements, F. E. 1916. Plant Succession: An analysis of the
development of vegetation. Publ. No. 424. Carnegie Inst.,
Washington, D.C.

Collins, J. P., C. Young, J. Howell and W. L. Minckley. 1981.
Impact of flooding in a Sonoran Desert stream, including
elimination of an endangered fish population (Poeciliopsis
o. occidentalis, Poeciliidae). Southwest. Nat. 26:415-423.

Connell, J. P. and R. O. Slatyer. 1977. Mechanisms of
succession in natural communities and their role in
community stability and organization. Amer. Nat.
111:1119-1144.

Cowles, H. C. 1901. The physiographic ecology of Chicago and
vicinity: a study of the origin, development, and
classification of plant societies. Bot. Gaz. 31:73-108,
145-182.

Crisp, D. T. 1970. Input and output of minerals for a small
watercress bed by chalk water. J. Appl. Ecol. 7:117-140.

Cummins, K. W. 1974. Structure and function of stream
ecosystems. BioScience 24:631-641.

Dickman, M. 1974. Changes in periphyton community structure
following diatom inhibition. Oikos 25:187-193.

Dimond, J. B. 1967. Pesticides and stream insects. Bull. 23.
Marine Forest Service. 21 pp.

Drury, W. H. and I. C. T. Nisbet. 1973. Succession. J. Arnold
Arboretum 54: 331-368.

Edwards, A. M. C. 1969. Silicon depletion in some Norfolk
waters. Freshwat. Biol. 4:267-274.

Egler, F. E. 1954. Vegetation science concepts I. Initial
floristics composition, a factor in old-field vegetation
development. Vegetatio. 4:412-417.

Eichenberger, E. and K. Wuhrmann. 1975. Growth and
photosynthesis during the formation of a benthic algal
community. Verh. Internat. Verein. Limnol. 19:2035-2042.

Fisher, S. G. 1977. Organic matter processing by a
stream-segment ecosystem: Fort River, Massachusetts, U.S.A.
Internat. Rev. Ges. Hydrobiol. 62:701-727.

Fisher, S. G. and G. E. Likens. 1973. Energy flow in Bear
Brook, New Hampshire: an integrative approach to stream
ecosystem metabolism. Ecol. Monogr. 43:421-439.

Fisher, S. G., L. J. Gray, N. B. Grimm and D. E. Busch. 1982.
Temporal succession in a desert stream ecosystem following
flash flooding. Ecol. Monogr. 52:93-110.

Fraleigh, P. C. and R. G. Wiegert. 1975. A model explaining
successional change in standing crop of thermal blue-green
algae. Ecology 56:656-664.

Gale, W. F. and A. J. Gurzynski. 1979. Colonization and
standing crops of epilithic algae in the Susquehanna River,
Pennsylvania. J. Phycol. 15:117-123.

Gleason, H. A. 1926. The individualistic concept of the plant association. Bull. Torrey Bot. Club 44:1-20.

Gray, L. J. 1981. Species composition and life histories of aquatic insects in a lowland Sonoran Desert stream. Amer. Midl. Nat 106:229-242.

Gray, L. J. and S. G. Fisher. 1981. Postflood recolonization pathways of macroinvertebrates in a lowland Sonoran Desert stream. Amer. Midl. Nat 106:249-257.

Grimm, N. B., S. G. Fisher and W. L. Minckley. 1981. Nitrogen and phosphorus dynamics in hot desert streams of southwestern U.S.A. Hydrobiologia 83:303-312.

Hall, C. A. S. 1972. Migration and metabolism in a temperate stream ecosystem. Ecology 53:585-604.

Hannan, H. H. and T. C. Dorris. 1970. Succession of a macrophyte community in a constant temperature river. Limnol. Oceanogr. 15:442-453.

Hanson, D. L. and T. F. Waters. 1974. Recovery of standing crop and production rate of a brook trout population in a flood-damaged stream. Trans. Amer. Fish Soc. 103:431-439.

Harrison, A. D. 1966. Recolonization of a Rhodesian stream after drought. Arch. Hydrobiol. 62:405-421.

Hastings, J. R. 1959. Vegetation change and arroyo cutting in southeastern Arizona. J. Ariz. Acad. Sci. 1:60-67.

Holmes, N. T. H. and B. A. Whitton. 1977. Macrophytic vegetation of the River Swale, Yorkshire. Freshwat. Biol. 7:545-558.

Hoopes, R. L. 1974. Flooding, as a result of hurricane Agnes, and its effect on a macrobenthic community in an infertile headwater stream in central Pennsylvania. Limnol. Oceanogr. 19:853-857.

Horn, H. S. 1974. Ecology of secondary succession. Ann. Rev. Ecol. Syst. 5:25-37.

Horn, H. S. 1975. Markovian properties of forest succession. pp. 196-211. In: M. Cody and J. Diamond (eds.), Ecology and evolution of communities. Harvard Univ. Press, Cambridge.

Huet, M. 1959. Profiles and biology of western European streams as related to fish management. Trans. Amer. Fish. Soc. 88:155-163.

Hynes, H. B. N. 1958. The effect of drought on the fauna of a small mountain stream in Wales. Verh. Internat. Ver. Limnol. 13:836-833.

Hynes, H. B. N. 1963. Imported organic matter and secondary productivity in streams. Proc. 16th Internat. Cong. Zool. 16:324-329.

Hynes, H. B. N. and N. K. Kaushik. 1969. The relationship between dissolved nutrient salts and protein production in submerged autumnal leaves. Verh. Internat. Ver. Limnol. 17:95-103.

Illies, J. and L. Botosaneanu. 1963. Problems et methods de la classification et de la zonation ecologique de eaux

courantes, considerees surtout du point de vue faunistique.
Mitt. Internat. Ver. Theor. Angew. Limnol. 12:1-57.

Jones, R. C. 1978. Algal biomass dynamics during colonization
of artificial islands: experimental results and a model.
Hydrobiologia 59:165-180.

Kaufman, L. H. 1980. Stream aufwuchs accumulation processes:
effects of ecosystem depopulation. Hydrobiologia 70:75-81.

Kaushik, N. K. and H. B. N. Hynes. 1971. The fate of dead
leaves that fall into streams. Arch. Hydrobiol. 68:465-515.

Kaushik, N. K. and J. B. Robinson. 1976. Preliminary
observations on nitrogen transport during summer in a small
spring-fed Ontario stream. Hydrobiologia 49:59-63.

Keever, C. 1950. Causes of succession on old fields of the
Piedmont, North Carolina. Ecol. Monogr. 20:231-250.

Kennedy, H. D. 1955. Colonization of a previously barren stream
section by aquatic invertebrates and trout. Prog.
Fish-Culturist 17:119-122.

Kuehne, R. A. 1962. A classification of streams, illustrated by
fish distribution in an eastern Kentucky creek. Ecology
43:608-614.

Larimore, R. W., W. F. Childers and C. Heckrotte. 1959.
Destruction and re-establishment of stream fish and
invertebrates affected by drought. Trans. Amer. Fish. Soc.
88:261-285.

Lewis, W. M. 1978. Analysis of succession in a tropical
plankton community and a new measure of succession rate.
Amer. Nat. 112: 401-414.

Margalef, R. 1960. Ideas for a synthetic approach to the
ecology of running waters. Internat. Rev. Ges. Hydrobiol.
45:133-153.

Margalef, R. 1968. Perspectives in ecological theory.
University of Chicago Press. Chicago. 111 pp.

McIntosh, R. P. 1980. The relationship between succession and
the recovery process in ecosystems. pp. 11-62. In: J.
Cairns (ed.), The recovery process in damaged ecosystems.
Ann Arbor Science, Ann Arbor, Mich.

Moore, J. W. 1976. Seasonal succession of algae in rivers. I.
Examples from the Avon, a large slow-flowing river. J.
Phycol. 12:342-349.

Naiman, R. J. and J. R. Sedell. 1980. Relationships between
metabolic parameters and stream order in Oregon. Can. J.
Fish. Aquatic Sci. 37:834-847.

Newbold, J. D., J. W. Elwood, R. V. O'Neill and W. Van Winkle.
1981. Measuring nutrient spiralling in streams. Can. J.
Fish. Aquat. Sci. 38:860-863.

Odum, E. P. 1969. The strategy of ecosystem development.
Science 164:262-270.

Odum, H. T. 1971. Environment, power and society.
Wiley-Interscience. New York. 331 pp.

Peters, R. H. 1976. Tautology in evolution and ecology. Amer. Nat. 110:1–12.

Petersen, R. C. and K. W. Cummins. 1974. Leaf processing in a woodland stream. Freshwat. Biol. 4:343–368.

Pickett, S. T. A. 1976. Succession: an evolutionary interpretation. Amer. Nat. 110:107–119.

Reice, S. R. 1980. The role of substratum in benthic macroinvertebrate microdistribution and litter decomposition in a woodland stream. Ecology 61:580–590.

Reisen, W. K. and D. J. Spencer. 1970. Succession and current demand relationships of diatoms on artificial substrates in Prater's Creek, South Carolina. J. Phycol. 6:117–121.

Sedell, J. R., F. J. Triska, J. D. Hall, N. H. Anderson and J. H. Lyford. 1974. Sources and fates of organic inputs in confierous forest streams. In: R. H. Waring (ed.), Symposium: Synthesis of Coniferous Forest Biome Research. IBP Analysis of Ecosystems.

Seegrist, D. W. and R. Gard. 1972. Effects of floods on trout in Sagehen Creek, California. Trans. Amer. Fish. Soc. 101:478–482.

Sheldon, A. L. 1968. Species diversity and longitudinal succession in stream fishes. Ecology 49:193–198.

Siegfried, C. A. and A. W. Knight. 1977. The effects of washout in a Sierra foothill stream. Amer. Midl. Nat. 98:200–207.

Suberkropp, K. and M. J. Klug. 1976. Fungi and bacteria associated with leaves during processing in a woodland stream. Ecology 57:707–719.

Swanson, F. J. and G. W. Lienkaemper. 1978. Physical consequences of large organic debris in Pacific Northwest streams. Forest Service General Technical Report PNW–69. Pacific Northwest Forest and Range Experiment Station. Portland, Oregon, U.S.A.

Tansley, A. G. 1929. Succession: the concept and its value. Internat. Congr. Plant. Sci., Ithaca, Proc. 1926. 1:677–686.

Thorup, J. 1970. The influence of a short-termed flood on a springbrook community. Arch. Hydrobiol. 66:447–457.

Vannote, R. L., G. W. Minshall, K. W. Cummins, J. R. Sedell and C. E. Cushing. 1980. The river continuum concept. Can. J. Fish. Aquat. Sci. 37:130–137.

Wallace, J. B. 1977. The role of filter feeders in flowing waters. Arch. Hydrobiol. 79:506–532.

Vincent, W. F. and M. T. Downs. 1980. Variation in nutrient removal from a stream by watercress (Nasturtium officinale R. Br.). Aquat. Bot. 9:221–235.

Wiegert, R. G. and P. C. Fraleigh. 1972. Ecology of Yellowstone thermal effluent systems: net primary production and species diversity of a successional blue-green algal mat. Limnol. Oceanogr. 17:215–228.

Williams, W. T., G. N. Lance, L. J. Webb, J. B. Tracey and J. H.
     Connell. 1969. Studies in the numerical analysis of
     complex rain forest communities. IV. A method for the
     elucidation of small-scale forest pattern. J. Ecol.
     57:635-654.
Wuhrmann, K. 1974. Some problems and perspectives in applied
     limnology. Mitt. Internat. Verein. Limnol. 20:324-402.

PRIMARY PRODUCTIVITY IN STREAMS

Thomas L. Bott

Stroud Water Research Center
Academy of Natural Sciences, R.D. 1, Box 512
Avondale, Pennsylvania 19311

INTRODUCTION

The most widely cited studies concerning the energetics of
lotic systems have been conducted on headwater streams in
temperate deciduous forested regions. These studies showed the
importance of allochthonous organic matter inputs supporting
consumer organisms and gave rise to the generalization that
stream ecosystems were heterotrophic (i.e., that respiration
exceeded photosynthesis annually, Hynes, 1963; Cummins, 1974).
Fisher and Likens (1973) expanded the conceptualization of stream
system energetics to include consideration of import and export
properties. When photosynthesis and import exceeded respiration
and export the system was considered accretive and when the
reverse occurred the system was considered remissive. Cushing
and Wolf (1982) discuss communications of Minshall and Fisher and
point out that a system may shift from being annually accretive
to remissive depending on flood flows and channel conditions that
affect storage capability.

The generalization concerning stream heterotrophy has been
reexamined by some. Wetzel (1975a) stated that the "assumptions
that in most streams autochthonous primary productivity is
negligible is unwarranted." McIntire (1973) had considered
earlier that rapid turnover of primary producer biomass may be
sufficient to support considerable consumer biomass even when
algal standing crops are not particularly large. Minshall (1978)
pointed out that most studies have been done on small, shaded
stream reaches and that measurements in larger sized reaches and
on systems arising on non-forested watershed were needed. He
further emphasized that annual photosynthesis/respiration (P/R)

29

ratios may be misleading because they obscure seasonally
important peaks of primary productivity. Busch and Fisher
(1981), after study of a desert stream, defined an autotrophic
stream ecosystem as one in which total gross primary production
exceeded community respiration and in which accretion of excess
photosynthate and/or export of biomass occured.

The River Continuum concept (Vannote et al., 1980) predicts
that in unperturbed stream systems the energy sources supporting
biotic processes change in downstream direction. It was
hypothesized that community metabolism would shift from a
dominance of respiration in headwater reaches to a dominance of
photosynthesis in wider but still shallow mid-sized reaches at
least at some seasons of the year, if not annually. Further
downstream a return to dominance of respiration was expected as
river depth and turbidity increased. Here, I briefly summarize
some of our findings concerning primary productivity in first
through eighth order stream reaches which are discussed in detail
elsewhere (Minshall et al., in press; Bott et al., in review).
Some of the important factors regulating primary productivity in
terrestrial and aquatic systems will then be discussed. This
treatment, while cursory, is intended to stimulate recognition of
analogous or contrasting features in different habitats and
encourage the transfer of ideas across system boundaries.
Lastly, some areas of research need are highlighted.

COMPARISON OF PRIMARY PRODUCTIVITY IN STREAMS WITH OTHER SYSTEMS

Primary productivity is defined as the rate of conversion of
solar energy to reduced chemical energy through photosynthesis.
Gross primary productivity (GPP) is the total quantity of energy
fixed (although data is more commonly reported in carbon or
oxygen units) including that used subsequently in respiration and
net primary productivity (NPP) is the amount stored in biomass
(i.e., GPP - respiration by autotrophs). The ^{14}C uptake and
harvest procedures employed to measure primary productivity focus
on the metabolism of autotrophs. Although the length of
incubation influences results, it generally is considered that
the ^{14}C procedure measures something close to NPP. Excretion of
DOM by algae is recognized as an important loss of photosynthate,
and some data suggests that the ^{14}C method inaccurately estimates
the quality and quantity of excreted DOM (Storch and Saunders,
1975; Steinberg, 1978; Kaplan and Bott; in press). Harvest
techniques (which ideally should include below ground material)
also yield a net estimate since losses through respiration,
grazing, fragmentation, or sloughing usually are not accounted

for. On the other hand, as usually employed, gas change
procedures yield primary productivity estimates from measures of
total community metabolism, which includes both photosynthesis
and community respiration. Since respiration cannot be related
to autotrophs, estimates of net productivity are termed net
community primary productivity (NCPP). Estimates of GPP are
imprecise because they are based on respiration of the community
in the dark. However, it is possible to calculate a term
reported here as net daily metabolism (NDM) from NCPP and dark
community respiration (NDM = NCPP - CR_d) as well as from GPP and
community respiration for 24 hr (NDM = GPP - CR_{24}) because the
questionable respiration component of GPP is subtracted in CR_{24}.
NDM is equivalent to quantities reported by others as net
photosynthesis (Marker, 1976), net primary productivity (Pennak
and LaVelle, 1979), net community productivity (Odum, 1971) and
net ecosystem production (Brock, 1981).

In aquatic systems, gas change or ^{14}C uptake procedures
usually are employed for measures with plankton and attached
communities and sequential harvesting is used for macrophytes
(Hall and Moll, 1975). We compared ^{14}C uptake and dissolved
oxygen procedures simultaneously with benthic communities in two
stream reaches. In one, NPP estimates from ^{14}C uptake were
84-93% of estimates of NCPP from dissolved oxygen change
(converted to carbon for comparison) in four of five experiments;
in the other, estimates based on oxygen production were twofold
higher in two experiments (Bott and Ritter, 1981).   Hunding and
Hargrave (1973) reported reasonable agreement between methods
but others have reported considerable differences (Schindler et
al., 1973). Revsbech et al. (1981) found that oxygen and ^{14}C
uptake methods agreed well in porous marine sediments at low
light intensities; at higher intensities or in highly reduced
sediments they did not. However, unless the specific activity of
bicarbonate was based on the bicarbonate content of sediment
interstitial water, as opposed to that of the overlying water,
primary productivity would have been underestimated fivefold.
This consideration may account for the results of our procedure
comparison and perhaps for those of Schindler et al. (1973)
because it is unexpected that NCPP which includes the respiration
of heterotrophs in the communities should exceed NPP estimates.
In terrestrial environments net primary productivity is most
often measured. The methods usually employed are harvest
techniques or measures of specific tissue growth and to a lesser
extent gas change (Whittaker and Marks, 1975).

Assessment of ecosystem primary productivity also requires
consideration of several communities (e.g., benthos, plankton,

periphyton in aquatic habitats; soil, understory and canopy in
terrestrial habitats) comprised usually of many primary producer
species. Most ecosystems contain a species mix of small
organisms with rapid reproduction and larger organisms with
slower reproduction which serves to insure an autotrophic energy
base via alternative pathways (Reichle et. al., 1975). In small
and shallow intermediate size streams the benthic community
(comprised of algae and perhaps mosses and liverworts) is
important but with increasing stream size other communities
(vascular macrophytes and phytoplankton) may also assume
importance (Wetzel, 1975a; Minshall, 1978).

Comparisons of primary productivity in diverse systems are
complicated by differences in methods and in sampling intensity
used to obtain original data as well as by various assumptions
and conversions used to generate final values. The major sites
of primary productivity in the biosphere have been identified as
the forests and oceans (Woodwell et al., 1978). Freshwater
systems (lakes and streams) are thought to make a relatively
minor contribution to global primary production because inland
waters occupy a minor portion of the global surface. Primary
productivity of inland waters expressed on an areal basis ranked
14th out of 19 habitat types and 5th out of 7 aquatic habitat
types if the transitional swamp and marsh systems are included
(Table 1). Likens (1975) noted that some lakes (particularly
those subjected to cultural eutrophication) have elevated primary
productivity levels that may be similar to those of highly
productive tropical forests, coral reefs, swamps, and marshlands.
Data for streams (expressed on an areal basis) yield a wide range
of values depending on climatic and habitat influences. A few
flowing water systems are highly productive with mean annual GPP
values in excess of $10 \text{ g } O_2 \text{ m}^{-2} \text{ day}^{-1}$; e.g., Silver Springs, FL
(Odum, 1957), reaches on the Blue River, OK (Duffer and Dorris,
1966) and the Raritan River, NJ (Flemer, 1970). These data are
roughly comparable to data for tropical rain forests (8.67 g
$O_2 \text{ m}^{-2} \text{ day}^{-1}$, calculated from Table 1, assuming a Photosynthetic
Quotient of 1.2). However, most reported data is for temperate
or cold water systems with lower productivity.

For the River Continuum studies we measured primary
productivity in 19 first through eighth order (Strahler
classification, 1957) relatively unperturbed reaches on stream
systems located in four biomes. Study locations were:
Pennsylvania (PA, eastern deciduous forest, coastal climate),
Michigan (MI, mesic hardwood forest, continental climate), Idaho
(ID, coniferous vegetation on north facing slopes, sagebrush on
south facing slopes, cool arid climate with most precipitation as
snowfall), Oregon (OR, coniferous vegetation, coastal climate).
Benthic community metabolism was measured at stations 1-4 at all

Table 1. Net primary productivity of major habitats in the
          biosphere, after Woodwell et al. (1978).

Ecosystem Type	Net Primary Productivity $gC\ m^{-2}\ year^{-1}$
Terrestrial	
Tropical rain forest	988.2
Tropical seasonal forest	720.2
Temperate evergreen forest	580.0
Temperate deciduous forest	542.9
Savanna	406.7
Boreal forest	358.3
Woodland and shrubland	317.7
Cultivated land	292.9
Temperate grassland	266.7
Tundra and alpine meadow	62.5
Desert scrub	38.9
Rock, ice, and sand	1.3
Aquatic	
Swamp and marsh	1350.0
Algal bed and reef	1166.7
Estuaries	714.3
Upwelling zones	250.0
Lake and stream	200.0
Continental shelf	161.7
Open ocean	56.3

sites during 1976. Measures made on the large Salmon River,
Idaho reaches (4-8) during 1977 included both benthic communities
and communities in transport to provide a total system estimate.
Measurements were made by placing intact colonized streambed
sediments contained in trays or separate rock, detritus or
transport samples into Plexiglas chambers with stream water. The
chambers were incubated in situ, the water recirculated, and
dissolved oxygen concentration concentrations were continuously
monitored. Water in the chambers with benthic communities was
changed periodically to circumvent dissolved oxygen
supersaturation or extreme depletion, and this exchange also
served to replenish nutrients.

Annual NDM for each station was estimated from seasonal
means (dervied from 6 to 16 measurements). The data are plotted
against stream width in Fig. 1. Other measures of stream size
could have been used but width is particularly well linked with
separation of the tree canopy. At all sites, station 1 had a
negative NDM value, i.e., metabolism was dominated by
respiration. Photosynthesis was predominant in mid-sized reaches
everywhere but at the Pennsylvania stations. Local influences
modified the point along the river system where metabolism
shifted from annual heterotrophy to autotrophy. This occurred by
station 2 at Idaho, station 3 at Oregon and station 4 at
Michigan. Higher light inputs and lesser retention of detritus
at Idaho favored photosynthesis over respiration. While annual
autotrophic status was not achieved at any Pennsylvania station

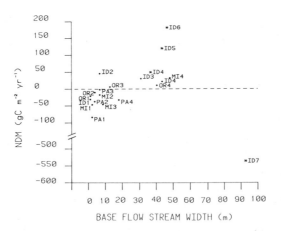

Fig. 1. Net daily metabolism in study reaches on four stream
systems: (PA, White Clay Creek, Buck Run; MI, Augusta
Creek, Kalamazoo River; ID, Salmon River; OR, Devils
Club Creek, Mack Creek, Lookout Creek, McKenzie River).
Idaho data from 1976 (·) and 1977 (*). All 1976 data is
for benthic communities; 1977 data at Idaho includes
metabolism of communities in transport.

during these studies, extensive experimentation at PA 3 conducted
between 1970 and 1974 indicated that either autotrophy or
heterotrophy could predominate annual metabolism there. Stations
5 and 6 at Idaho were autotrophic to an extreme degree. Both of
these reaches were wide and relatively shallow so that solar
radiation reaching the benthos was high. On the other hand,
ecosystem metabolism at ID 7 showed strong heterotrophy which
resulted from respiration of communities in transport at this
large, deep station ($\bar{X}$ depth = 7.37 m vs. 0.06 - 1.60 m for all
other stations). Benthic community metabolism at ID 7 was
reduced compared to that at ID 5 and 6 but still favored autotrophy. These data confirmed the hypothesis concerning a predictable shift in community metabolism with increasing stream size.

Even in the low order reaches dominated annually by
respiration, photosynthesis was predominant during at least one
season. Photosynthetic dominance occurred at OR 1 in the spring,
PA 2 and OR 2 in the spring, MI 2 in the autumn, PA 3 and MI 3 in
the spring and at PA 4 in the winter and summer. However, this
did not occur at ID 7, where respiration always dominated
ecosystem metabolism.

Net community primary productivity data at River Continuum
study stations are compared with net primary productivity data
for lakes in each geographical region in Table 2. Most studies
in these north temperate lakes were of phytoplankton; littoral
periphyton and macrophyte productivity were considered of lesser
importance. However, where the lake was shallow this may make
for a considerable underestimate. For Lake Lawrence an
integrated value is available. For that system, Wetzel (1975b)
reported that macrophytes contributed 51% of the primary
productivity. Fisher and Carpenter (1976) summarized data
showing that the percent contribution of macrophytes to total
primary productivity of five stream systems ranged from 9.2 -
30%. Lakes used for comparison possessed a wide range of
nutrient loadings. While the relative contribution of primary
productivity to the system energy budget may be lower for streams
than for lakes because of greater allochthonous inputs to streams
(Fisher and Likens, 1973; Cummins, 1974; Marker, 1976) the data
in Table 2 show that primary productivity in the mid-sized and
larger downstream reaches at all sites can approach the
productivity in some of the lesser impacted lakes in the region.

REGULATION OF PRIMARY PRODUCTIVITY IN STREAMS AND OTHER SYSTEMS

Geology and Geomorphology

Geological and geomorphological factors affect the
distribution of plant communities and primary productivity in

Table 2. Net community primary productivity of River Continuum study reaches and net primary productivity of lakes in each geographical region.

Reach	Stream Reaches	Lakes in Area (comments)
	Mean daily net primary productivity for year (mgC m^{-2} day^{-1})	
PA 1	-27.0	197.3 Lake George, NY (Ferris and Clesceri, 1975; Rast and Lee, 1978)
2	84.6	517.0 Lake Erie, east basin, US-CANADA (Glooschenko et. al., 1974)
3	151.3	548.0 Cayuga Lake, NY (Stuart and Stanford, pers. comm.; from Oglesby, 1978)
4	246.8	517.8 Canadarago Lake, NY (alkaline; Hetling et. al., 1975; Rast and Lee, 1978)
MI 1	-55.9	280.3 Gull Lake, MI (alkaline, phytoplankton, 1972–74; Moss, Wetzel, and Lauff, 1980)
2	242.3	
3	156.5	469.0 Lawrence Lake, MI (alkaline, phytoplankton, macrophytes, periphyton; Wetzel, 1975b; Devol and Wissmar, 1978)
4	486.4	1012.0 Wintergreen Lake, MI (shallow, extensive nutrient loading. Phyto-plankton data. Extensive macrophytes and epiphytes would increase primary productivity at least threefold. Wetzel, 1975b, pers. comm.)
ID 1*	48.2	337.0 Flathead Lake, MT (Phytoplankton data; Periphyton and macrophytes probably insignificant, 2 years, Stuart and Stanford, pers. comm.)
2	236.1	
3	246.1	
4	136.7, 230.5	
5	816.9	
6	897.7	
7	524.7	
OR 1	-21.0	3 Waldo Lake, OR (Powers et. al., 1975; Rast and Lee, 1978)
2	24.2	234.3 Lake Tahoe, CA-NV (1978–1980; C. R. Goldman, pers. comm.)
3	83.5	98 Castle Lake, CA (deep alpine, Wetzel, 1975b)
4	93.8	438 Clear Lake, CA (large shallow, Wetzel, 1975b)
		652 Lake Sammamish, WA (Welch et. al., 1975; Rast and Lee, 1978)
		969.9 Lake Washington, WA (Edmondson, 1975; Rast and Lee, 1978; data from 1971)

*Data for Stations 1-3 are from 1976, for Station 4 from 1976 and 1977, and Station 5-7 from 1977.

Table 3. Influences of geology, geomorphology, and climate on terrestrial and aquatic primary productivity.

| FACTOR | PRIMARY MODE OF INFLUENCE ON PLANT COMMUNITIES AND PRIMARY PRODUCTIVITY | |
	IN TERRESTRIAL SYSTEMS	IN AQUATIC SYSTEMS
GEOLOGY	Soil fertility and drainage	Water chemistry, benthic substrate characteristics
GEOMORPHOLOGY	Development of moisture and shading gradients, erosion potential of soil	Shading, benthic contours, flow pattern, current velocity, stream power
CLIMATE	Precipitation, temperature, wind disturbance	Temperature, precipitation effects on hydrologic regime, wind generated turbulence

both terrestrial and aquatic systems as shown in Table 3.
Geology influences entire regions whereas geomorphology (e.g.,
slope or aspect) have more immediate effects but both control in
part the development of terrestrial communities (Swanson, 1979).
Concerning aquatic systems, regional geology is primarily
responsible for concentrations of dissolved inorganic chemical
species and geomorphology (e.g., slope) establishes benthic
contours and thus the proportion of illuminated benthos. In
streams, slope also partially regulates flow patterns, water
velocity, and stream power (a measure of the ability of flowing
water to do work, e.g., transport a sediment load) thereby
affecting sediment deposition. Sediment size and stability as
well as current influence the distribution of plant communities
in streams, e.g., macrophytes occur primarily on fine sediments
deposited in mid-sized reaches of moderate flow or in alcove
areas. Geomorphology has been invoked as a regulatory factor of
primary productivity in streams. For example, stream orientation
in a north - south vs. east - west direction (Naiman and Sedell,
1980) and shading from steep, wooded side slopes (Bott and
Ritter, 1981) or from canyon walls (Kobayasi, 1961) have been
related to levels of productivity primarily through effects on
the light regime. MLR analyses of site specific or combined data
sets from the River Continuum studies showed that primary
productivity was related strongly to the inverse of slope (which
decreases with downstream distance). It is likely that the
influence of slope on lotic primary productivity is mediated
through the direct effect of current velocity on the metabolic
rates of autrotrophs (Whitford and Schumacher, 1964; Horner and
Welch, 1981) and through stream power which affects the
transparency/turbidity properties of the system and silt
deposition onto the benthic flora.

Climate

Climatic features such as the temperature decrease from the
tropics to the polar regions and variation in precipitation are
major factors regulating terrestrial primary productivity and
vegetation type (Ricklefs, 1979). In arid regions, precipitation
also may be an important direct source of dissolved nutrients.
In aquatic systems, temperature is the more important influence.
A positive correlation between temperature and primary
productivity or community metabolism has been shown in many
studies (e.g., Malone, 1976; Phinney and McIntire, 1965; Hameedi,
1976; Summer and Fisher, 1979). MLR analyses of River Continuum
data showed temperature was important in regulating GPP and
especially benthic community respiration.

Other climatic factors such as fog (Triska et. al., in
press) and prevailing winds may also influence both aquatic and
terrestrial primary productivity. Fog alters light regimes and

moisture gradients. Prevailing winds influence the distribution
of airborne particulate and dissolved compounds (Swan, 1963) and
is especially well illustrated by the acid rain patterns which
have received recent widespread attention. In one specialized
situation, absorption of ammonia volatilized from cattle feedlots
was shown to enrich nearby surface waters in nitrogen (Hutchinson
and Viets, 1969). Winds also generate internal currents in
lakes which increase nutrient supply and stimulate primary
productivity especially in the littoral zone (Wetzel, 1975b).

Light

The effects of light and temperature on primary productivity
are difficult to separate in field studies since both vary with
solar radiation. Experimental manipulations have demonstrated
the regulation of primary productivity by light in small or
mid-sized streams (Sumner and Fisher, 1979; Triska et. al., in
press) and in larger systems (Malone, 1977). In other studies,
correlation and regression techniques have revealed its
importance. Brylinsky and Mann (1973) considered that solar
radiation, determined by latitude, exerted principal control over
lentic primary productivity, although within a narrow range of
latitude nutrients were important. MLR analysis of data obtained
during the River Continuum studies demonstrated that light
explained the largest amount of the variance in NDM data. Light
was second to a geomorphological variable in explaining variance
in the GPP and NCPP data sets. In addition to geomorphological
features, riparian vegetation at the study sites strongly
influenced the light regime. Hameedi (1976) used MLR analysis to
examine primary productivity data for populations in the plume of
the Columbia River and in ambient northern Pacific seawater. In
the plume, depth of the photic zone was one of the factors
explaining a large proportion of variance in the data. In
ambient seawater, solar radiation of the previous day was
important. Path analysis was used to assess factors affecting
primary productivity in Lake Monroe, Indiana and revealed that
nutrients and solar radiation exerted independent affects, and
that low light during the winter months may have been an
important variable (Chang, 1981).

Nutrients

Ricklefs (1979) and Likens (1975) among others, however,
have considered that nutrients are the primary controlling
mechanism for aquatic primary productivity. Nutrients are
supplied exogenously and endogenously to aquatic systems.
Watershed studies have shown that a tight linkage exists between
the terrestrial system and stream drainage networks, that
terrestrial vegetation controls the supply of some dissolved
nutrients (e.g., nitrate) to streams and that disturbance (e.g.,

through logging) increases nitrate loss from the watershed
(Likens et. al., 1970; Gosz, 1978) although the extent of loss is
modified in part by vegetation type (Vitousek et al., 1979;
Gessel and Cole, 1965). While the supply of dissolved nutrients
supporting production is important, particulate inputs also
affect the net metabolism of aquatic communities primarily by
increasing community respiration. This effect is particularly
pronounced in small streams draining deciduous forested
watersheds.

Concern over cultural eutrophication has generated much
interest in exogenously supplied anthropogenic nutrients,
primarily phosphorus to lakes. Nutrient additions to some lakes
have revealed convincingly the principal importance of phosphorus
in stimulating phytoplankton productivity (Schindler and Fee,
1974), as has the long term monitoring of phosphorus, nitrogen,
and chlorophyll a concentrations in another lake (Edmondson,
1970) and numerous bioassays (Rast and Lee, 1978). While the
limiting nutrient has been shown to change through time in lakes
(Storch and Dietrich, 1979) and varies between systems, the
control of phosphorus is of demonstrated practical value.

In streams, nutrient supply to periphyton communities is
controlled by the interaction of concentration and current
velocity. In one study (Horner and Welch, 1981) there was
continuous periphyton accrual when the o-PO_4 concentration
exceeded 40 - 50 ug/1 at velocities   50 cm/sec. At lower
phosphate concentrations, increasing the velocity reduced the
accrual rate. Velocities   50 cm/sec eroded algae. Wong and
Clark (1976) found a critical phosphorus concentration of 60 ug/1
(coupled with 1.6 mg P/g algal tissue) for algae in an Ontario
stream. Triska et. al. (in press) discuss studies showing
nitrogen limitation in some Pacific northwest streams. Manuel
and Minshall (1980) have shown stimulation of algal productivity
through nitrate and phosphate additions. Wuhrmann (1974)
however, reported that inorganic N and P additions to groundwater up to 3.6 and 0.1 mg/1 respectively caused little
stimulation of attached algal growth in experimental streams but
that amendment with inorganics plus minute concentrations of
sewage effluent enhanced phototrophic productivity.

Nutrients for primary productivity also are supplied by
recycling within the system. Both physical transfer of water and
biological aspects are involved. In oceans, upwelling tends to
counteract the sinking of plankton and, in lakes and oceans,
other currents distribute both non-vagile organisms and
nutrients. Consumer organisms regulate nutrient turnover rates
in lakes through two general phenomena: translocation by neuston
or sediment dwelling burrowing organisms and transformations that
alter nutrient cycling rates primarily through change of the
surface area/volume ratio such as size selective predation,

agglomeration of particles, or invasion of particulates by
burrowing organisms (Kitchell et al., 1979). Protozoan or
amphipod grazing on bacteria has been shown to accelerate
phosphorus recycling (Johannes, 1965; Barsdate et al., 1974;
Morrison and White, 1980). In the open ocean, nutrient recycling
via zooplankton grazing in the photic zone is considered the
major source of nutrients, particularly nitrogen, supporting
primary productivity (Eppley, 1981). Regeneration of ammonia
nitrogen also supports phytoplankton production in both
productive (Liao and Lean, 1978) and meso-oligotophic (Axler et
al., 1981) lakes. Even in terrestrial systems, grazing affects
nutrient recycling rates by accelerating release of nutrient ions
through leaching.

    In streams, downstream displacement of nutrients adds an
important dimension to nutrient cycling concepts. Downstream
communities in natural system have evolved to capitalize on the
inefficiencies of utilization by upstream communities (Vannote
et al., 1980; Minshall et al., 1983) and the concept of nutrient
spiralling has emerged to characterize nutrient flux in streams
(Elwood et al., in press; Newbold et al., 1981).

Biomass

    Primary productivity usually has been closely correlated
with chlorophyll a or some other measure of plant biomass,
although there have been some exceptions (Wetzel, 1964; Naiman,
1976). One control of plant biomass is through herbivory.
Consumption of periphyton biomass by macroinvertebrates was
estimated as < 10% of net periphyton production in the Fort
River, MA (Sumner and Fisher, 1979). However, other experiments
suggested a significant impact on primary producer biomass
through herbivory (Wurhmann, 1974) as have the studies of
McAuliffe (this volume) and by inference those of Elwood et al.
(1981). Gregory (this volume) has reviewed many features of
herbivory in streams. Grazing is in part dependent on plant
type. For example, little aquatic macrophyte biomass is consumed
directly and most stored energy is either released at the end of
the growing season through lysis and decomposition or is utilized
in the food web as detritus (Koslucher and Minshall, 1973; Fisher
and Carpenter, 1976). Cladophora is a poorly utilized algal food
source compared to other periphyton (Moore, 1975; R. Patrick,
pers. comm.). Microalgae with filamentous growth habits or loose
adhesion were ingested whereas more tightly adhering species were
not in both intertidal (Nicotri, 1977) and freshwater communities
(Patrick, 1970; Sumner and McIntire, 1982) grazed by gastropods.
Several species of blue-green algae are considered a poor food
source compared to other algae (Porter, 1978) which may be due to
toxic properties of some (Collins, 1978) or perhaps cell
structure (Hargrave, 1970). Selective grazing may alter
community species composition but this may also occur through

differential digestibility and enhanced productivity of undigested forms. The green alga, Sphaerocystis schroeteri, exhibited enhanced growth after passage through the gut tract of the grazing zooplankter, Daphnia magna (Porter, 1976), thereby compensating for losses through grazing.

Data collected on Walker Branch, TN suggested that grazing limited periphyton productivity through control of the periphyton standing crop (Elwood and Nelson, 1972). Flint and Goldman (1975) reported that, depending on animal biomass, crayfish grazing could either stimulate or inhibit periphyton productivity in Lake Tahoe. Grazing by snails reduced attached algal primary productivity in a small Michigan pond (Hunter, 1980). O'Neill (1976) has considered that the regulation of autotrophic productivity by heterotrophs is particularly important in aquatic environments.

Microbial pathogens also have been shown to affect the biomass of algal populations in lakes (viruses on blue-green algae; Granhall, 1972; Safferman and Morris, 1963; fungi on diatoms, Canter and Lund, 1948; Masters, 1971; fungi on blue-green algae, Redhead and Wright, 1978) and in microcosms (Bott and Rogenmuser, 1980). In streams, little is known of such interactions and occasional reports of rapid dieback of plant growth have been attributed to other causes, e.g., temperature (Wong et al., 1978).

It appears thus far that in streams high discharge is the major mechanism affecting the biomass of benthic algae. Tett et al. (1978) concluded it was impossible to predict benthic algal biomass (chlorophyll a) in an Appalachian stream because of the unpredictability of high flows. Stockner and Shortreed (1976) considered losses of algae through grazing were negligable compared to those through scour in a Pacific northwest stream. Inadequate measurement of algal losses through herbivory and sloughing may help explain why allochthonous detritus standing crops appear to be so important in streams (Minshall, 1978). Effects on flora also will be influenced by sediment stability under various flow conditions.

Disturbance may be less important in plant biomass regulation in terrestrial than in aquatic systems. However, Bormann and Likens (1979) stated that minor disturbance was essential for forest ecosystem persistence. Forests are returned to a less mature successional state by natural mechanisms of windthrow (the potential for which is affected by geomorphology) and fire and through logging. They characterized the mature presettlement New England hardwood forest as a shifting mosaic of patches of vegetation of different ages. Martin (1979) has pointed out the potential for greater nutrient retention of such systems than of old, even aged growths.

While primary productivity is directly related to biomass, studies at the River Continuum sites and elsewhere (Marker, 1976; Pfeifer and McDiffett, 1975) have shown an inverse relationship between chlorophyll-specific periphyton primary productivity and chlorophyll levels. Thus, there is an apparent compensatory response to biomass reduction mediated by factors such as more favorable nutrient supply, removal of metabolic wastes, habitat spatial features, or light conditions. Phytoplankton density also has been inversely related to the assimilation ratio (Findenegg, 1965).

Summary

The principal controls of primary productivity differ between systems. In terrestrial systems, moisture and temperature are most important, whereas in oceans and lakes, nutrients and light (where nutrients are adequate) are of primary importance. In some small-to-mid-sized relatively unperturbed streams, nutrient additions have been shown to stimulate productivity but control through grazing, spates, and in particular, light, may have even greater importance. Where a factor influences productivity in both terrestrial and aquatic habitats its relative importance and mode of influence can vary. The role of slope in terrestrial systems, for example, is mediated through potential for erosion, climatic zonation, or disturbance, whereas in streams effects are mediated primarily through current.

SOME RESEARCH NEEDS

Many studies of lotic periphyton productivity have been conducted with natural communities that were transferred to flowing water microcosms where current effects were provided. Such an experimental approach includes many of the physical and chemical constraints and population interactions operative in nature. Other researchers have measured total system metabolism directly in the open system. The most common fault with studies of periphyton productivity has been the failure to encompass either the spatial or temporal heterogeneity or both present in stream systems. Continued work incorporating these aspects will establish more firmly the range of primary productivity values to be expected in streams of differing character.

In addition, studies which focus on the dynamics of community response to fluctuating environmental conditions (e.g., light, nutrients, temperature) are needed to advance our knowledge of the physiological ecology of lotic algae. These studies should emphasize the degree of environmental change required to elicit a biological response and what constitutes

background variation to which organisms show no response.
Oceanographers have elucidated some of the responses of marine
plankton algae to changing envrionmental factors in studies which
may guide stream ecologists into productive research areas (see
papers in Falkowski, 1980). Factorial experimental design, work
with natural communities and manipulation experiments conducted
in the field whenever possible should yield greatest significant
progress in our understanding of factors regulating primary
productivity in streams. Others have discussed the difficulty of
extrapolating the results of nutrient enrichment bioassays with
single species of plankton algae conducted in the laboratory to
the field (Lane and Levins, 1977). The problem of scale noted by
Allen (1977), i.e., that measures are often made at a level too
coarse to be meaningful to the cell should be considered in data
interpretation. Some aspects of this problem may be
insurmountable with available methods and instrumentation.
However, a microenvironmental approach should be used where
feasible.

O'Neill and Reichle (1979) have succinctly delineated the
theoretical elements concerning ecosystem function and
persistence as: (1) primary producers or the energy base; (2)
storage capability (detritus), nutrient cycling, and ecosystem
resistence; (3) rate regulation carried out by heterotrophic
organisms; (4) population interactions and ecosystem dynamics;
and (5) heterogeneity as a mechanism for ecosystem persistence.
The discussion of factors affecting primary productivity has
demonstrated many linkages to other features of ecosystem
function deserving study, e.g., the role of herbivory or other
population interactions. It is clear that a wholistic
perspective is needed wherein experimental design and
interpretation of results are related to other aspects of system
structure and function.

Studies linking watershed characteristics and uses with
instream primary productivity are also needed. In particular,
alterations of nutrient supply and/or the flow regime resulting
from agriculture, clearing, or development deserve study.
Considerable attention has been given to the nutrient loading of
lakes and their eutrophication (e.g., Rast and Lee, 1978).
Interpretation of responses of lotic primary producers to given
nutrient concentration should include consideration of current
velocity as affecting nutrient supply (Horner and Welch, 1981).
Concerning watershed manipulations, it has been suggested that
regulation of the light regime through the planting of riparian
vegetation would encourage the development of desirable aquatic
plant communities (Dawson and Kern-Hansen, 1979). The potential
practical use of streamside vegetation as a management tool for
controlling light, litter, and nutrient inputs to streams should
be more thoroughly investigated.

The relative importance of autotrophic production and allochthonous inputs to streams has been evaluated through system level energy budgets and through P/R ratios. Further substantiation of the role of detrital inputs in the system has been gained from feeding studies with detritivores. Less attention has been given to the fate of benthic autotrophic biomass in streams than to the fate of detritus. Alternative fates include losses to transport through sloughing, death and lysis, decomposition, burial, or use as a food resource. Material in transport has seldom been evaluated with respect to the autotrophic component although Swanson and Bachmann (1976) related algal export to upstream benthic standing crop in one study. Similarily, the influence of benthic algae on transported DOC which contains excreted photosynthetic products has received scant attention in streams (Kaplan and Bott, in press) although algal excretion in plankton systems has received much study. Studies of algal decomposition (DePinto and Veroff, 1977; Mills and Alexander, 1974; Sudo et al., 1978) and macrophyte decay (Godshalk and Wetzel, 1978) have been done but more work with periphyton and more in situ studies are needed to quantitatively assess this fate in an ecosystem framework. Feeding studies to evaluate the relative importance of different dietary items for consumers are required as exemplified in the studies of Benke and Wallace (1980; see also Cummins and Klug, 1979). Studies of the autotroph-meiofaunal linkage may reveal an important route by which primary productivity is linked with macrofaunal productivity. Greater consideration of the fates of autotrophic biomass will enhance our appraisal of the importance of autotrophy in streams.

ACKNOWLEDGMENTS

The River Continuum data was gathered under NSF Grant No. BMS-75-0733. The Boyer Research Endowment Fund provided additional support for T.L.B. Reviews of earlier drafts by C.D. McIntire, G.W. Minshall and R.G. Wetzel are gratefully acknowledged. This is contribution number 21 of the River Continuum project.

REFERENCES

Allen, T. F. H. 1979. Scale in microscopic algal ecology: a neglected dimension. Phycol. 16:253-257.
Axler, R. P., G. W. Redfield and C. R. Goldman. 1981. The importance of regenerated nitrogen in a subalpine lake. Ecology 62:345-354.
Barsdate, R. J., R. T. Prentki and T. Fenchel. 1974. Phosphorus cycle of model ecosystems: significance for decomposer food chains and effect of bacterial grazers. Oikos 25:239-251.

Benke, A. C. and J. B. Wallace. 1980. Trophic basis of
    production among net-spinning caddisflies in a southern
    Appalachian stream. Ecology 61:108–118.
Bormann, F. H. and G. E. Likens. 1979. Catastrophic disturbance
    and the steady state in northern hardwood forests. Am. Sci.
    67:660–669.
Bott, T. L. and F. P. Ritter. 1981. Benthic algal productivity
    in a Piedmont stream measured by ^{14}C and dissolved oxygen
    change procedures. J. Freshwater Ecol. 1:267–278.
Bott, T. L. and K. Rogenmuser. 1980. Fungal pathogen of
    Cladophora glomerata. Appl. Environ. Microbiol. 40:977–980.
Brock, J. T. 1981. Annual metabolism of a desert stream-segment
    ecosystem: Rock Creek, Idaho. M.S. Thesis. Idaho State
    Univ., Pocatello.
Brylinsky, M. and K. H. Mann. 1973. An analysis of factors
    governing productivity in lakes and reservoirs. Limnol.
    Oceanogr. 18:1–14.
Busch, D. E. and S. G. Fisher. 1981. Metabolism of a desert
    stream. Freshwat. Biol. 11:301–307.
Canter, H. M. and J. W. G. Lund. 1948. Studies on plankton
    parasites. I. Fluctuations in the number of Asterionella
    formosa Hass. in relation to fungal epidemics. New Phytol.
    47:238–261.
Chang, W. Y. B. 1981. Path analysis of factors affecting
    primary productivity. J. Freshwater Ecol. 1:113–120.
Collins, M. 1978. Algal toxins. Microbiol. Rev. 42:725–746.
Cummins, K. W. 1974. Structure and function of stream
    ecosystems. BioScience 24:631–641.
Cummins, K.W. and M.S. Klug. 1979. Feeding ecology of stream
    invertebrates. Ann. Rev. Ecol. Systematics 10:149–172.
Cushing, C. E. and E. G. Wolf. 1982. Organic energy budget
    of Rattlesnake Springs, Washington. Am. Mid. Natur.
    107:404–407.
Dawson, F. H. and V. Kern-Hansen. 1979. The effect of natural
    and artificial shade on the macrophytes of lowland streams
    and the use of shade as a management technique. Int. Rev.
    ges. Hydrobiol. 64:437–455.
DePinto, J. V. and F. H. Veroff. 1977. Nutrient regeneration
    from aerobic decomposition of green algae. Environ. Sci.
    Tech. 11:371–377.
Devol, A. H. and R. C. Wissmar. 1978. Analyses of five North
    American lake ecosystems. V. Primary production and
    community structure. Verh. Internat. Verein. Limnol.
    20:581–586.
Duffer, W. and T. C. Dorris. 1966. Primary productivity in a
    southern Great Plains stream. Limnol. Oceanogr. 11:143–151.
Edmondson, W. T. 1970. Phosphorus, nitrogen, and algae in Lake
    Washington after diversion of sewage. Science 169:690–691.
Edmondson, W. T. 1975. Lake Washington. Report to U.S. E.P.A.,
    Environmental Research Laboratory, Corvallis.

Elwood, J. W. and D. J. Nelson. 1972. Periphyton production and grazing rates in a stream measured with ^{32}P material balance method. Oikos 23:295-303.

Elwood, J. W., J. D. Newbold, R. V. O'Neill and W. VanWinkle. In press. Resource spiralling: an operational paradigm for analyzing lotic ecosystems. In: T. D. Fontaine, III, and S. M. Bartell (eds.), The dynamics of lotic ecosystems. Ann Arbor Science, Ann Arbor, Michigan.

Elwood, J. W., J. D. Newbold, A. F. Trimble, and R. W. Stark. 1981. The limiting role of phosphorus in a woodland stream ecosystem: Effects of P enrichment on leaf decomposition and primary procedures. Ecology 62:146-158.

Eppley, R. W. 1981. Autotrophic production of particulate matter, pp. 343-361. In: A. R. Longhurst (ed.), Analysis of marine ecosystems. Academic Press, London.

Falkowski, P. G. (ed.). 1980. Primary productivity in the sea. Plenum Press, New York.

Ferris, J. J. and N. L. Clesceri. 1975. A description of the trophic status and nutrient loading for Lake George, New York. Report to U.S. E.P.A., Environmental Laboratory, Corvallis.

Findenegg, I. 1965. Relationship between standing crop and primary productivity, pp. 271-289. In: C. R. Goldman (ed.), Primary production in aquatic environments. Mem. Ist. Ital. Idrobiol., 18 Suppl., Univ. California Press, Berkeley.

Fisher, S. G. and S. R. Carpenter. 1976. Ecosystem and macrophyte primary production of the Fort River, Massachusetts. Hydrobiologia 47:175-187.

Fisher, S. G. and G. E. Likens. 1973. Energy flow in Bear Brook, New Hampshire: An integrative approach to stream ecosystem metabolism. Ecol. Monogr. 43:421-439.

Flemer, D. A. 1970. Primary productivity of the north branch of the Raritan River, New Jersey. Hydrobiologia 35:273-296.

Flint, R. W. and C. R. Goldman. 1975. The effects of a benthic grazer on the primary productivity of the littoral zone of Lake Tahoe. Limnol. Oceanogr. 20:935-944.

Gessel, S. P. and D. W. Cole. 1965. Movement of elements through a forest soil as influenced by tree removal and fertilizer addition, pp. 95-105. In: C. T. Youngberg (ed.), Forest soil relationships in North America. Second N. Am. Forest Soils Conf.

Glooschenko, W. A., J. E. Moore, M. Munawar, and R. A. Vollenweider. 1974. Primary production in Lakes Ontario and Erie: a comparative study. J. Fish. Res. Board Can. 31:253-263.

Godshalk, G. L. and R. C. Wetzel. 1978. Decomposition of aquatic angiosperms. II. Particulate components. Aquatic Bot. 5:301-327.

Goldman, C. R. 1975. Trophic status and nutrient loading for
    Lake Tahoe, California-Nevada. Report to U.S. E.P.A.,
    Environmental Research Laboratory, Corvallis.
Gosz, J. R. 1978. Nitrogen inputs to stream water from forests
    along an elevational gradient in New Mexico. Water Res.
    12:725-734.
Granhall, U. 1972. Aphanizomenon flos-aquae: Infection by
    cyanophages. Physiol. Plant. 26:332-335.
Hall, C. A. S. and R. Moll. 1975. Methods of assessing aquatic
    primary productivity, pp. 19-53. In: H. Lieth and R. H.
    Whittaker (eds.), Primary productivity of the biosphere.
    Springer-Verlag, New York.
Hameedi, M. J. 1976. An evaluation of the effects of
    environmental variables on marine plankton primary
    productivity by multivariate regression. Int. Rev. ges.
    Hydrobiol. 61:529-550.
Hargrave, B. T. 1970. The utilization of benthic microflora by
    Hyalella azteca (Amphipoda). J. Anim. Ecol. 39:427-437.
Hetling, L. J., T. E. Harr, G. W. Fuhs and S. P. Allen. 1975.
    North American Project Trophic status and nutrient budget of
    Canadarago Lake. New York State Dept. of Environ.
    Conservation Tech. Report. No. 34.
Horner, R. R. and E. B. Welch. 1981. Stream periphyton
    development in relation to current velocity and nutrients.
    Can. J. Fish. Aquatic Sci. 38:449-457.
Hunding, C. and B. T. Hargrave. 1973. A comparison of benthic
    microalgal production measured by ^{14}C and oxygen methods.
    J. Fish. Res. Bd. Can. 30:309-312.
Hunter, R. D. 1980. Effects of grazing on the quantity and
    quality of freshwater aufwuchs. Hydrobiologia 69:251-259.
Hutchinson, G. L. and F. G. Viets, Jr. 1969. Nitrogen
    enrichment of surface water by absorption of ammonia
    volatilized from cattle feedlots. Science 166:514-515.
Hynes, H. B. N. 1963. Imported organic matter and secondary
    productivity in streams. Proc. XVI Internat. Congr. Zool.
    3:324-329.
Johannes, R. E. 1965. Influence of marine protozoa on nutrient
    regeneration. Limnol. Oceanogr. 10:434-442.
Kaplan, L. A. and T. L. Bott. In press. Diel fluctuations of
    DOC generated by algae in a piedmont stream. Limnol. and
    Oceanogr.
Kitchell, J. F., R. V. O'Neill, D. Webb, G. W. Gallepp, S. M.
    Bartell, J. F. Koonce, and B. S. Ausmus. 1979. Consumer
    regulation of nutrient cycling. BioScience 29:28-34.
Kobayasi, H. 1961. Productivity in sessile algal community of
    Japanese mountain river. Bot. Mag. Tokyo 74:331-341.

Koslucher, D. G. and G. W. Minshall. 1973. Food habits of some
    benthic invertebrates in a northern cool-desert stream (Deep
    Creek, Curlew Valley, Idaho-Utah). Trans. Am. Micros. Soc.
    92:441-452.
Liao, C. F.-H. and D. R. S. Lean. 1978. Nitrogen transformation
    within the trophogenic zone of lakes. J. Fish. Res. Board
    Can. 35:1102-1108.
Likens, G. E., F. H. Bormann, N. M. Johnson, D. W. Fisher, and R.
    S. Pierce. 1970. Effects of forest cutting and herbicide
    treatment on nutrient budgets in the Hubbard Brook watershed
    ecosystem. Ecol. Monogr. 40:23-47.
Likens, G. E. 1975. Primary productivity of inland aquatic
    ecosystems, pp. 185-202. In: H. Lieth and R. H. Whittaker
    (eds.), Primary productivity of the biosphere.
    Springer-Verlag, New York.
Malone, T. C. 1977. Environmental regulation of phytoplankton
    productivity in the lower Hudson estuary. Est. Coast. Mar.
    Sci. 5:157-171.
Manuel, C. Y. and G. W. Minshall. 1980. Limitations on the use
    of microcosms for predicting algal response to nutrient
    enrichment in lotic systems, pp. 645-667. In: J. P. Giesy,
    Jr. (ed.), Microcosms in ecological research. U.S. Dept. of
    Energy, Washington, D.C.
Marker, A. F. H. 1976. The benthic algae of some streams in
    southern England. II. The primary production of the
    epilithon in a small chalk-stream. J. Ecol. 64:359-373.
Marra, J. 1978. Effect of short-term variation in light
    intensity on photosynthesis of a marine phytoplankter: a
    laboratory simulation study. Mar. Biol. 46:191-202.
Martin, C. W. 1979. Precipitation and streamwater chemistry in
    an undisturbed forested watershed in New Hampshire. Ecology
    60:36-42.
Masters, M. J. 1971. The ecology of Chytridium deltanum and
    other fungus parasites on Oocystis spp. Can. J. Bot
    49:75-87.
McIntire, C. D. 1973. Periphyton dynamics in laboratory
    streams: a simulation model and its implications. Ecol.
    Monogr. 43:399-420.
Mills, A.L. and M. Alexander. 1974. Microbial decomposition of
    species of freshwater planktonic algae. J. Environ. Qual.
    3:423-428.
Minshall, G. W. 1978. Autotrophy in stream ecosystems.
    BioScience 28:767-771.
Minshall, G. W., R. C. Petersen, K. W. Cummins, T. L. Bott, J. R.
    Sedell, C. E. Cushing, and R. L. Vannote. 1983. Interbiome
    comparison of stream ecosystem dynamics. Ecol. Monogr.
    53:1-25.
Moore, J. W. 1975. The role of algae in the diet of Asellus
    aquaticus L. and Gammarus pulex L. J. Animal Ecol.
    44:714-730.

Morrison, S. J. and D. C. White. 1980. Effects of grazing by
    estuarine gammaridean amphipods on the microbiota of
    allochthonous detritus. Appl. Environ. Microbiol.
    40:569-671.

Moss, B., R. G. Wetzel, and G. H. Lauff. 1980. Annual
    productivity and phytoplankton changes between 1969 and 1974
    in Gull Lake, Michigan. Freshwat. Biol. 10:113-121.

Naiman, R. J. 1976. Primary production, standing stock and
    export of organic matter in a Mohave Desert thermal stream.
    Limnol. Oceanogr. 21:60-73.

Naiman, R. J. and J. R. Sedell. 1980. Relationships between
    metabolic parameters and stream order in Oregon. Can. J.
    Fish. Aquat. Sci. 37:834-847.

Newbold, J. D., J. W. Elwood, R. V. O'Neill, and W. VanWinkle.
    1981. Nutrient spiralling in streams: The concept and its
    field measurement. Can. J. Fish. Aquat. Sci. 38:860-863.

Nicotri, M. E. 1977. Grazing effects of four marine intertidal
    herbivores on the microflora. Ecology 58:1020-1032.

Odum, E. G. 1971. Fundamentals of Ecology. W. B. Saunders,
    Philadelphia.

Odum, H. T. 1957. Trophic structure and productivity of Silver
    Springs, Florida. Ecol. Monogr. 27:55-112.

Oglesby, R. T. 1978. The limnology of Cayuga Lake, p. 1-120.
    In: J. A. Bloomfield (ed.), Lakes of New York State, V. 1.
    Ecology of the Finger Lakes, Academic Press. New York.

O'Neill, R. V. 1976. Ecosystem persistence and heterotrophic
    regulation. Ecology 57:1244-1253.

O'Neill, R. V. and D. E. Reichle. 1979. Dimensions of ecosystem
    theory, pp. 11-26. In: R. H. Waring (ed.), Forests: Fresh
    perspectives from ecosystem analysis. Proc. 40th An. Biol.
    Colloquium. Oregon State Univ. Press, Corvallis.

Patrick, R. 1970. Benthic stream communities. Am. Sci.
    58:546-549.

Pennak, R. W. and J. W. Lavelle. 1979. In situ measurements of
    net primary productivity in a Colorado mountain stream.
    Hydrobiologia 66:227-235.

Pfeifer, R. F. and W. F. McDiffett. 1975. Some factors
    affecting primary productivity of stream riffle communities.
    Arch. Hydrobiol. 75:306-317.

Phinney, H. K. and C. D. McIntire. 1965. Effect of temperature
    on metabolism of periphyton communities developed in
    laboratory streams. Limnol. Oceanogr. 10:341-34.

Porter, K. G. 1976. Enhancement of algal growth and
    productivity by grazing zooplankton. Science 192:1332-1334.

Porter, K. G. 1978. The plant-animal interface in freshwater
    systems. Am. Sci. 65:159-170.

Powers, C. F., W. D. Sanville, and F. S. Stay. 1975. Waldo
    Lake, Oregon. Report to U.S. E.P.A., Environmental Research
    Laboratory, Corvallis.

Rast, W. and G. F. Lee. 1978. Summary analysis of the North American (U.S. Portion) OECD Eutrophication Project: Nutrient loading – Lake response relationships and trophic state indices. E.P.A. Rept. No. 600/3-78-008. U.S. E.P.A., Corvallis.

Redhead, K. and S. J. L. Wright. 1978. Isolation and properties of fungi that lyse blue green algae. Appl. Environ. Micorbiol. 35:962-969.

Reichle, D. E., R. V. O'Neill, and W. F. Harris. 1975. Principles of energy and material exchange in ecosystems, pp. 27-43. In: W. H. vanDobben and R. H. Lowe – McConnell (eds.), Unifying concepts in ecology. W. Junk, The Hague.

Revsbech, N. P., B. B. Jorgensen, and O. Brix. 1981. Primary productivity of microalgae in sediments measured by oxygen microprofile, $H^{14}CO_3$-fixation and oxygen exchange methods. Limnol. Oceanogr. 26:717-730.

Ricklefs, R. E. 1979. Ecology, 2nd edition. Chiron, New York.

Safferman, R. S. and M. E. Morris. 1967. Observations on the occurrence, distribution, and seasonal incidence of blue-green algal viruses. Appl. Microbiol. 15:1219-1222.

Schindler, D. W. and E. J. Fee. 1974. Experimental lakes area: Whole-lake experiments in eutrophication. J. Fish. Res. Bd. Can. 31:937-953.

Schindler, D. W., V. E. Frost and R. V. Schmidt. 1973. Production of epilithophyton in two lakes of the Experimental Lakes Area, northwestern Ontario. J. Fish. Res. Bd. Can. 30:1511-1524.

Steinberg, C. 1978. Freitsetzung gelosten organischen Kohlenstoffs (DOE) verschiedener Molekulgroben in Planktongesellachaften. Arch. Hydrobiol. 82:155-165.

Stockner, J. A. and K. R. S. Shortreed. 1976. Autotrophic production in Carnation Creek, a coastal rainforest stream on Vancouver Island, British Columbia. J. Fish. Res. Board Can. 33:1553-1563.

Storch, T. A. and G. A. Dietrich. 1979. Seasonal cycling of algal nutrient limitation in Chatauqua Lake, New York. J. Phycol. 15:399-405.

Storch, T. A. and G. W. Saunders. 1975. Estimating daily rates of extracellular dissolved organic carbon release by phytoplankton populations. Verh. Internat. Verein. Limnol. 19:952-958.

Strahler, A. N. 1957. Quantitative analysis of watershed geomorphology. Trans. Am. Geophys. Union 38:913-920.

Sudo, R., H. Ohtake, S. Aiba and T. Mori. 1978. Some ecological observations on the decomposition of periphytic algae and aquatic plants. Water Res. 12:179-184.

Sumner, W. T. and S. G. Fisher. 1979. Periphyton production in Fort River, Massachusetts. Freshwat. Biol. 9:205-212.

Sumner, W. T. and C. D. McIntire. 1982. Grazer - periphyton interactions in laboratory streams. Arch. Hydrobiol. 93:135-157.

Swan, L. A. 1963. Aeolian zone. Science 140:77-78.

Swanson, C. D. and R. W. Bachmann. 1976. A model of algal exports in some Iowa streams. Ecology 57:1076-1080.

Swanson, F. J. 1979. Geomorphology and ecosystems, pp. 159-170. In: R. H. Waring (ed.), Forests: Fresh perspectives from ecosystem analysis. Proc. 40th Annual Biol. Colloquium. Oregon State Univ. Press, Corvallis.

Tett, P., C. Gallegos, M. G. Kelly, G. M. Hornberger and B. J. Cosby. 1978. Relationships among substrate, flow, and benthic microalgal pigment density in the Mechums River, Virginia. Limnol. Oceanogr. 23:785-797.

Triska, F. J., V. C. Kennedy, R. J. Avanzino, and B. N. Reilly. In press. Effect of simulated canopy cover on regulation of nitrate uptake and primary production by natural periphyton assemblages. In: T. D. Fontaine, III and S. M. Bartell (eds.), The dynamics of lotic systems. Ann Arbor Press, Ann Arbor, Michigan.

Vannote, R. L., G. W. Minshall, K. W. Cummins, J. R. Sedell and C. E. Cushing. 1980. The river continuum concept. Can. J. Fish. Aquat. Sci. 37:130-137.

Vitousek, P. M., J. R. Gosz, C. C. Grier, J. M. Melillo, W. A. Reiners and R. L. Todd. 1979. Nitrate losses from disturbed ecosystems. Science 204:469-474.

Webster, J. R., J. B. Waide and B. C. Patten. 1975. Nutrient cycling and the stability of ecosystems, pp. 1-27. In: F. G. Howell, J. B. Gentry and M. H. Smith (eds.), Mineral cycling in Southeastern ecosystems. ERDA Symp. Ser. CONF-740513. Techn. Information Center, Office of Public Affairs.

Welch, E. B., T. Wiederholm, D. E. Spyrdakis and C. A. Rock. 1975. Nutrient loading and trophic state of Lake Sammamish, Washington. Report to U.S. E.P.A., Environmental Research Laboratory, Corvallis.

Wetzel, R. G. 1964. A comparative study of the primary productivity, higher aquatic plants, periphyton, and phytoplankton in a large shallow lake. Internat. Rev. ges Hydrobiol. 49:1-61.

Wetzel, R. G. 1975a. Primary production, pp. 230-247. In: B. A. Whitton (ed.), River ecology. Univ. California Press, Berkeley.

Wetzel, R. G. 1975b. Limnology. Saunders, Philadelphia.

Whitford, L. A. and G. L. Schumacher. 1964. Effect of a current on respiration and mineral uptake in Spirogyra and Oedogonium. Ecology 45:168-170.

Whittaker, R. H. and G. E. Likens. 1973. Carbon in the biota, pp. 281-301. In: G. M. Woodwell and E. V. Pecan (eds.), Carbon in the biosphere. A.E.C. Techn. Information Center, Washington.

Whittaker, R. H. and P. L. Marks. 1975. Methods of assessing terrestrial productivity, pp. 55-118. In: H. Lieth and R. H. Whittaker (eds.), Primary productivity in the biosphere. Springer-Verlag, New York.

Wong, S. L. and B. Clark. 1976. Field determination of the critical nutrient concentration for Cladophora in streams. J. Fish. Res. Bd. Can. 33:85-92.

Wong, S. L., B. Clark, M. Kirby and R. F. Kosciew. 1978. Water temperature fluctuations and seasonal periodicity of Cladophora and Potomogeton in shallow rivers. J. Fish. Res. Bd. Can. 35:866-870.

Woodwell, G. M., R. H. Whittaker, W. A. Reiners, G. E. Likens, C. C. Delwiche and D. B. Botkin. 1978. The biota and the world carbon budget. Science 199:141-146.

Wuhrmann, K. 1974. Some problems and perspectives in applied limnology. Mitt. Internat. Verein. Limnol. 20:324-402.

DENSITY INDEPENDENCE VERSUS DENSITY DEPENDENCE IN STREAMS

Dennis K. Shiozawa

Department of Zoology
Brigham Young University
Provo, Utah 84602

INTRODUCTION

In the past 40 years stream ecology has been largely a
descriptive science. Taxonomy has been a major thrust. Stream
researchers have described the basic structure of stream
communities and the life histories of some component species have
been worked out. Enough information has now been amassed for
researchers to examine functional relationships within the stream
community. The formalization of the functional group concept
(Cummins, 1974), the River Continuum Concept (Vannote et al.,
1980), the recognition of the importance of detritus (Cummins,
1974; Fisher and Likens, 1972), and the importance of instream
autotrophy (McIntire, 1973; Minshall, 1978) all represent major
advances in our understanding of stream community structure.
Yet, few studies have attempted to evaluate density dependence in
structuring the organismal associations in stream communities.

Stream systems are characterized by annual floods,
occasional catastrophic floods, pulsed periods of detrital input
(Fisher and Likens, 1972, 1973; Minshall, 1967) and seasonal
temperature changes. Streams are open systems. Material at a
given site can be lost downstream while additional material can
be imported from upstream sources. The bed of the stream itself
is subject to both erosion and deposition, for instance, rocks
located on a given riffle may move downstream to other riffles
during flood periods (Leopold et. al., 1964). Are the
communities existing in such conditions, equilibrium communities,
controlled and structured to a large extent by biotic
interactions? Or are they simply concentrations of opportunistic
species existing only because the appropriate physical conditions

are present, and perturbations have not yet eliminated them?
This question is basically one concerning the relative roles of
density dependent versus density independent mechanisms in
structuring lotic communities (Shiozawa, 1978).

The objective of this paper is not to advocate density
dependence or independence, but to discuss the concepts in light
of what is now known of stream systems. I also will elucidate
some promising avenues where testing for the existence of density
dependence is possible. We need to know whether density
dependent factors, such as competition, exist, and also whether
such factors are intense enough to have a significant influence
on population levels and community structure.

DENSITY INDEPENDENCE AND DENSITY DEPENDENCE DEFINED

Density dependence and density independence are two major
classes of population regulating mechanisms (Smith, 1935).
Density independent mechanisms are catastrophic mechanisms which
act independently of the density of the population being
affected. For instance, a spate or sudden freeze may eliminate a
high proportion of a population regardless of whether the initial
population density was 100 or 100,000. Under such regulation,
the population could be pictured as growing exponentially until
some catastrophic event reduces its numbers. Once the event has
passed, the population again resumes exponential growth. The
population is never controlled by density dependent mechanisms
because frequent catastrophies prevent it from obtaining a level
where such mechanisms could operate.

Density dependent mechanisms are those which change in
intensity as a function of population density. As the population
increases, the influence of density dependent mechanisms also
increases. At some point, the reduction in population growth due
to density dependent factors equals the population's intrinsic
rate of growth. The net population growth is zero. This point
is an equilibrium. The carrying capacity, K, in the logistic
growth curve, is an example of such an equilibrium condition.
Density dependent mechanisms include factors such as competition,
predation, and parasitism. Even physical factors can be included
if the existence of suitable refuges are important. The concept
that density dependence could regulate populations has been
repeatedly attacked (e.g., Andrewartha and Birch, 1954) and
defended (e.g., Soloman, 1958; Tanner, 1966; see also Horn,
1968). Its status is still controversial (Schoener, 1982),
although many ecologists support its existence (e.g., Krebs,
1978; Ricklefs, 1979).

The implications of the mode of density determination in
streams extend beyond whether a population is controlled by food

supplies or floods. It has connotations concerning reproductive strategies, diversity, and the general applicability of equilibrium based ecological theory. On a more applied scale, density dependence has implications for management strategies, and impact assessment and prediction. Assemblages of opportunists should react very differently to the extinction of various species than would equilibrium communities, especially if those species eliminated played a key role (e.g., Paine, 1966, 1969) in structuring the community itself.

The existence of density dependent factors in stream systems has been inferred by several investigators. Macan (1962) noted the possible impact of predaceous flatworms in restructuring a stream community. Edington (1965) discussed the factors influencing the distribution of lotic Trichoptera. While physical factors set the general distributional template, he felt territorial behavior was the ultimate regulating mechanism. Similar mechanisms may maintain the uniform spacing of larval Simuliidae on the upper surface of stones (Wiley and Kohler, 1981). Allan (1975) in a study of Cement Creek, Colorado noted that cogeneric species with similar microhabitat requirements were separated vertically (elevationally), while those with similar elevational distributions had different microhabitats. He concluded that this evidence supported the competitive exclusion principle. Microhabitat segregation also was found to occur between two cogeneric harpacticoid copepods, Bryocamptus vejdorskyi and B. zschokkei in Valley Creek, Minnesota (Shiozawa, 1978).

A few investigators have attempted to experimentally evaluate various density dependent mechanisms in stream systems. Peckarsky (1979) manipulated invertebrate densities in cages and concluded that movements were density dependent (see also Peckarsky, 1981; Sell, 1981). McAuliffe (this volume) and Hart (this volume) both present evidence of competitive interactions among benthic grazers through the use of field experiments. Oberndorfer et al. (unpublished manuscript), using caged leaf packs with varied predator densities, found significant differences in leaf pack decomposition rates. Perhaps the best known working hypothesis concerning density dependence in streams is the excess production hypothesis (Waters 1961, 1965, 1966) which relates behavioral drift to assimilation and secondary production above the carrying capacity of the environment. This concept, as we will see later, has some important implications and sets some key guidelines for future research on density dependence (see also Muller, 1954; Waters, 1972). Hildebrand (1974) manipulated food level in artificial streams and found significantly more drift with lower food concentrations. This tends to support the excess production hypothesis, but other experimental studies have failed to produce conclusive results.

Allan (this volume) did not detect any significant change in
stream benthos after trout were excluded from a section of a
Colorado stream. A similar study (Reice, 1980) excluded fish
from sections of a North Carolina stream, and no significant
effect of exclusion was noted.

Many stream researchers support density independent
regulation as the major factor regulating organisms in stream
systems. Hynes (1970) discusses factors controlling benthic
invertebrates in running waters. About 90% of his discussion
relates to density independent mechanisms. Margalef (1960)
states that "no community of running water may be considered as
climax, but as transitory or more or less permanent stages of
seres...." A climax community implies equilibrium conditions
controlled by biotic interactions. He emphasizes physical
factors, such as the "violence of flow," as the main regulatory
mechanisms. Patrick (1972) portrays headwater stream communities
as assemblages of opportunists. Stout and Vandermeer (1975)
proposed opportunists as being the inhabitants of headwater
streams in temperate areas (but see Fox, 1977), although they did
portray tropical stream systems as being equilibrium communities.
These and other investigators support density independence
largely due to the seasonal scouring in temperate streams.
Perturbations are assumed to occur frequently enough to force the
community into a constant state of immaturity.

THE DOMINANCE OF DENSITY INDEPENDENCE--IS IT VALID?

While neither density dependence nor density independence
have been unequivocally proven to predominate in stream systems,
most published information (as noted above) implies that density
independence is the dominant regulating factor. Four possible
reasons for this tendency will be discussed.

First, density independence may be the major factor
regulating stream community structure. Many examples of density
independent events exist (see Hynes, 1972). In my studies of
stream systems, I have observed the effects of three catastrophic
floods which caused major restructuring of the stream ecosystems.
One was a snow melt flood associated with an ice jam. Its
effects were localized, and the community several kilometers
downstream had minimal disturbance (Shiozawa, personal
observation). The other two were caused by summer cloud bursts.
Undoubtedly, these perturbations caused catastrophic changes in
community structure, but the results were very different from the
usual spring flood associated with snow melt. What happens when
communities are not disturbed by massive spring perturbations
over several years time? Do populations continue to increase in
numbers or do density dependent mechanisms intervene? Our
knowledge of the long-term dynamics of a single stream community

is essentially nonexistent and density dependence, like density independence, need not operate continually (MacArthur, 1972; Wiens, 1977). There is no valid basis on which to reject the operation of density dependent mechanisms outright.

A second reason favoring the explanation of density independent regulatory mechanisms is the high diversity that exists in temperate streams. Diversity may develop from a patchy environment induced by density independent factors. The high diversity in temperate streams was one of the few major contradictions noted by MacArthur (1972) when comparing the diversity of the tropics to temperate areas. This has been attributed to the opening of new habitats by perturbations. These openings would give competitively inferior species refuges in a community, thus increasing overall community diversity. The disturbance must act at an intermediate level, since intense disturbances would completely remove all members of the community (e.g., Menge and Sutherland, 1976). Connell (1978, 1979) has hypothesized that intermediate disturbances are important factors in controlling diversity in both tropical coral reefs and tropical forests (see Stanford and Ward, this volume for application of this idea to streams). If intermediate level disturbances create conditions that result in the high diversity of stream systems, several questions should be considered. A basic precept of Connell (1979) is that such systems are in the process of moving towards (undergoing succession) a lower diversity climax, but disturbances are continually setting succession back by opening new, uncolonized cells or patches.

At what level should the intermediate disturbance operate in stream systems? Is spring flooding the disturbance which sets the community back? Does the cell or patch size extend down to a single stone, such that within a riffle, randomly moving or disturbed stones constitute the opening of new cells? If the most influential cell size is the entire stream, then researchers must address the impact of various degrees of perturbation in the recolonization of the stream and its subsequent succession towards a low diversity climax. If this scale is the most influential one, then streams should exist in many degrees of succession even within a given drainage since not all streams will be subject to perturbations of equivalent intensity. If the predominantly perturbed cell size is a single stone, then rapid recolonization of disturbed stones is possible from surrounding unperturbed cells. At this resolution, patches should continually be opening or the system will rapidly decline in diversity. Will stabilization of a stone (protection from perturbation) result in a loss of diversity? Adjustments must also be made for the life history and growth of cohorts of aquatic insects. Do insects which always hatch after major seasonal (= annual) perturbations (such as spring runoff) represent species which have adjusted to the frequency of those perturbations, and if so, are the events really perturbations?

A third reason for the support of density independence is the highly variable distribution of stream benthos. This line of reasoning contends that the variability will render experimental data useless, so by default only correlative (sampling) field relationships are usable. Correlative relationships tend to support the role of physical factors (discussed below). The study by Needham and Usinger (1956) in Prosser Creek, California is the most familiar example. They sampled a "relatively uniform riffle" and found that to estimate community density with 95% confidence, 73 surber samples would be required. Processing this number of quantitative samples would be beyond the limits of all but the best budgeted aquatic ecologists. Later investigators have improved the quantitative nature of sampling gear (e.g., Waters and Knapp, 1961; Mundie, 1971), and others have refined our statistical methodology (e.g., Elliott, 1977). Yet, contagious distributions remain a problem in both sampling and the application of parametric statistics to the analysis of data. In short, answers to many questions are not obtainable because we cannot obtain precise estimates of numbers, etc., with a reasonable sampling effort.

Lotic sampling concepts and methodology must be reevaluated. Benthic sampling relies heavily on plot type sampling techniques. The standard area of the Surber sampler, one square foot, has little biological basis, at least as far as the benthos is concerned. Stream researchers should consider changing the size and shape of sampling gear as plant ecologists have done (e.g., Muller-Dombois and Ellenberg, 1974). Selecting sampling gear types which will best answer a particular question (e.g., Gray, 1971) should be the rule rather than simply utilizing a standard sampling gear type (Mundie, 1971). For instance, increasing the area sampled will tend to average heterogeneous (patchy) conditions and the precision of the data may improve substantially. The processing of samples in the laboratory can be simplified by subsampling since both the statistics (e.g., Cochran, 1963; Green, 1979; Stewart, 1968; Sudman, 1976) and equipment (e.g., Waters, 1969; Mundie, 1971) have been developed. Conversely, the options of plotless techniques (Shiozawa, 1978) or of reduced sampler area so as to improve resolution of habitat grain should be considered (Elliott, 1977). These are just a few examples of how sample variance or noise in stream sampling might be reduced. The high variance in stream systems may not be as important as it currently seems to be.

A fourth reason for support of density independence as the dominant community regulatory mechanism is that stream ecology has been predominantly a descriptive science. The basic approach to the study stream systems has been the sample survey. Sample numbers are converted to densities or biomass and from this a picture of stream community structure is obtained. Such a picture is useful but often leaves the investigator with little more than data lists. To remedy this problem, the researcher attempts to find determinants of the structure. The advent of computers and powerful multivariate statistical packages has solved part of the problem. Enormous amounts of data can be collected and analyzed. The researcher thus obtains statistical results from which to base his inferences. With the backing of complex statistical methodology, these data sets gain more validity, and interpretations obtain more legitimacy.

Two basic problems arise with this methodology. First, we select the parameters to measure through intuition, which is biased by our own limited ability in perceiving all potential causal factors. We may have the ability to discern physical variations on a microscale, but we often cannot connect these variations with species assemblages at that same time. Instead, we usually take samples and analyze them later when the time and equipment is available. This results in a tendency to segregate stream ecosystems into abiotic (often field observed) and biotic (often laboratory derived) components. When the data are analyzed, this separation is retained. The abiotic data sets are then treated as the independent variables, while the biotic data sets are treated as dependent variables. This aids interpretations of causality since it is easily stated for instance, that a species is commonly found in fast current with large substrate sizes. Thus, the physical habitat becomes the factor to which the organisms are assumed to be adjusted. If such studies are included when reviewing the literature for support of either density independent or density dependent mechanisms, the review will strongly favor density independence. This is based on the prediction of an organisms presence or abundance with physical factors and not species associations.

The second problem is that data from sampling surveys cannot prove causality. Regardless of the complexity of the statistical technique used, when the data are from field samples, only correlative associations can be determined. Even high degrees of association do not prove causality. For example, if the density of a mayfly was regressed with rock size and a linear relationship resulted, it implies that rock size is the determining factor for mayfly abundance. Water velocity could be the actual factor, or any of a number of other variables or combinations of variables that were not measured. Our interpretation is limited to the factors measured and entered into the regression. Alternatively, given the high association between mayfly abundance and rock size, one could regress rock size as the dependent variable with mayfly abundance as the

independent variable. The resulting equation would say that the
size of the rock is determined by the abundance of mayflies.
Statistically, this relationship is just as valid as the first
although logically it is unreasonable. Basically we tend to
transpose predictability with causality. Thus, while rock size
in the above example allows an accurate prediction of mayfly
abundance, it does not necessarily prove that rock size causes
the abundance of mayflies observed. Statistics based on sample
surveys do not allow us to make definitive conclusions about
causality.

TESTING STREAM SYSTEMS FOR DENSITY REGULATING MECHANISMS

How does one go about testing for the existence of density
dependent regulation? Basically, an unbiased evaluation of
stream systems for these interactions require the formulation of
logical hypotheses which can then be tested in the stream system.
Initially, our solution appears simple: formulate a hypothesis,
test it, and make a conclusion based on the result of the test.
A hypothesis may be based on field observations or on precepts
about the behavior of a system or parts of the system. The
formulation of the hypothesis itself can be done in several ways.
Most biologists simply ask the question (hypothesis) and then
test it under experimental conditions. An alternative method is
the use of mathematical modeling. This is an extension of the
previous method, except that the logic has been placed in mathematical terms rather than verbal terms. These models, however,
can give precise predictions and the opportunity of formulating
biological relationships and observing "noiseless" system changes
according to these relationships. Statistics can then be used to
test the adherence of the model.

These hypotheses can be tested either by very extensive
sampling-type surveys or through experimental manipulation
(Amant, 1970). Experimental manipulation is probably the best
approach for stream researchers. In the case of density
dependence, such a hypothesis would usually involve the
manipulation of density, food, or space among the experimental
units. Unfortunately, stream systems have many complex factors.
Because of this, the outcome of research designed to test
hypotheses in streams depends upon the experimenter being able to
discern areas where discrete, non-confounded results can be
obtained. In many attempts at experimental manipulations these
areas have not been clearly defined. This approach requires
controls, replicates, and careful selection of the appropriate
experimental units. If properly conducted, these experiments
will result in definitive conclusions free of most, if not all,
of the objections raised above. Here causality can be
demonstrated.

Formulation of a good hypothesis will be futile unless other conditions are considered. Attempts at evaluating the effects of density dependence at a community level have a high probability of failure, even if density dependence exists. Not all members of the benthic community are regulated by the same factors. Some may be regulated by predation, others by food, space, or density independent factors. By definition, the community will include organisms of different trophic levels as well as different functional groups (Cummins, 1974; Cummins and Klug, 1979) within or across these levels. Thus, in a community the manipulation of one factor may only influence a few species. The other species might show no effect, and their numbers or biomass could mask any actual community changes due to density dependence.

The functional group concept establishes an excellent base for the initial selection of organisms to work with. However, functional designations (e.g., Merritt and Cummins, 1978) are mutable. Organisms may specialize differentially within separate communities (Fox and Morrow, 1981) and thus have different functional roles in those communities. Within a community a given species may perform several functions. For instance, a generalist may feed on algae, which is capable of regeneration quite rapidly (McIntire, 1973) and detritus which is not regenerated as rapidly. To set K in this situation would be difficult if not impossible. Related to this problem is switching (Murdock, 1969) which allows an organism the option of changing diet, often in response to decreasing food supply or changing physiological needs. Such changes in foraging behavior have been recorded in stream systems (Fuller and Mackay, 1980a, 1980b; Rhame and Stewart, 1976; Anderson and Cummins, 1979).

Grazers are among the groups in which the search for density dependence can be most easily conducted. In the Raft River, Idaho, high densities of mountain suckers (<u>Pantosteus platyrhynchus</u>) occurred in some pools and were in low densities in others. The riffles adjacent to pools containing high sucker densities had low filamentous algal cover. However, a highly developed filamentous algal cover occurred in areas adjacent to low fish density pools (Barnes and Shiozawa, 1980). Mountain suckers are grazers (Scott and Crossman, 1973), and it is unlikely that the high biomass residing in some pools could be supported by autotrophy occurring within the pools. It is probable that the fish move into riffle areas to feed and their grazing pressure is sufficient to crop the algae. This grazing is probably a significant component in structuring the local algal communities. In a study of a Northern California stream, Lamberti and Resh (1980) reported the impact of invertebrate grazers on tiles placed in the stream. The exclusion of grazers from the tiles resulted in an increase in algal biomass. Hart (this volume) and McAuliffe (this volume) also utilized grazers when examining competitive interaction.

Grazing studies can be conducted over relatively short periods of time and many confounding factors can be controlled. But what about similar studies made over an entire year or several years? Several problems arise that should be considered. First is the influence of seasonality on density dependent interactions. Factors such as canopy cover, angle of the sun, turbidity, and water temperature will vary seasonally. These can influence the autotroph production rate and instream primary production will not remain constant (Hickman, 1974). If K is set by the primary producers, then it also will change with seasons. These seasonally induced changes in K must be addressed in long-term studies. A second problem is that many aquatic organisms grow in cohorts. The young emerge from eggs at approximately the same time and at any given instant the population will consist of organisms that are close to the same size and age. The cohorts will begin with high numbers of small individuals. With time the numbers diminish owing to various mortality factors, but concurrently the organisms that survive increase in biomass. Over long time periods these cohorts cannot be treated the way continually reproducing populations are treated, at least with the conventional methods of evaluation (numbers and biomass). Simple numerical evaluations of population changes are confounded with biomass changes as the population grows. Biomass is confounded with time. Some of the biomass (standing stock) of a population represents materials assimilated and stored during earlier life stages, which could be a year or more before.

Use of production rates would avoid these complications. This is the significance of the excess production hypothesis (Waters, 1961, 1965, 1966). This hypothesis not only invokes density dependence as a regulating mechanism, but also converts the mechanism into one related to energy flow and production. Respiratory costs of organisms must be met if they are to maintain their weight, and these costs must be exceeded if the organisms are to grow. As individuals produce more biomass the respiratory costs to maintain that biomass can increase. They also grow in weight exponentially. Four first instar larvae will not weigh the same as one fourth instar larva. Thus, the need to produce additional biomass increases with size. To achieve this, the population will increase its assimilative demands on the food base. If these demands exceed the food replacement rate or capacity (K) of the environment (in the case of grazers upon periphyton, this is net primary productivity), then part of the population will be forced to drift in search of additional food sources. This is the excess production. The remaining organisms would then be at or near K. Drift acts as a mortality factor. Not all drifting organisms will die, but the probability of mortality is higher (Allan, 1978; Waters, 1972) for drifters than it is for sedentary benthos. Since the population will continually be growing (assimilating and producing), excess production will continually drift as long as the population is at or approaching K. The diel nature of behavioral drift strongly implies that the total biomass drifting should be a function of the daily production rates when the above conditions are met and K is constant. Continual monitoring of both K and secondary production rates would be necessary in long- term studies. The measurement of primary and secondary production (Edmondson and Winberg, 1971; Vollenweider, 1969; Waters, 1977; Winberg, 1971) is not easy and is normally conducted in unmanipulated field conditions. This area has potential for further research both on theoretical grounds and in terms of making measurements under manipulated field conditions.

The problems of changing K levels will be common in many food limited groups. Most detritus in temperate climates is introduced into the stream during autumn leaf fall. The leaves are colonized by microbial organisms which in turn form the food base for detritivores (Cummins, 1974). The annual pulsing of leaf litter input should also cause the carrying capacities of detritus based food webs to be pulsed. If detrital input is idealized or smoothed over time, a sequential hierarchy of detrital particles in the stream can be hypothesized. The first and most cohesive pulse would represent the fall input of unconditioned leaves. Following this would be a pulse of conditioned leaves, resulting from leaching and the growth of the microflora and fauna. This pulse would have less cohesiveness and less biomass, since it represents leaf material minus the leachate. It would be present for a longer period of time (thus the lower pulse cohesion), disappearing from the system in a negative exponential fashion. With the breakdown of the conditioned leaves a subsequent increase in coarse particulate organic matter (CPOM) would be expected. The CPOM pulse should be lower in magnitude than that of the conditioned leaves, and it should be represented by a greater curve breadth. Beyond this level the sequential series will be less discrete. When CPOM is formed, so too will be medium and fine particulate matter (MPOM and FPOM, respectively) and dissolved organic matter (DOM). The leachate and other DOM materials can floculate (Lock and Hynes, 1975, 1976) and make up a significant portion of the fines in the system (see also Boling et al., 1975; Short and Ward, 1980).

The abundance of finer particulate materials may fluctuate annually (Hynes et al., 1974), but since their sources are varied, they may be difficult to evaluate. The fines may be so abundant that they play no role in population regulation. Anyone who has placed a fine mesh drift net into a stream would agree that the leakage of fine organics out of stream systems appears to be much greater than that captured by filter feeders. Filter feeders may be more limited by competition for attachment sites

(Edington, 1965; Boon, 1979) than by food. Alternatively the
high degree of age and species specificity in the filter net size
(Wallace and Merritt, 1980; Wallace et al., 1977; Malas and
Wallace, 1977) implies some degree of specialization to the food
resources being exploited. Is this specialization induced by
competition for food, or is it merely a byproduct of instar size
and interspecific variation?

It appears, therefore, that the cohesively pulsed leaf
materials are more predictable than the fines. This makes them
an excellent resource to investigate for regulatory influence.
Periods of food scarcity (e.g., Weins, 1977) are most likely to
occur during the periods of decline of the pulses. Organisms
utilizing the pulsed conditioned leaves resource would be the
shredder functional group (Cummins 1974). Here, as discussed
earlier, the cohort growth patterns of stream invertebrates
should be recognized. Since later instars must assimilate more
energy (for the resultant secondary production) than the earlier
instars, selection pressures should favor cohorts where the
larger instars are reached during periods of maximal food
standing stock. At some point, food may be reduced below levels
required to sustain the population's growth, and density
dependence would begin to operate.

But how do we develop a hypothesis that will allow a
correction for a changing K? Mathematical models work well in
this case. For instance, if K is set by leaf pack size, then
with the subsequent decomposition of leaf material, K must also
decline. Simple models can be formulated for such changing
levels of leaf materials, and the associated changes in shredder
community can also be predicted. A series of differential and
difference equation models are currently being developed in my
laboratory with the objective of forming more definitive
hypotheses for testing in stream systems. I will discuss one
model based on whole stream detrital dynamics, but the
mathematics involved in the formulation and testing of it is
beyond the scope of this paper and will not be included.

The model was formulated to predict changes in leaf standing
stock with different storage capacities of the stream system.
The rate of leaf decomposition in a stream is related to factors
such as leaf species, shredder abundance, water temperature, and
location of the leaf material in the stream (Cummins et al.,
1973; Hynes et al., 1974; Kaushik and Hynes, 1971; Petersen and
Cummins, 1974; Reice 1974, 1977; Welton, 1980). If detritus
decomposition and processing is location specific (Meyer, 1980)
then the standing stock of leaf material will be a function of
the diversity of locations. One factor which appears relevant in
determining locational effects is whether the environment is in
an oxidizing or a reducing state. Decomposition will be

significantly slower in reducing environments. Within the stream
these locations can be considered as depositional regions (=
reducing), which tend to be low in oxygen, and erosional regions
(= oxidizing), which tend to be high in oxygen because of
turbulence. For simplicity, I have designated these as pools and
riffles, respectively. While these classifications are
simplistic, the reader must be aware that the rate for pools is
assumed to be the average rate within depositional areas. Some
will have higher rates, others lower rates of breakdown. The
same logic prevails for riffles. A leaf falling into the stream
may either enter a pool or a riffle. Material in either category
can "leak" out of the system, or into the other category. While
the leaf material is in a given location, it will be processed at
the rate appropriate for that category. Storage capacities for
both locations were set with logistic type functions. The model
was then used to simulate streams with differing pool-riffle
ratios.

For this discussion I will center on the two extremes of the
model, 90% riffle and 10% riffle. Initially the simulation was
run with continuous litter input, and the results behave as
anticipated (see Figures 1 and 2). The low riffle ratio (high
pool) stored more material than the high riffle system. However,
litter input is pulsed, not continuous, and our interests lie in
the levels of material resulting in streams with differing
storage capacities. A simple function (Dixon, 1976) was used
which allowed a changing rate of leaf fall (Figure 3). The
simulations were then run with this function incorporated
(Figures 4 and 5). Systems with high proportions of riffles tend
to have a more pulsed leaf litter standing stock in the riffle
areas. That is, the input of leaves from the terrestrial system
and subsequent leakage out of the system tends to dominate.
Systems with low riffle ratios and high pool ratios have less
pulsed leaf litter standing stocks. Here the impact of leakage
from the system is buffered by leakage to the riffle from the
pool storage (Figure 6). This model is still being modified but
data collected from Valley Creek, Minnesota, concurrent with the
initial formulation of the model, support some of the assumptions
of the model. While general predictions are still accommodated,
storage capacities and leakage rates differ more than originally
assumed. The results of adjusting for these variations will
probably increase the differences between different riffle-pool
ratios rather than diminish them. Even without complete
verification of the model, we can begin to discuss some of the
implications relative to density dependence.

Density dependence (relative to shredders) will be more
likely in the high riffle ratio systems. Here leaves for
shredding should become limiting rapidly. Life histories of the
shredding components should be adjusted to quickly taking

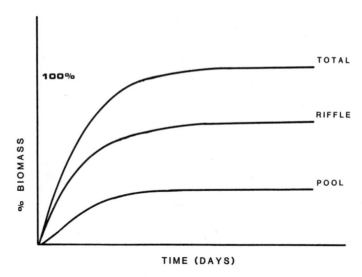

Fig. 1.   Comparison of riffle, pool and total leaf standing crop
          in a stream with 90% riffle area.

advantage of the short-lived resource. High storage capacity
streams will have a more gradual leakage of leaves to the
riffles. In these streams the riffle based shredders will not be
forced to key their life cycles as tightly to the autumn litter
pulse as those in riffle dominated systems. Another implication
of this model includes the possibility of determining optimum
riffle-pool ratios for streams. Since most assimilation appears
to occur in riffles, the ratio of riffles to pools may be related
to total secondary production rates. At some point the increased
secondary production from increased riffle area may be countered
by the concurrently reduced storage capacity of the stream. Thus
certain riffle-pool ratios may actually maximize production. Yet
the entire senerio discussed above is based upon one factor—
density dependence—where food stored in the stream may act to
limit populations and thus productivity.

     Another model series on which I am working utilizes the leaf
pack as the unit which sets K. Leaf packs were selected because
they represent easily manipulated units which have been used
experimentally in many past studies. It is known that leaf pack
decomposition approximates a negative exponential curve (Petersen
and Cummins, 1974) and if leaf pack size sets K, K will
eventually approach zero. Classical stability analysis then
would be expected to indicate that no stable equilibria other
than zero exist. Evaluation of the initial models confirms this.
The shredder curve (either biomass or numbers depending on the
assumptions) shows an initial colonization pulse and,
subsequently, smoothly tracks the declining K. This model fits

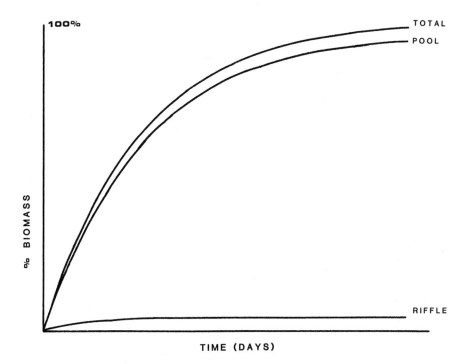

Fig. 2.   Comparison of riffle, pool and total leaf standing crop
in a stream with 10% riffle area.

data gathered for shredder and leaf pack biomass with box elder
leaves in a Rocky Mountain stream (Barnes, unpublished), although
variation exists in the field data.   Whether this variation is
random noise due to factors such as individual leaf pack
difference, locational effects, etc., is not known, but some
additional possibilities exist.   Since drift occurs in a diel
pattern, the feeding activity patterns of drifting organisms are
likely to be diel also.   This implies that events occur in
discrete time periods rather than continuously.   Therefore
difference equations may be more appropriate for modeling leaf
pack-shredder interactions.   Likewise the roles of growth of the
associated microflora and fauna should be taken into account,
especially if several days are required for cropped decomposers
to rebuild their biomass.   This introduces a potential time delay
and by so doing may induce oscillations about K rather than the
smooth convergence seen in the differential model.   Other factors
also should be considered, such as changes in microorganism
growth rates with time, since nutrients and easily assimilated
materials should be removed first.   Will this have an influence
on the shredding community?   Oscillations about K, while theoret-
ically interesting, would also increase the variance in field

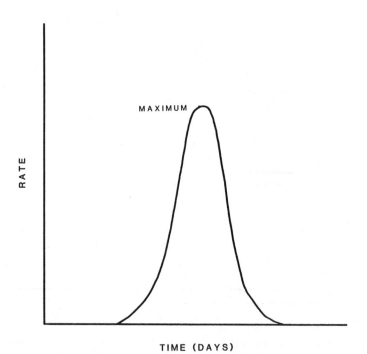

Fig. 3.   Leaf fall rate.

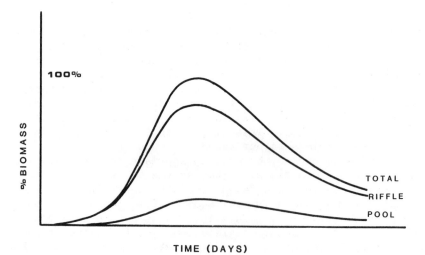

Fig. 4.   Comparison of pool, riffle and total standing crop in a
stream with 90% riffle area.

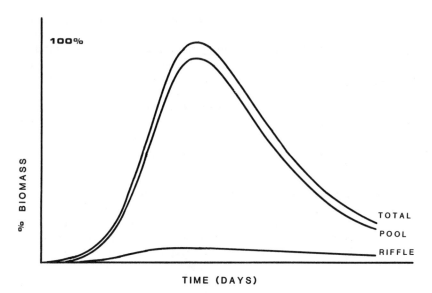

Fig. 5.   Comparison of pool, riffle and total standing crop in a
stream with 10% riffle area.

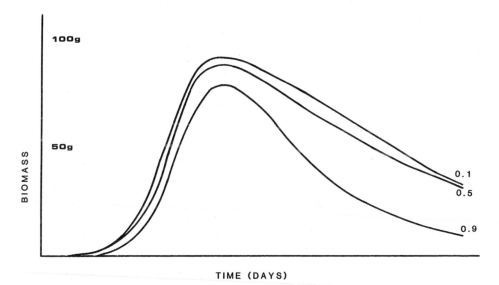

Fig. 6.   Comparison of riffle standing crops at 3 different
riffle/pool ratios.

experiments with leaf packs.   Could such behavior be an explaina-
tion for some of the noise typically found in such studies?

CONCLUSION

I have outlined some of the problems stream ecologists face
in interpreting past studies and, paradoxically, problems that
are arising with the advent of complex statistical packages. I
have also noted areas where stream experiments will benefit
greatly from the field investigators techniques. In many cases,
primary and secondary production rates may be the only methods
giving the resolution necessary to answer abundance regulating
type questions. The careful formulation of hypotheses is also
important. Studies often are conducted with only a slight notion
of what the hypothesis is, and this results in less than
conclusive answers. Streams are complex systems but their
complexity does not make them impossible subjects for study of
density dependent and independent factors. Such studies, based
on sound hypotheses and experimental design, should be where the
next series of major advances in understanding stream systems
will develop.

REFERENCES

Allan, J.D. 1975. The distributional ecology and diversity of
    benthic insects in Cement Creek, Colorado. Ecology
    56:1040-1053.
Allan, J.D. 1978. Trout predation and the size composition of
    stream drift. Limnol. Oceanogr. 23:1231-1237.
Amant, J.L. 1970. The detection of regulation in animal
    populations. Ecology 51:823-828.
Anderson, N.H. and K.W. Cummins. 1979. Influences of diet on
    the life histories of aquatic insects. J. Fish. Res. Board
    Can. 36:335-342.
Andrewartha, H.G. and L.C. Birch. 1954. The distribution and
    abundance of animals. Univ. Chicago Press, Chicago, 782 pp.
Barnes, J.R. and D.K. Shiozawa. 1980. A description and
    assessment of the Raft River lotic system in the vicinity of
    the Raft River Geothermal area. Annual Report, Department
    of Energy, Idaho Operations Office, Idaho Falls, Idaho.
Boling R.H., E.D. Goodman, J.A. Van Sickle, J.O. Zimmer, K.W.
    Cummins, R.C. Petersen and S.R. Reice. 1975. Toward a
    model of detritus processing in a woodland stream. Ecology
    56:141-151.
Boon, P.J. 1979. Adaptive strategies of Amphipsyche larvae
    (Trichoptera: Hydropsychidae) downstream of a tropical
    impoundment. pp. 237-255. In: The ecology of regulated
    streams. J. V. Ward and J.A. Stanford (eds.), Plenum Press,
    New York.
Cochran, W.G. 1963. Sampling Techniques. John Wiley and Sons
    Inc., New York, 413 pp.

Connell, J.H. 1978. Diversity in tropical rain forests and coral reefs. Science 199:1302–1310.

Connell, J.H. 1979. Tropical rainforests and coral reefs as open non-equilibrium systems. pp. 141–163. In: Population dynamics. R.M. Anderson, B.D. Turner and L.R. Taylor (eds.), Blackwell Scientific Publications, Oxford.

Cummins, K.W.. 1974. Structure and function of stream ecosystems. BioScience 24:631–641.

Cummins, K.W. and M.J. Klug. 1979. Feeding ecology of stream invertebrates. Ann. Rev. Ecol. Syst. 10:147–172.

Cummins, K.W., R.C. Petersen, F.O. Howard, J.C. Wuycheck and V.I. Holt. 1973. The utilization of leaf litter by stream detritivores. Ecology 54:336–345.

Dixon, K.R. 1976. Analysis of seasonal leaf fall in north temperate deciduous forests. Oikos 27:300–306.

Edington, J.M. 1965. The effect of water flows on populations of net spinning trichoptera. Mitt. Internat. Verein. Limnol. 13:40–48.

Edmondson, W.T. and G.G. Winberg, Eds. 1971. A manual on methods for the assessment of secondary productivity in freshwaters. Blackwell Sci. Pub. Co., Oxford, 358 pp.

Elliott, J.M. 1977. Some methods for the statistical analysis of samples of benthic invertebrates. Freshwater Biological Assoc. Sci. Pub. 25, 159 pp.

Fisher, S.G. and G. Likens. 1972. Stream ecosystem: Organic energy budget. BioScience 22:33–35.

Fisher, S.G. and G. Likens. 1973. Energy flow in Bear Brook, New Hampshire: an integrative approach to stream ecosystem metabolism. Ecol. Monogr. 43:421–439.

Fox, L.R. 1977. Species richness in streams--an alternative mechanism. Amer. Nat. 111:1017–1021.

Fox, L.R. and P.A. Morrow. 1981. Specialization: species property or local phenomenon? Science 211:887–893.

Fuller, R.L. and R.J. Mackay. 1980a. Field and laboratory studies of net spinning activity by Hydropsyche larvae (Trichoptera: Hydropsychidae) Can. J. Zool. 58:2006–2014.

Fuller, R.L. and R.J. Mackay. 1980b. Feeding ecology of three species of Hydropsyche (Trichoptera: Hydropsychidae) in Southern Ontario. Can. J. Zool. 58:2239–2251.

Gray, J.J. 1971. Sample size and sample frequency in relation to the quantitative sampling of sand meiofauna. pp. 191–197. In: Proceedings of the First International Conference on Meiofauna. N.C. Hullings (ed.), Smithsonian Contributions to Zoology, Washington D.C.

Green, R.H. 1979. Sampling design and statistical methods for environmental biologists. John Wiley & Sons, New York, 257 pp.

Hickman, M. 1974. The standing crop and primary productivity of the epiphyton attached to Chiloscyphus polyanthus (L.) and Chorda rivularis (Schrad.) Nees in a spring fed stream. Arch. Hydrobiol. 73:464-469.

Hildebrand, S.G. 1974. The relation of drift to benthos density and food level in an artificial stream. Limnol. Oceanogr. 19:951-957.

Horn, H.S. 1968. Regulation of animal numbers: a model counter-example. Ecology 49:776-778.

Hynes, H.B.N. 1970. The ecology of running waters. U. of Toronto Press, Toronto, 555 pp.

Hynes, H.B.N., N.K. Kaushik, M.A. Lock, D.L. Lush, Z.S.J. Stocker, R.R. Wallace and D.D. Williams. 1974. Benthos and allochthonous organic matter in streams. J. Fish Res. Board Can. 31:545-553.

Kaushik, N.K. and H.B.N. Hynes. 1971. The fate of the dead leaves that fall into streams. Arch. Hydrobiol. 68:465-515.

Krebs, C.J. 1978. Ecology: The experimental analysis of distribution and abundance, 2nd Ed. Harper & Row Pub. Inc., New York, 678 pp.

Lamberti G.D. and V.H. Resh. 1980. Geothermal influences on the interactions of benthic algae, bacteria, and herbivorous insects in a Northern California stream. Paper presented at 28th Annual Meeting, North American Benthological Society, Savannah, Georgia

Leopold, L.B., M.G. Wolman and J.P. Miller. 1964. Fluvial processes in geomorphology. W.H. Freeman & Co., San Francisco, 522 pp.

Lock, M.A. and H.B.N. Hynes. 1975. The disappearance of four leaf leachates in hard and soft water streams in Southwestern Ontario, Canada. Int. Rev. ges. Hydrobiol. 60:847-855.

Lock, M.A. and H.B.N. Hynes. 1976. The fate of 'dissolved' organic carbon derived from autumn shed maple leaves (Acer saccharum). Limnol. Oceanogr. 21:436-443.

Macan, T.T. 1962. Biotic factors in running water. Schweiz. Z. Hydrol. 24:386-407.

MacArthur, R.M. 1972. Geogrpahical ecology. Harper & Row Pub., New York, 269 pp.

Malas, D. and J.B. Wallace 1977. Strategies for coexistence in three species of net spinning caddisflies (Trichoptera) in second order Southern Appalachia streams. Can. J. Zool. 55:1829-1840.

Margalef, R. 1960. Ideas for a synthetic approach to the ecology of running waters. Int. Rev. ges. Hydrobiol. 45:133-153.

May, R.M. 1973. Stability and complexity in model ecosystems. Princeton U. Press, New Jersey, 265 pp.

McIntire, C.D. 1973. Periphyton dynamics in laboratory streams:
    a simulation model and its implications. Ecol. Monogr.
    43:399-420.
Menge, B.A. and J.P. Sutherland. 1976. Species diversity
    gradients: synthesis of the role of predation, competition,
    and temporal heterogeneity. Amer. Natur. 110:351-369.
Merritt, R.E. and K.W. Cummins, Eds. 1978. An introduction to
    the aquatic insects of North America. Kendall/Hunt Pub.
    Co., Dubuque, 441 pp.
Meyer, J.L. 1980. Dynamics of phosphorus and organic matter
    during leaf decomposition in a forest stream. Oikos
    34:44-53.
Minshall, G.W. 1967. Role of allochthonous detritus in the
    trophic structure of a woodland springbrook community.
    Ecology 48:139-149.
Minshall, G.W. 1978. Autotrophy in stream ecosystems.
    BioScience 28:767-771.
Muller, K. 1954. Investigations on the organic drift in North
    Swedish streams. Rep. Inst. Freshwat. Res. Drottningholm
    35:133-148.
Muller-Dombois, D. and H. Ellenberg. 1974. Aims and methods of
    vegetation ecology. John Wiley & Sons, New York, 547 pp.
Mundie, J.H. 1971. Sampling benthos and substrate materials
    down to 50 microns in size in shallow streams. J. Fish.
    Res. Bd. Canada 28:849-860.
Murdock, W.W. 1969. Switching in general predators: experiments
    on predator specificity and stability of prey populations.
    Ecol. Monogr. 39:335-354.
Needham, P.R. and R.L. Usinger. 1956. Variability in the
    macrofauna of a single riffle in Prosser Creek, California,
    as indicated by the Surber samples. Hilgardia 24:383-409.
Paine, R.T. 1966. Food web complexity and species diversity.
    Amer. Nat 100:65-75.
Paine, R.T. 1969. A note on trophic complexity and community
    stability. Amer. Nat. 103:91-94.
Patrick, R. 1972. Benthic communities in streams. In: Growth
    by Intussusception. E.S. Devey (ed). Conn. Acad. Arts and
    Sciences 44:169-184.
Peckarsky, B.L. 1979. Biological interactions as determinants
    of distributions of benthic invertebrates within the
    substrate of stony streams. Limnol. Oceanogr. 24:59-68.
Peckarsky, B.L. 1981. Reply to comment by Sell. Limnol.
    Oceanogr. 26:982-987.
Petersen, R.C. and K.W. Cummins. 1974. Leaf processing in a
    woodland stream. Freshwat. Biol. 4:343-368.
Reice, S.R. 1974. Environmental patchiness and the break down
    of leaf litter in a woodland stream. Ecology 55:1271-1282.
Reice, S.R. 1977. The role of animal associations and current
    velocity in sediment specific leaf litter decomposition.
    Oikos 29:357-365.

Reice, S.R. 1980. The effect of fish exclusions and detritus
     levels on a woodland stream ecosystem. Paper presented at
     28th Annual Meeting, North American Benthological Society.
     Savannah, Georgia.

Rhame, R.E. and K.W. Stewart. 1976. Life cycles and food habits
     of three Hydropsychidae (Trichoptera) species in the Brazos
     River, Texas. Trans. Am. Entomol. Soc. 102:65-99.

Ricklefs, R.E. 1979. Ecology, 2nd Ed. Chiron. Press Inc. New
     York, 966 pp.

Schoener, T. W. 1982. The controversy over interspecific
     competition. Amer. Sci. 70:586-595.

Scott, W.B. and E.J. Crossman. 1973. Freshwater fishes of
     Canada. Bull. 184. Fisheries Research Board Can. Ottawa,
     966 pp.

Sell, D.W. 1981. Comment on "Biological interactions as
     determinants of distributions of benthic invertebrates
     within the substrate of stony streams." Limnol. Oceanogr.
     26:981-982.

Shiozawa, D.K. 1978. The habitat preferences, seasonal drift
     and abundance of stream microcrustacea. Unpub. Ph.D.
     thesis, U. of Minnesota, 151 pp.

Short, R.A. and J.V. Ward. 1981. Benthic detritus dynamics in a
     mountain stream. Holarctic Ecology 4:32-35.

Smith, H.S. 1935. The role of biotic factors in determining
     population densities. J. Econ. Entom. 28:873-898.

Solomon, M.R. 1958. Meaning of density dependence and related
     terms in population dynamics. Nature 181:1778-1780.

Stewart, A. 1968. Basic ideas of scientific sampling. Hafner
     Pub. Co., New York, 99 pp.

Stout, J. and J. Vandermeer. 1975. Comparison of species
     richness for stream inhabiting insects in tropical and mid
     -latitude streams. Amer. Nat. 109:263-280.

Sudman, S. 1976. Applied sampling. Academic Press, New York,
     249 pp.

Tanner J.T. 1966. Effects of population density on growth rates
     of animal populations. Ecology 47:734-740.

Vannote, R.L., G.W. Minshall, K.W. Cummins, J.R. Sedell and C.E.
     Cushing. 1980. The river continuum concept. Can. J. Fish.
     Aquat. Sci. 37:130-137.

Vollenweider, R.A., Ed. 1969. A manual on methods for measuring
     primary production in aquatic environments. Blackwell Sci.
     Pub., Oxford, 213 pp.

Wallace, J.B. and R.W. Merritt. 1980. Filter feeding ecology of
     aquatic insects. Ann. Rev. Entomol. 25:103-132.

Wallace, J.B., J.R. Webster and W.R. Woodall. 1977. The role of
     filter feeders in flowing waters. Arch. Hydrobiol.
     79:506-532.

Waters, T.F. 1961. Standing crop and drift of stream bottom
     organisms. Ecology 42:532-537.

Waters, T.F. 1965. Interpretation of invertebrate drift in
     streams. Ecology 46:327-334.

Waters, T.F.   1966.   Production rate, population density, and
      drift of a stream invertebrate.   Ecology 47:595-604.
Waters, T.F.   1969.   Subsampler for dividing large samples of
      stream invertebrate drift.   Limnol. Oceanogr. 14:813-815.
Waters, T.F.   1972.   The drift of stream insects.   Ann. Rev.
      Entomol. 17:253-272.
Waters, T.F.   1977.   Secondary production in inland waters.   Adv.
      Ecol. Res. 10:91-164.
Waters, T.F. and R.J. Knapp.   1961.   An improved stream bottom
      fauna sampler.   Trans. Amer. Fish. Soc. 90:225-226.
Welton J.S.   1980.   Dynamics of sediment and organic detritus in
      a small chalk stream.   Arch. Hydrobiol. 90:162-181.
Wiens J.A.   1977.   On competition and variable environments.
      Amer. Sci. 65:590-597.
Wiley M.J. and S.L. Kohler.   1981.   An assessment of biological
      interactions in an epilithic stream community using
      time-lapse cinematography.   Hydrobiologia 78:183-188.
Winberg, G.G., Ed.   1971.   Methods for the estimation of
      production of aquatic animals.   Academic Press, New York,
      175 pp.

USE OF BEHAVIORAL EXPERIMENTS TO

TEST ECOLOGICAL THEORY IN STREAMS

Barbara L. Peckarsky

Department of Entomology
Cornell University
Ithaca, New York 14853

INTRODUCTION

Direct behavioral observation of stream invertebrates was practiced extensively by aquatic entomologists in the early part of this century. Early descriptions of stream insect behavior were usually qualitative but of masterful literary style. Needham et. al. (1935) indulged in countless flowery descriptions of mayfly behavior, for example, this description of activities of close-clinging, stone-loving mayfly nymphs: "Under the stones we may find representatives of other ecological groups. Some dwell there; others are stranded there; others run in for shelter there. Every creature has a right to such shelter as he can find from the vicissitudes of life. In havens of refuge many strangers meet" (p. 182). Neave (1930) reported directly observing mass upstream migrations of mayfly nymphs along the banks of rivers, a behavior later reinvestigated by Hayden and Clifford (1974). Behavioral reports have an even older legacy as exemplified by writings of the habits of burrowing mayfly nymphs by the Dutch naturalist, Swammerdam (1737): "Of all species of insects, I never saw any one more mild, gentle, or innocent. For, in whatever manner it is treated, it is always calm and peaceful; and when left to itself, it immediately goes to work and begins to dig a cell for its habitation" (Needham et. al., 1935, p. 6).

The standards of the ecologists of the 1980's do not allow such qualitative indulgence, although much of the natural history of stream insects, written in this style, forms the basis for modern stream ecology. Unfortunately, the increasing emphasis on obtaining detailed quantitative distributional information has caused stream ecologists virtually to abandon behavioral work.

Not until recently have quantitative techniques and an
experimental approach been utilized to study the behavior of
stream insects. My objective in this paper is to show that a
combination of direct behavioral observation, with quantification
of insect activity and controlled experimentation, can provide
rigorous data that can be interpreted in the context of general
ecological theory. The primary advantage of behavioral
experiments over other methods is that they can allow
determination of mechanisms for observed phenomena.

Stream insect behavior can be observed in situ or in the
laboratory. Observations of subjects in their natural habitat
with minimal disturbance provide the most reliable data, but are
often difficult because many stream insects are nocturnal or
crepuscular. Field manipulations or laboratory experiments must
be carefully designed and cautiously interpreted to avoid drawing
erroneous conclusions from unnatural behavior stimulated by
artificial conditions.

Altmann (1974) developed an observer's guide to behavioral
sampling methods. This guide describes and lists recommended uses
of all qualitative and quantitative techniques known at the time
of publication. I will use her classification of quantitative
methods in discussing behavioral studies addressing questions on a
broad range of ecological concepts.

1. Ad libitum sampling - recording as much information as
    possible; typical field notes;

2. Sociometric matrix completion - supplementing
    ad lib sampling with additional observations on
    particular pairs of individuals;

3. Focal-animal sampling - recording all activities of one
    individual or a specified group;

4. Sampling all occurrences of some behavior(s);

5. Sequence sampling - recording the sequence of a
    particular interaction or occurrence;

6. One-zero sampling - occurrence or non-occurrence;

7. Instantaneous and scan sampling - preselecting a sampling
    frequency and recording current activities.

INFLUENCE OF ABIOTIC FACTORS ON DISTRIBUTION

Species of plants and animals are restricted to certain
ecological ranges by constraints imposed by their abiotic

surroundings. Microdistributions of organisms are influenced
further within prescribed ranges by variations in physical and
chemical parameters. Stream invertebrate distributions are
affected by substrate, current velocity, temperature, oxygen,
detritus, and other variables. This generality has been supported
primarily by data from traditional invertebrate survey techniques
with simultaneous measurements of abiotic factors, or by
manipulation of abiotic factors and assessment of its effect on
invertebrate colonization. A few behavioral studies have provided
convincing corroboration of mechanisms producing the observed
field associations between benthic distributions and abiotic
factors.

To test the orientation of mayfly nymphs to light, Hughes
(1966a) counted the number of four instar stages of Baetis
harrisoni (Baetidae) resting on their dorsal or ventral surfaces,
given top or bottom illumination (focal animal sampling).
Experiments were carried out in the laboratory in still water;
thus, behavior may be an artifact of unnatural conditions. Nymphs
were shown to maintain their dorso-ventral orientation using light
as a proximal cue. Hughes also observed B. harrisoni repeatedly
to release their hold of the substrate at low current velocity,
swim upward, and fall back in a "somersaulting" behavior in the
light beam. His interpretation of the importance of this response
was that at low stream current velocities B. harrisoni nymphs thus
created a current over their integument for respiratory purposes.
However, he conducted no experiments to test this interpretation.
Hughes (1966b) also tested the behavioral responses of B.
harrisoni and another mayfly, Tricorythodes discolor
(Tricorythidae), to light in an outdoor artificial stream (focal
animal sampling). Both species were shown to use light as a
proximal cue in the selection and maintenance of microhabitat.

Gallepp (1977) manipulated temperature, food availability
(brine shrimp), and current velocity singly and interactively, and
recorded the resultant behavior of the caddisflies Brachycentrus
americanus and B. occidentalis (Brachycentridae) in a plexiglas
insert in an artificial stream tank. He emphasized the importance
of certain criteria in the choice of species for observation.
Animals should be easy to observe (which eliminates many strongly
photonegative or highly mobile species), should exhibit
identifiable behavior patterns that allow ecological
interpretation, should be adapted to life in the laboratory (in
his case), and should be easy to collect in large numbers.
Gallepp observed the percent of the larvae in his systems
filtering (feeding), case-building, withdrawn inside cases, and
unattached (crawling, drifting, or stationary) at established
intervals (instantaneous and scan sampling). Results showed that
Brachycentrus behavior was significantly affected by the abiotic
variables, and by food availability, and that temperature and food

were of primary importance. These data can be used to identify
the behavioral mechanisms mediating associations between the
animal's environment and its distribution pattern. In other
words, adaptive behavior categories, such as filtering and
case-building, are restricted to certain ranges of food and
temperature, with these variables overriding the effects of
current velocity. Such behavioral experiments can be used to
interpret more clearly patterns of caddisfly distributions
observed in the field.

Mackay (1977) performed laboratory experiments designed to
determine substrate choice by larvae of the caddisfly genus
Pycnopsyche (Limnephilidae) under approximately natural
conditions. She introduced specimens, evenly spaced, into
plexiglas trays containing different substrate types, and noted
their positions at designated intervals (instantaneous and scan
sampling). She showed that substrate particle size may be an
important abiotic factor limiting the distribution of P.
scabripennis. As in the previous example, these behavioral data
may be applied to identify the mechanisms producing observed
associations between caddisfly distributions and substrate types
in the field. However, they do not elucidate the relative
importance of this variable, since other factors were not tested
simultaneously.

Edington (1968) conducted laboratory experiments to
corroborate field experiments and correlations suggesting a
cause-effect relationship between current speed and the
distribution of two net-spinning caddisfly species. Hydropsyche
instabilis (Hydropsychidae) and Plectrocnemia conspersa
(Polycentropodidae) larvae were exposed to various current
velocities in a laboratory tank for 24 hours, after which the
number of larvae having constructed nets was counted
(instantaneous and scan sampling). Net spinning activity of H.
instabilis, the species with a larger mesh-size net, increased at
higher current velocities; whereas P. conspersa net spinning
decreased at higher flow rate. Edington concluded that current
speed is an important variable limiting the distribution of these
two species of caddisflies.

Wiley and Kohler (1980) observed the positioning behavior of
four mayfly species in the laboratory in a small circular
flowing-water chamber under several combinations of dissolved
oxygen (DO) concentration and current velocity. Dissolved oxygen
was reduced from saturation in 2 ppm decrements; nymphs were
exposed to each DO level at three different current velocities for
15 minutes, after which the number of individuals on current-
exposed surfaces of stone substrates was recorded (instantaneous
and scan sampling). The design of their chamber allowed viewing
of individuals on all stone surfaces. All four species moved to

exposed substrate surfaces under respiratory stress, when gill
ventilation was insufficient to meet respiratory needs.

DISPERSAL MECHANISMS (DRIFT)

     Wiley and Kohler's (1980) study not only provided information
on proximal cues for microhabitat choice by stream insects, but
also can be interpreted in the context of mechanisms for dispersal
by stream insects. They observed active entry of all four mayfly
species into the drift when dissolved oxygen (DO) fell below a
certain critical level at which substrate-position and
gill-ventilation behavior could no longer satisfy oxygen
requirements (sampling all occurrences of some behavior). A
Baetidae species (Pseudocloeon sp.) drifted at the highest DO,
followed by Stenonema pulchellum (Heptageniidae), Ephemerella lata
(Ephemerellidae), and Stenacron interpunctata (Heptageniidae).
These responses were consistent with field distributional data and
respiratory requirements, and reflected relative abundances of
these taxa in traditional studies of stream drift. This study
confirmed that an increase in exposure, and an active behavior
might account for the entry of large numbers of individuals into
the drift.

     The active-passive drift controversy has been addressed by
numerous authors (Waters, 1972). The adaptive significance of
drift behavior is dependent, in part, upon whether it is an active
dispersal behavior or an accidental removal phenomenon, and on the
effect of drift on stream invertebrate community structure.
Behavioral studies can effectively document mechanisms controlling
entry into and resettling from the drift. Several papers have
described experiments conducted within an elliptical artificial
stream (e.g., Ciborowski et. al., 1977). Ciborowski and Corkum
(1980) observed the settling ability of four mayfly species by
introducing 50 live nymphs into the stream and recording the
number suspended within the water column at one-minute intervals
until 90% of the individuals had settled (instantaneous and scan
sampling). Insects were then killed and the experiment repeated.
An index of settling capacity (SC) was calculated for live and
dead individuals of each species. Comparison of such indices
allowed determination of the portion of the SC due to
morphological characters (MC) and that due to the behavior of the
individual (BC).

     Ciborowski and Corkum showed that the relative contribution
of behavior to the SC of an organism (the behavioral index, BI =
BC - SC) was highest in Baetis vagans and intermediate in
Ephemerella spp. and Paraleptophlebia mollis (Leptophlebiidae).
These data and similar calculations from data of Elliott (1971)
showed high behavioral indices for those individuals with a high

tendency toward diel periodicity (Ephemeroptera, Amphipoda) and low BI for some Plecoptera, Simuliidae, Chironomidae, and Elmidae, which showed less diel periodicity. The authors concluded that those species most prone to drift are best able to remove themselves behaviorally from the water column. They suggested that where the BI was minimal (Chironomidae, Simuliidae), capture in drift nets may be due to a low SC as well as a high propensity to drift. The behavior of these taxa should be studied under more controlled conditions before such conclusions be accepted without question.

Behavioral studies by Corkum and Clifford (1980) have documented effects of food, substrate type, light levels, and predators on the settling capacity of two mayflies. As in Ciborowski and Corkum (1980), the number of introduced nymphs of Baetis tricaudatus and Leptophlebia cupida suspended in the water column were counted at given time intervals (instantaneous and scan sampling) under different light, substrate, predator, and food regimes. Nymphs resettled faster during the day, larger nymphs resettled faster than smaller, and small Baetis settled more quickly when food or a predaceous stonefly (Isogenoides elongatus) were present. Such studies on resettling motivation provide more complete information on the role of drift in the redistribution and habitat choice of stream insects than do data from traditional drift net studies. Drift net contents only approximate numbers of organisms present in the water column over a given time interval. They do not allow identification of how far an insect drifted, why it resettled, or why it released hold of the substrate.

Other studies have implemented behavioral techniques to document release motivation. Corkum (1978) tested the departure rates of nymphs of Paraleptophlebia mollis and Baetis vagans under still water conditions in response to photoperiod and population density. She conducted nighttime observations of nymphs marked with luminous paint using a 40-W red bulb for illumination, which was shown not to affect nymphal activity. Nymphs were pipetted or poured through a funnel onto bricks, and total numbers leaving the brick were tallied at 10-sec intervals (instantaneous and scan sampling); or verbal tallies were made on a tape recorder of all departures from the brick (sampling all occurrences of some behavior). These experiments revealed that an increase in density did not result in an increased proportion of nymphs leaving the substrate but that behavioral type (B. vagans, swimmers; P. mollis, crawlers) was the most important determinant of drift behavior.

Peckarsky (1980) recorded the results of encounters between several mayfly species and stonefly predators within plexiglas observation boxes in a Colorado and a Wisconsin stream (sampling

all occurrences of some behavior). Baetis bicaudatus (Colorado)
and B. phoebus (Wisconsin) responded to encounters with predators
by entering the water column actively, either drifting, swimming,
or drift-swimming (releasing hold of the substrate, and drifting
instantaneously, followed by active swimming behavior). Other
mayfly types (Heptageniidae, Ephemerellidae) did not drift or swim
in significant numbers in response to the predators. This simple
behavioral experiment clearly documented that predator-avoidance
is a motivation for active drift in some mayflies. Corkum and
Clifford's (1980) study, described above, also showed that the
presence of stonefly predators increased drift of mature Baetis
and Leptophlebia.

BIOLOGICAL INTERACTIONS

        The studies of Corkum (1978), Corkum and Clifford (1980), and
Peckarsky (1980) added an interesting dimension to the testing of
ecological theory in stream ecosystems. Data on such biological
interactions as competition and predation in streams are
conspicuously absent from the literature. Perhaps this lack of
information is due to the difficulty of developing appropriate
experimental designs to test such questions (see Hart, this
volume). Most studies on predation or competition have been based
on stomach-content analyses or interpretation of data on spatial
or temporal overlap between closely related species. Controlled
experiments, including those of a behavioral design, have been
conducted only recently. These experiments have contributed
significantly to the knowledge of predation and competition
mechanisms in streams.

Competition

        The frequency and importance of interference and exploitation
competition in streams is not well understood. Few studies have
investigated the role of competition in determining spacing
patterns of stream invertebrates. (See chapters by Hart and
McAuliffe in this volume for a thorough discussion of the
application of competition theory to streams.) The behavioral
experiments described below provide insight into the role of
competition in structuring stream invertebrate communities.

        Corkum (1978) examined the question of spacing of mayfly
nymphs within the laboratory system described above. She observed
nymphal movement on brick surfaces for Baetis vagans and
Paraleptophlebia mollis during the first hour of every experiment
(ad lib sampling). Contrasting responses to high density were
observed between these two species (sociometric matrix
completion). Baetis nymphs remained stationary until they swam or
drifted from the brick; P. mollis, on the other hand, actively

crawled about the brick exhibiting intraspecific aggressiveness.
Nymphs bent their abdomens in a horizontal plane from side to
side, stricking adjacent individuals; or when facing each other,
they raised and lowered forelegs and antennae repeatedly in direct
contact with another nymph. These interactions were followed by
withdrawal behavior or by one nymph "chasing" the other over the
brick. Drift rarely occurred from such encounters. Since this
aggressive behavior was recorded under still-water conditions, it
is possible that such interactions do not occur in the
flowing-water habitat.

Others have recorded competitive or aggressive interactions
among stream insects under still-water conditions. Glass and
Bovbjerg (1969) studied aggressive behavior among groups of four
caddisfly larvae (Cheumatopsyche sp., Hydropsychidae) forced into
close contact in glass bowls. Encounters were classified as (1)
avoidance (10%) - a head-on encounter, no contact; one retreats;
(2) strike (15%) - mouth contact, but no biting; attacked larva
retreats; (3) bite (50%) - mandibles close on attacked larva,
which then retreats; and (4) fight (25%) - both animals strike or
bite, one retreats. The percentage of each type of tension
contact was recorded with and without the presence of
pebble-refuges (sociometric matrix completion, sequence sampling).
The authors also conducted spacing experiments by placing large
numbers of larvae in a clump in a glass bowl. After a few hours,
the population had become uniformly dispersed. Aggressive
encounters were shown to mediate the spacing behavior, and were
reduced by the presence of refuges.

Hildrew and Townsend (1980) quantified aggressive behavior
between larvae of the predatory net-spinning caddisfly
Plectrocnemia conspersa (Polycentropodidae) in an artificial
laboratory stream with a 5 cm/s current velocity. Intruder larvae
were introduced on substrate patches where a resident had
established a catch net. If the intruder touched the resident's
net, a confrontation was triggered. A "rearing up" behavior
initiated the encounter, after which repeated mandibular striking
and "rearing up" occurred. Severe biting occasionally resulted.
The contest was terminated when one of the larvae retreated
(sociometric matrix completion, sequence sampling). The resident
retreated through its living tube; the intruder actively drifted
in a characteristic movement pattern. Such mutual interference
generally resulted in the larger of the two individuals "winning"
the contest and a weak prior residency effect. Hildrew and
Townsend interpreted these results in the context of a motive for
active drift behavior, and as an explanation for the failure of P.
conspersa to exhibit optimal foraging behavior in some seasons
(see below).

Jansson and Vuoristo (1979) provoked fights between individuals of four Hydropsyche spp. (Hydropsychidae) within laboratory containers (40 X 25 X 12 cm) in which the water was aerated vigorously. They removed one larva, the "intruder," from its retreat, and introduced it into the net of a "defender." Observation vessels were placed on top of smaller containers so that larvae building retreats under rocks could be observed from underneath with mirrors. Fights were initiated when the intruder attempted to enter the defender's retreat. Biting behavior ensued, and the defender often backed into the retreat and began to stridulate. The authors recorded stridulation behavior with a hydrophone (sampling all instances of some behavior), and general behavior during fights (sociometric matrix completion). Larvae produced stridulation sounds by rubbing ventrolateral striations of the head (files) against specialized tubercles on the anterodorsal sides of the fore femora (scrapers). Upon the initiation of stridulation by the defender, the intruder either gave up or made further attempts to enter the retreat. Fighting continued with alternating bouts of biting and stridulation until one of the antagonists won. Stridulation by defenders increased the probability that they would win the fight, especially with larger intruders. These observations suggest that aggressive competition among stream-dwelling caddisflies may be an important density-regulating mechanism where sites for building catch-nets are limited.

Black fly (Simuliidae) larvae have been observed by Hart (1979) to bite at neighboring conspecifics, a behavior that may mediate observed spacing patterns. Wiley and Kohler (1981) devised a simple cinematographic technique with which they have recorded numerous interactions among epilithic species in a Michigan stream (ad lib sampling). This technique has tremendous potential for documenting biotic interactions among stream insects, although it is not an experimental manipulation. Analysis of 78 hours of film shot at 30-60 seconds/frame showed aggressive interactions between simuliids, and between simuliids and other taxa, the majority of which resulted in local displacement or migration of one of the antagonists (sociometric matrix completion).

Wilson et. al. (1978) conducted a simple field study on interference competition in a population of the tropical ripple bug Rhagovelia scabra (Hemiptera:Veliidae) that feeds primarily on soft-bodied terrestrial insects floating on the surface of streams. The optimal location for food capture was at the head of pools where the current was swift. The authors noticed that 95% of the ripple bugs in this microhabitat were adult females. They conducted behavioral experiments to determine: (1) whether food capture was highest at the heads of pools, (2) whether the dominance of adult females occupying the optimal habitat was due

to their exclusion of adult males and juveniles, and (3) the
mechanisms whereby interference among age and sex classes
occurred. Their techniques involved recording capture rate of
introduced fruit fly food (sampling all occurrences of some
behavior), releasing individuals of adult males, adult females,
and juveniles into a pool cleared of all R. scabra and observing
the redistribution of each group alone and in combination with
others at a designated time interval (instantaneous and scan
sampling), and recording interactions between pairs of bugs for a
five-minute period (focal animal sampling, sequence sampling).
The results of these behavioral experiments showed clearly that
the population of ripple bugs was structured by interference
competition.

This convincing documentation of interference competition
among stream insects suggests that such behavior is, perhaps, more
prevalent in stream systems than had been previously suspected.
In many hours observing the behavior of perlid and perlodid
stonefly predators, I have often observed aggressive interactions
between conspecifics or confamilials. Preliminary observations of
the perlodid predators Megarcys signata and Kogotus modestus
within observation boxes in a Colorado stream indicate that active
avoidance of inter-individual contacts may be the mechanism
mediating spacing behavior observed in experiments with substrate
cages (Peckarsky, 1979; Peckarsky and Dodson, 1980). However, I
cannot be certain that these experimental conditions produce
"normal" aggressive behavior.

The activity of M. signata and K. modestus was recorded
continuously for 10-minute periods for single individuals, those
in the presence of potential (conspecific or confamilial)
competitors, of prey, and of prey and potential competitors
(sequence sampling). The repertoire of behavior recorded included
crawling with antennae searching the substrate, stationary with
antennae searching, crawling, swimming, drifting, and stationary.
The data showed that M. signata behavior was significantly
affected by the presence of potential competitors, K. modestus and
Pteronarcella badia, and that K. modestus behavior was altered by
the presence of Baetis bicaudatus prey (Figs. 1 and 2). The
presence of the potential competitor M. signata also significantly
reduced the proportion of time that K. modestus spent attacking
and eating prey. Murdoch and Sih (1978) recorded a 90% decrease
in feeding rate by pool-dwelling Notonecta hoffmanni immatures in
the presence of conspecific adults (focal animal sampling) in the
laboratory. These observations suggest that "interference" with
competitors may reduce the feeding efficiency of some
stream-dwelling invertebrate predators.

Hart (1981) introduced the first behavioral evidence, to my
knowledge, suggestive of exploitative competition among a guild of

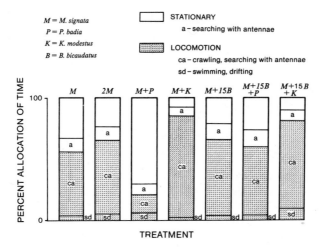

Fig. 1. Activity of <u>Megarcys signata</u>.

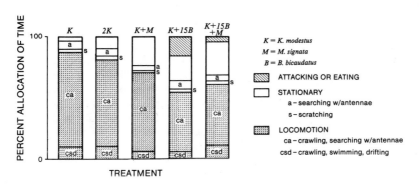

Fig. 2. Activity of <u>Kogotus modestus</u>.

grazing stream insects. He chose 12 focal animals of the
caddisfly <u>Dicosmoecus gilvipes</u> (Limnephilidae) and quantified
their activity by recording their behavioral state for 80
sequential 5-s intervals (instantaneous and scan sampling).
Observations were made in pools of a California stream with
facemask and snorkel on a mosaic of quarry tiles on which
periphyton cover was manipulated. Grazers were shown to depress
substantially food available for subsequent grazers, although
grazer densities may have been artificially high. Individuals
spent significantly more time on ungrazed food patches than on
grazed areas. These data suggest that grazers can actually cause
resource patchiness, and reduce the quantity of food available to

competitors, thereby generating a potentially negative effect on
their fitness. These experiments, and those of McAuliffe (see
elsewhere in this volume), suggest a resource-scramble or
exploitative competition among stream grazers for a limited
periphyton resource.

Predator-prey interactions

     Little is known about the role ot predator-prey interactions
in determining patterns of species abundance and distribution in
streams. Predator hunting strategies, prey preferences, prey
escape mechanisms, and responses of predators to varying prey
densities are poorly understood. (See chapter by Allan, this
volume, for discussion of the application of predator-prey theory
to streams.) The study of predator-prey interactions can be
accomplished effectively by behavioral experimentation if the
appropriate cautions are excercised in design of techniques such
that observed predation events are not artifacts of unnatural
conditions. Below is a summary of a few behavioral studies of
stream invertebrates that have been conducted to address questions
related to predator-prey theory.

     Wiley and Kohler's (1981) photographic observation technique
allowed the enumeration of predator-prey encounters, although
experiments were not conducted. The predatory caddisfly,
Rhyacophila acropedes grp (Rhyacophilidae) was observed capturing
some simuliid larvae and initiating the emigration of others.
Peckarsky (1980) also quantified the responses of mayfly species
in a Wisconsin and a Colorado stream to encounters with stonefly
predators within plexiglas observation boxes (sampling all
occurrences of some behavior). Baetidae species, as mentioned
earlier, drifted, swam, or drift-swam in response to encounters
with large stonefly predators in both streams. Heptageniid
species crawled primarily in response to encounters with
predators. Such avoidance behavior may be an important mechanism
explaining biological patchiness of stream invertebrates.

     Ephemerellid species in both streams exhibited a "scorpion
posture," which may serve as a display behavior in response to
encounters with predators. This posture acted as a deterrent to
the tactile predators, presumably creating the "image" of a spiny
creature of an unpreferred size and shape; it may also be a
generalized stress response. Behavioral experiments are presently
being conducted to determine the nature of stimuli producing this
response, and its role in reducing the effectiveness of the
predators.

     Peckarsky (1980) also tested the hypothesis that mayfly
species detect and avoid stonefly predators given non-contact
cues. The positions of 15 mayflies of each test species were

recorded in the observation boxes before, during, and after the
presence of a stonefly predator in a screen tube or a test tube in
the center of the box (instantaneous scan sampling). Results of
these experiments showed that some species, including those that
postured and the Colorado Baetis species, avoided a region of
presumed high chemical stimulus downstream from the predator in
the screen tube. This result has interesting implications for
search and escape strategies for stream invertebrate predators and
prey. If chemical cues can be detected indicating presence of a
predator or prey, the predator should search in an upstream
direction, and prey should escape downstream (such as in Baetis
drifting). Also, Ephemerella species may use a chemical sense to
detect predators and assume their defensive posture before the
predator comes into physical contact with them. These questions
can be pursued by using in situ behavioral experiments.

     Predation strategies of net-spinning caddisflies
(Plectrocnemia conspersa) have been studied by several authors.
Tachet (1977) designed an apparatus that vibrated an artificial
lure (a bristle) so that the frequency and amplitude of the
net-vibrations could be varied independently while the larva was
in its prey-catching net (in still water). A sophisticated coding
system was implemented for recording from a large repertoire of
behavior (focal animal sampling). A number of behaviors related
to prey-capture were recorded in sequence (sequence sampling).
Larvae of P. conspersa exhibited a characteristic sequence of
responses to vibrations of their nets, including waiting,
awakening of interest, orientation and movement toward prey,
mandibular capture, withdrawal of prey, examination of prey,
ingestion of prey, and egestion of feces. Tactile stimulation of
the nets provided sensory cues for prey capture by this predator.

OPTIMAL FORAGING THEORY

     Hildrew and Townsend (1980) and Townsend and Hildrew (1980)
examined predatory behavior of P. conspersa using behavioral
techniques. They recorded net-building behavior, widespread
movement (wandering), ventilation behavior, incidence of
stationary behavior, and agonistic behavior of one or a pair of
larvae (focal animal sampling, sociometric matrix completion,
sequence sampling). The authors found that the availability of
prey significantly affected the length of time a larva remained at
an established retreat. Larvae abandoned sites after a certain
threshold time during which they had not captured a prey
(giving-up time, GUT). The authors interpreted the results of
these experiments in the context of optimal foraging theory, which
predicts that predators should maximize their net energy gain per
unit time (Krebs, 1978).

Optimal foraging theory assumes that animals have
sophisticated data-gathering and data-assimilation capacity.
Townsend and Hildrew (1980) showed that a fixed GUT could result
in approximately optimal behavior consistent with the prediction
that predators should leave a patch when the instantaneous rate of
energy gain drops to the equivalent of the average of the entire
habitat (Charnov, 1976). This marginal value theorem was derived
by assuming a variable GUT. However, Townsend and Hildrew's
(1980) observations suggest that a simpler behavior can
approximate an optimal solution. One possible problem with this
study is that the average field prey-capture rate was calculated
as a seasonal mean. High seasonal variability in capture rate
might cast doubt on the accuracy of this estimate.

Others have interpreted stream-insect behavior in the context
of optimal foraging theory. Hart and Resh (1980) recorded
behavior of the grazing caddisfly Dicosmoecus gilvipes in a pool
of a California stream by a number of methods involving
observations by SCUBA. This study did not involve experiments,
however. They mapped the distribution of tagged individuals twice
daily (instantaneous scan sampling), and recorded their behavior
each five seconds, for 3.5-minute intervals (sequence sampling),
from a repertoire including feeding on periphyton, feeding on
leaves, walking, withdrawing inside case, holding onto substrate,
and falling. The purpose was to determine how these larvae
partitioned their time budget among various activities. Results
showed that greater than two-thirds of the time was spent feeding.
The authors interpreted the pattern of movement of D. gilvipes to
be influenced by the spatial heterogeneity of the available
periphyton. Larvae remained in one area for several days, then
moved long distances, rarely feeding while on long walks, and
settled on another patch for an extended period of time. The
amount of movement across small-scale food patches was negatively
correlated with the amount of periphyton per patch (determined
subjectively). This "area-restricted search" behavior is
consistent with optimal foraging theory (Krebs, 1978).

In an elegant study of the predatory bug Notonecta hoffmanni
(Notonectidae) that inhabits stream pools, Sih (1980) demonstrated
that conflicting demands, such as prey-search and predator-
avoidance, may require predators to assume a suboptimal strategy
as far as energy intake is concerned. Adult N. hoffmanni are
cannibalistic on immatures, both of whom consume dipteran prey.
Sih recorded the amount of time spent in regions of high and low
prey (Drosophila) density by sampling at 30-minute intervals
(instantaneous and scan sampling). After 24 hours he recorded the
survival and feeding rates of the adults and nymphs. First-instar
Notonecta spent more time in the low prey-density area, but
actually consumed more prey in this region than in the region of
high prey density. Older instars and adult N. hoffmanni spent

more time and consumed more prey in the area of high prey density.
Sih concluded that juvenile N. hoffmanni balanced the conflicting
demands of predator-avoidance and prey-search by feeding in areas
apart from adults. Since optimality is maximizing fitness, a
behavior balancing energy intake and predator avoidance may be
optimal. Each instar preferred the microhabitat with higher
associated fitness. Sih suggested that this information should be
incorporated into optimal foraging models. Hildrew and Townsend
(1980) showed, similarly, that suboptimal foraging behavior by P.
conspersa larvae might be a result of mutual interference between
foragers. Aggressive interactions could have caused a lack of
aggregation of P. conspersa in regions of high prey density during
an August sampling period. Such a lack of a positive numerical
response by stonefly predators (Peckarsky and Dodson, 1980) could
also be attributed to interference with competitors that has been
observed in behavioral experiments.

     The question of relative allocation of time among various
activities, such as foraging and predator-avoidance, also has been
examined by Peckarsky (1980). The behavior of six mayfly nymphs
was recorded in observation boxes in the Colorado and Wisconsin
streams before and after the introduction of three stonefly
predators. Behavior was enumerated each minute for 15 minutes
from a repertoire including crawling, swimming, drifting,
drift-swimming, stationary, and posturing (instantaneous and scan
sampling). Prey that were relatively stationary in the absence of
predators became highly active in their presence (Baetis spp.,
Heptageniidae spp.), and vice versa (Ephemerellidae spp.) (Fig.
3). These responses suggest that different mayflies allocated
different energy resources to predator-avoidance. In addition,
such observations revealed an increase in certain behaviors in the
presence of predators, such as the "scorpion" posture in
Ephemerella and a "tail curl" behavior in B. bicaudatus, in which
the mayflies flexed their abdomens laterally, usually in the
direction of the predator. Further behavioral experiments can be
designed to identify the function of these responses in defense
against predation, and their role in maximizing fitness.

REPRODUCTIVE BEHAVIOR

     Few aquatic insects breed in the stream habitat. Only
representatives of the Hemiptera and Coleoptera have aquatic
adults. Most studies of mating behavior in aquatic insects
including those of courtship displays, acoustical signalling, and
paternity assurance have been conducted on lentic Hemiptera or
Coleoptera (see Jansson, 1973; Ryker, 1976; Smith, 1979).
However, one truly elaborate study of the reproductive behavior of
a semi-aquatic lotic insect deserves discussion here.

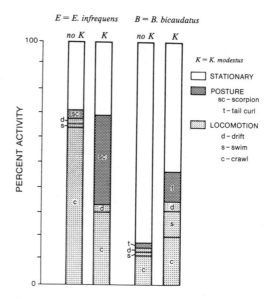

Fig. 3. Change in activity of <u>Ephemerella infrequens</u> and <u>Baetis</u>
<u>bicaudatus</u> in the presence of <u>Kogotus modestus</u>.

    Male riverine water striders (Gerridae, Hemiptera) produce
calling signals of a wide range of frequencies by generating
surface waves using their pro-, meso-, and metathoracic legs
(Wilcox, 1972, 1979). Females also signal, but only at low
frequencies using their forelegs. Wilcox conducted numerous
experiments to document the call sequence in courtship and mating
of two species of Gerridae. To do so, he used a variety of
ingenious gadgetry and techniques, including recording mating
behavior in total darkness by connecting a galvanometer and float
system to a Rika Denkl Model B-1 activity recorder, generating
artificial courtship signals with the galvanometer, blindfolding
males with custom-made rubber masks, and gluing to the foreleg of
a female a tiny samarium-cobalt magnet through which a signal
simulating the frequency and amplitude of the male signal could be
transmitted from an IMSAI 8080 microcomputer. His entire
computerized recording system could be converted to battery
operation for field use.

    Female or male responses were scored from a range of
behaviors. Female responses included approaching the male,
signalling, touching the male, spanning and moving laterally to
the mid- or hind-leg of the male, grasping a float that the male
inhabited, and ovipositing on the float (Wilcox, 1972). Male
responses included grasping an object (float), releasing the
object, attempts to copulate, backing away, and various signalling

responses. Behavior was recorded on videotape (<u>ad</u> <u>lib</u> sampling, sociometric matrix completion, sequence sampling). Results showed that surface waves produced by males and females entirely communicated the information required for sex-discrimination, courtship, mating behavior, and oviposition in Gerridae.

CONCLUSIONS

Simple to very complex behavioral experiments on stream insects have provided valuable data that can be interpreted in the context of general ecological theory. Careful quantification of behavior can be used to address questions of abiotic limitations to invertebrate distributions, the mechanism and role of dispersal within streams, competitive and predator-prey interactions, optimal foraging behavior, and reproductive behavior. This review demonstrates clearly the need for more rigorous behavioral experiments designed to test general ecological theory. Such data will allow us to evaluate the adequacy of the present conceptual framework for explaining phenomena occurring in streams.

ACKNOWLEDGMENTS

I'd like to thank the magic of rabbit-rabbit night for this manuscript and Jim Barnes and Wayne Minshall for not editing this acknowledgment. Steve Horn prepared the original illustrations. Beth French and Susan Pohl provided expert editing. Discussions with Dave Hart and reviews by Mike Wiley and Andy Sih provided valuable comments on earlier versions of this manuscript.

REFERENCES

Altmann, J. 1974. Observational study of behavior: sampling methods. Behaviour 49:2227-2267.
Charnov, E. L. 1976. Optimal foraging: the marginal value theorem. Theor. Popul. Biol. 9:129-136.
Ciborowski, J. J. H. and L. D. Corkum. 1980. Importance of behaviour to the re-establishment of drifting Ephemeroptera. pp. 321-330. <u>In</u>: J. F. Flannagan and K. E. Marshall (eds.), Advances in Ephemeroptera biology. Plenum Press, New York.
Ciborowski, J. J. H., P. J. Pointing, and L. D. Corkum. 1977. The effect of current velocity and sediment on the drift of the mayfly <u>Ephemerella</u> <u>subvaria</u> McDunnough. Freshwat. Biol. 7:567-572.
Corkum, L. D. 1978. The influence of density and behavioural type on active entry of two mayfly species (Ephemeroptera) into the water column. Can. J. Zool. 56:1201-1206.

Corkum, L. D. and H. F. Clifford. 1980. The importance of species associations and substrate types to behavioural drift. pp. 331-341. In: J. F. Flannagan and K. E. Marshall (eds.), Advances in Ephemeroptera biology. Plenum Press, New York.

Edington, J. M. 1968. Habitat preference in net-spinning caddis larvae with special reference to the influence of running water. J. Anim. Ecol. 37:675-692.

Elliott, J. M. 1971. The distances travelled by drifting invertebrates in a Lake District stream. Oecologia 6:350-379.

Gallepp, G. W. 1977. Responses of caddisfly larvae (Brachycentrus spp.) to temperature, food availability and current velocity. Amer. Mid. Natur. 98:59-84.

Glass, L. W. and R. B. Bovbjerg. 1969. Density and dispersion in laboratory populations of caddisfly larvae (Cheumatopsyche: Hydropsychidae). Ecology 50:1082-1084.

Hart, D. D. 1979. Patchiness and ecological organization: experimental studies within stream benthic communities. Ph.D. Thesis. University of California, Davis.

Hart, D. D. 1981. Foraging and resource patchiness: field experiments with a grazing stream insect. Oikos 73:46-52.

Hart, D. D. and V. H. Resh. 1980. Movement patterns and foraging ecology of a stream caddisfly larva. Can. J. Zool. 38:1174-1185.

Hayden, W. and H. F. Clifford. 1974. Seasonal movements of the mayfly Leptophlebia cupida (Say) in a brown-water stream of Alberta, Canada. Amer. Midl. Nat. 91:90-102.

Hildrew, A. G. and C. R. Townsend. 1980. Aggregation, interference and foraging by larvae of Plectrocnemia conspersa (Trichoptera:Polycentropodidae). Anim. Behav. 28:553-560.

Hughes, D. A. 1966a. On the dorsal light response in a mayfly nymph. Anim. Behav. 14:13-16.

Hughes, D. A. 1966b. The role of responses to light in the selection and maintenance of microhabitat by the nymphs of two species of mayfly. Anim. Behav. 14:17-33.

Jansson, A. 1973. Stridulation and its significance in the genus Cenocorixa (Hemiptera:Corixidae). Behaviour 46:1-36.

Jansson, A. and T. Vuoristo. 1979. Significance of stridulation in larval Hydropsychidae (Trichoptera). Behaviour 17:167-186.

Krebs, J. R. 1978. Optimal foraging decision rules for predators. pp. 26-63. In: J. R. Krebs and N. B. Davies (eds.), Behavioral ecology: An evolutionary approach. Sinauer, Sunderland, MA.

Mackay, R. J. 1977. Behavior of Pycnopsyche (Trichoptera: Limnephilidae) on mineral substrates in laboratory streams. Ecology 58:191-195.

Murdoch, W. W. and A. Sih. 1978. Age-dependent interference in a predatory insect. J. Anim. Ecol. 47:581-592.

Neave, F. 1930. Migratory habits of the mayfly Blasturus cupida Say. Ecology 11:568-576.

Needham, J. G., J. R. Traver, and Y-C Hsu. 1935. The biology of mayflies. Comstock Publish. Co., Inc., Ithaca, N.Y.

Peckarsky, B. L. 1979. Biological interactions as determinants of distributions of benthic invertebrates within the substrate of stony streams. Limnol. Oceanog. 24:59-68.

Peckarsky, B. L. 1980. Predator-prey interactions between stoneflies and mayflies: behavioral observations. Ecology 61:932-943.

Peckarsky, B. L. and S. I. Dodson. 1980. An experimental analysis of biological factors contributing to stream community structure. Ecology 61:1283-1290.

Ryker, L. C. 1976. Acoustic behavior of Tropisternus ellipticus, T. columbianus, and T. lateralis limbalis in western Oregon (Coleoptera:Hydrophilidae). Coleopt. Bull. 30:147-156.

Sih, A. 1980. Optimal behavior: can foragers balance two conflicting demands? Science 210:1041-1043.

Smith, R. L. 1979. Paternity assurance and altered roles in the mating behaviour of a great water bug, Abedus herberti (Heteroptera, Belastomatidae). Anim. Behav. 27:716-725.

Swammerdam, J. 1937. (Ephemera in) Biblia Naturae; sine Historia insectorum. Vols. 1 and 2, pls. 13-15. cited in Needham et al., 1935.

Tachet, H. 1977. Vibrations and predatory behaviours of Plectrocnemia conspersa larvae (Trichoptera). Ziet. Tierpsychol. 45:61-74.

Townsend, C. R. and A. G. Hildrew. 1980. Foraging in a patchy environment by a predatory net-spinning caddis larva; a test of optimal foraging theory. Oecologia 47:219-221.

Waters, T. F. 1972. The drift of stream insects. Ann. Rev. Entomol. 17:253-272.

Wilcox, R. S. 1972. Communication by surface waves. Mating behavior of a water strider (Gerridae). J. Comp. Physiol. 80:255-266.

Wilcox, R. S. 1979. Sex discrimination in Gerris remigis: role of a surface wave signal. Science 206:1325-1327.

Wiley, M. J. and S. L. Kohler. 1980. Positioning changes of mayfly nymphs due to behavioral regulation of oxygen consumption. Can. J. Zool. 58:618-622.

Wiley, M. J. and S. L. Kohler. 1981. An assessment of biological interactions in an epilithic stream community using time-lapse cinematography. Hydrobiologia 78:183-188.

Wilson, D. S., M. Leighton, and D. R. Leighton. 1978. Interference competition in a tropical ripple bug (Hemiptera: Veliidae). Biotropica 10:302-306.

THE IMPORTANCE OF COMPETITIVE INTERACTIONS WITHIN

STREAM POPULATIONS AND COMMUNITIES

David D. Hart[1]

W. K. Kellogg Biological Station
Michigan State University
Hickory Corners, Michigan   49060

INTRODUCTION

Of the ecological factors that shape populations and
communities, competition holds a prominent position. As ecology
seeks to gain predictive power as a growing science, it will
become increasingly important to know what role competitive
interactions play in producing patterns within natural
populations and communities. In this review, I consider some of
the evidence suggesting the relative importance of such
competitive interactions within stream ecosystems, emphasizing
studies of stream invertebrates. Though rigorous studies of such
processes are few, I will conclude that the role of competition
in these ecosystems has been underestimated. I suggest some of
the fruitful paths towards a better understanding of this
process, and consider the broader role of competition relative to
other processes which organize stream benthic communities.

Ecologists study the processes that govern the distribution
and abundance of organisms, and the patterns which result from
these processes. Probably the most fundamental dichotomy used to
categorize such processes is that of biotic vs. abiotic factors.
A broad view of ecological studies suggests that both factors can
operate to influence natural patterns in populations and
communities. Competition is one biotic factor that has received
a great deal of attention in many ecological studies. Yet the
prevailing view in stream ecology is that abiotic variables such
as temperature, water chemistry, substrate, and current velocity

---

[1]Present address: Academy of Natural Sciences, 19th and the
Parkway, Philadelphia, PA   19103.

have been of far greater importance than biotic factors such as
competition, predation, and other biological interactions. This
view is supported by a review of the literature, in that many
more studies demonstrate the clear cut role of physical-chemical
factors in affecting patterns (e.g. Hynes, 1970). Far fewer
studies even address the role of biotic factors, and only a
handful of these provide evidence demonstrating the importance of
these interactions in determining natural patterns. Indeed, as
recently as 1977, Minshall and Minshall indicated that "...there
is yet no concrete evidence of [competition's] importance in the
distribution of stream benthic invertebrates." In the next
section, I should like to ask whether this minimal amount
evidence for competitive interactions results from something
unique to the organization of stream ecosystems, or whether our
studies to date have fallen short in their attempts to fully
elucidate the role of these interactions.

Streams as Physically-Controlled, Heterogeneous Systems

Both theoreticians and empiricists have pointed out that the
potential importance of competition in natural systems can be
greatly reduced as a result of certain characteristics of
consumers and their environments. Streams are often considered
to be physically-controlled environments, in which flooding,
droughts, and associated changes in water chemistry, substrate
composition, and temperature act as strong sources of
density-independent mortality for stream inhabitants. There is
little question that such physical factors can and do influence
the distribution and abundance of stream organisms (for example,
see references in Hynes, 1970). However, this observation by
itself does not negate the potential importance of biotic
interactions in such systems. In terms of the ecological niche
(Hutchinson, 1957), it is clear that such abiotic factors do set
broad limits on the size and shape of an organism's fundamental
niche. No organism tolerates all environmental conditions. The
question left unanswered, however, is whether the niche occupied
by a species in the presence of other species (i.e. the realized
niche) is reduced in size below the wider limits resulting from
abiotic tolerances (cf. Birch, 1979).

Many authors have suggested that stream inhabitants have
been selected for a short-lived, fecund, highly dispersive (i.e.
r-selected) life history (e.g., Patrick, 1975) This life history
stereotype sacrifices competitive ability for rapid growth and
dispersal, in order to take advantage of temporally heterogeneous
environments (e.g., Southwood, 1981; cf. Parry, 1981). In such
habitats, conditions rarely remain constant long enough for
consumers to build up population densities at which competitive
interactions become important (e.g., Gray, 1981). There are a
number of problems with this simplified view (cf. Wilbur et al.,

1974; Stearns, 1976). For example, it makes the unreasonable assumption that life history characteristics require a one-for-one trade-off in abilities. In fact, however, it is entirely possible to find short-lived (e.g., univoltine) populations which are nonetheless adapted to relatively stable, high density conditions. In addition, calling stream organisms such as invertebrates "r-selected" is a bit like calling all vertebrates "K-selected." Although this may be a suitable perspective for a very broad overview of life histories, it is equally appropriate (and perhaps more interesting) to consider how consumers within a group of related taxa vary in life history features. Clearly, a wide range of tactics exist within any one of these groups (or even within a species - see Calow, 1981), which may be partially obscured by the application of coarse, life history stereotypes. Still, it is apparent that the habitats occupied by many stream organisms can be highly variable. The high seasonality of many streams is one component of this variability which may make it difficult for stream consumers to track resource availability in a changing environment. This scale of temporal heterogeneity, along with between-year variability, may be important in preventing competitive equilibria from being reached, as will be discussed later in more detail.

One of the most striking features of streams is their marked spatial heterogeneity at local scales (Hynes, 1970). Single stones can be extremely variable in size and shape, and substrate types in a given 10 m section of stream can include silt, sand, gravel, boulders, and macrophyte beds. There are clearly a number of microhabitats provided by the complex interaction between current and substrate within local areas, and a first impression is that the high local diversity of stream consumers may be a direct result of this wealth of microhabitats. To some, this pattern might suggest that there is plenty of "room" for coexistence of species, given such prominent spatial heterogeneity. Indeed, if all this diversity could be accounted for at the between-habitat scale (cf. MacArthur, 1965), species would not overlap in their utilization of resources, and thus, would not compete.

A final explanation for the apparently minor role played by competition in streams is that other biological interactions such as predation or parasitism may prevent many populations from reaching densities at which resources become limiting, thus eliminating competitive interactions within or between prey species (e.g. Patrick, 1972). The possibility that predators may mediate such interactions receives further attention below.

In summary, many workers have argued that competition is unlikely to play an important role in streams. They have suggested that a variety of factors may minimize its effect, including the harsh physical-chemical (abiotic) environment, the spatial and temporal heterogeneity of the environment, the non-competitive life history of the residents, and the density-reducing role played by predators. All of these views can only provide indirect evidence regarding the importance of competition in these systems, however. Descriptive and experimental studies of potential competitive processes themselves are required before any definitive perspective can be reached on this matter.

Earlier, I suggested that competitive interactions might be more important than previously supposed, and that the meager evidence for their influence might be more a result of insufficient and inadequate study than any inherent characteristic of stream ecosystems. In the sections to follow, I selectively review the evidence for such a position, after evaluating the necessary and sufficient forms of evidence required to demonstrate the role of competitive interactions.

Towards a Working Definition of Competition

A number of definitions and criteria have been proposed to demonstrate the existence of competitive interactions. Birch (1957) provided one of these: "Competition occurs when a number of animals (of the same or different species) utilize common resources the supply of which is short; or if the resources are not in short supply, competition occurs when the organisms seeking that resource nevertheless harm each other in the process." It is not my purpose to provide a detailed critique of this or any other definition that has been proposed. But at the core of most definitions lies the joint utilization of limiting resources, and the reduction in one or more components of an organism's absolute (as opposed to relative) fitness that results. There is some disagreement as to whether we should be measuring population-level effects of competition, such as reductions in density and rate of increase, or whether it is also necessary to demonstrate effects on individuals in terms of reduced growth, fecundity, or probability of survival. Although both levels unquestionably provide valuable forms of evidence, I believe that predictive models of community organization will be of limited utility until they incorporate information on the individual-level effects of competition, as I discuss later.

Two major types of competitive interactions have been identified by ecologists (Miller, 1967). The first, exploitation, occurs when competing individuals harm each other indirectly, by consuming and depleting a common, limiting resource. To demonstrate the operation of this mechanism, one must show both that the consumers deplete the supply of the

resource, and that they would be more successful if the resource
were in greater supply (i.e., if it were not limited).
Interference, on the other hand, involves a direct interaction
between consumers. Here, behavioral or allelochemical
"aggression" typically allows some consumers to prevent others
from gaining access to or using a resource. In this case, we
must show that some individuals (or species) interfere with the
ability of others to acquire resources, and that in the absence
of such interference, the former recipients of such aggression
would be more successful. Given these admittedly shortened
definitions, it should be possible to determine whether a
particular study provides sufficient evidence to justify the
conclusion that competition is operating.

EVIDENCE FOR COMPETITION IN STREAMS

     Many different approaches have been taken in studying the
role of competition in populations and communities. Undoubtedly
the most common approach involves descriptive analyses of the
resource utilization patterns of consumers. When practiced most
carefully, this method involves deducing patterns of resource
utilization given the assumption that competition is operating,
and then comparing these expected patterns with those observed in
nature. However, competition is but one of many factors which
act upon such patterns. The fact that many factors can interact
in their influence upon patterns of resource use suggests that it
will be difficult to evaluate the importance of any single factor
using descriptive approaches (cf. Dayton, 1973). Such
confounding variables are best controlled by the use of
experimental methods. Although laboratory experiments offer such
controls, they often fail to incorporate the realism of the
natural environment. Field experiments (cf. Connell, 1975)
involving the manipulation of consumers and their resources
frequently offer an especially rigorous means of evaluating the
effect of competition on resource utilization patterns.

Field Experiments

     The beauty of the field experimental approach is that it
allows us to work within the context of the complex natural
environment, yet systematically evaluate the role of particular
variables by comparing the response of organisms between
manipulated treatments and controls. Such in situ experiments
have already proven their usefulness in analyzing the importance
of particular features of the abiotic environment, such as
organism-substrate interactions (e.g., Rabeni and Minshall, 1977;
Hart, 1978; Reice, 1980). However, only a handful of field
experiments have been published which are designed to assess the
importance of competition in nature. In the section to follow, I
review some of these experimental studies.

    Wilson et al. (1978) carried out a series of field
experiments with a species of tropical ripple bug, Rhagovelia
scabra. They observed that individuals belonging to certain age
and sex classes were under-represented in habitats with the
highest availability of food. To determine whether intraspecific
competition contributed to this pattern, they removed the adult
females which were most abundant in the preferred food-rich
microhabitats, and noted that the remaining individuals moved
into these same preferred microhabitats in which they had
previously been under-represented. Thus, it was shown that the
distribution of these smaller individuals in lower quality
habitats resulted not from preference, but from active exclusion
by adult females, which chased the sub-ordinates out of these
preferred sites. Indeed, the ability of individuals to exclude
conspecifics from preferred microhabitats followed a type of
dominance hierarchy, with adult females superior to adult males,
which were superior to juveniles. Juvenile members often failed
to forage at all, and the authors suggested that this severe form
of interference might act to regulate population sizes much as
has been suggested for vertebrate populations (e.g., Brown,
1969).

    Hart (1982a) identified a similar phenomenon with a tunnel-
dwelling caddis larva, Psychomyia flavida. Here the contested
resource was silken tunnels in which larvae lived, and animals
were found wandering over the stone surface searching for vacant
tunnels. Most tunnels were already occupied, however, and fights
ensued between tunnel residents and these intruders. The outcome
of these fights depended on the relative size of the contestants,
with larger larvae (later instars) more successful at either
defending against or evicting smaller larvae (earlier instars).
It also was shown that smaller larvae often are under-represented
in preferred microhabitats, in which the larger larvae are
common. Upon removing these larger individuals however, small
instar larvae were quick to colonize the vacated tunnels. These
and other data were used to evaluate contrasting models of
habitat selection under the constraint of intraspecific
competition.

    Interference behavior can also be directed at individuals
that are not conspecifics. Hart (1982b) carried out descriptive
and experimental field studies of Leucotrichia pictipes. This
caddis larva lives in a fixed shelter and actively defends the
surrounding territory in which it grazes upon algae. Such
behavior not only resulted in a uniform spacing of conspecifics
on the surface of stones, but also prevented co-occurring baetid
mayfly nymphs from entering territories to feed, which they were
shown to utilize once territory residents were experimentally
prevented from defending these areas. A similar alteration in
the use of habitats by brook trout was observed when territorial

brown trout were experimentally removed from a stream (Fausch and White, 1981).

In the previous four studies, the competitive interaction has involved interference mechanisms. In many ways, the presence of interference competition is more easily identified than exploitation, since interference is so often manifested through visible agonistic encounters. Few other field studies have explicitly identified the competitive mechanism, but they are still useful for determining whether competition operates in these systems. During a series of field experiments which focused on the foraging behavior of grazing caddisfly larvae, Hart (1981) quantified natural rates of resource renewal ($\simeq$ supply) and resource depletion ($\simeq$ demand). The relative size of these rates, in relation to the density of conspecifics, suggested that these larvae were competing (probably exploitatively) for a limiting resource. A more direct approach to the evaluation of such intraspecific competitive interactions in grazers was taken in a subsequent study. There, Hart (1982c) noted that larval densities of a glossosomatid caddisfly reached values as high as one individual$\cdot$ cm^{-2}, and he hypothesized that these larvae might deplete their algal food supply at such densities. He carried out a field experiment in which larvae were confined to algal-covered tiles within enclosures at densities less than, equal to, and greater than those on the stream bottom. Algal biomass remaining on the tiles after several weeks was negatively correlated with grazer density, demonstrating resource depletion. The experiment also provided growth rate data indicating that this resource depletion resulted in a reduction in larval growth rates with increased densities. In this system, intraspecific competition appears to be mediated purely by resource exploitation, since interference behaviors directed towards competitors have never been observed in the field or laboratory.

Autotrophs, including both microalgae and macrophytes, are also known to deplete resources (e.g., nutrients and light) at natural densities (e.g., Gregory, 1978; Westlake, 1975). This fact, when combined with the results of field experiments in which growth rates can sometimes be stimulated by the addition of nutrients (e.g., Stockner and Shortreed, 1978; Gregory, 1980), suggests that competition for limiting resources may occur in various primary producers. In spite of the likely importance of such interactions in affecting patterns of seasonal and longitudinal succession in algae (Whitton, 1975), very little work has been done on this problem. Whether such interactions are mediated primarily by interference (cf. Patrick, 1978) or exploitation also remains to be determined.

Lock and Reynoldson (1976) experimentally reduced the density of two species of stream-dwelling triclads to determine whether competition might be affecting their "success". Under reduced densities, both species contained higher concentrations of an energy storage product (lyo-glycogen) compared with the pre-manipulation values. Although these experiments were flawed due to the inability to utilize experimental controls, they are nonetheless suggestive of resource limitation. Otto and Svensson (1976) performed a similar density reduction experiment with a detritivorous limnephilid caddis larva. Their primary interest was in the ability of animals to repopulate the area of reduction, but their results also indicated that larval survivorship in the low density region was significantly greater than would be expected from prior years of study. Warren et al. (1964) conducted a field experiment involving the enrichment of stream sections with sucrose. The enrichment stimulated the growth of the bacterium Sphaerotilus, which enhanced the productivity of invertebrates, thus leading to significant increases in the food consumption rate of coastal cutthroat trout. This result suggests that the trout may have been competing for food under natural circumstances.

Peckarsky (1979) carried out field experiments designed to determine whether dispersal to and from cage enclosures filled with substrate is dependent on the density of animals initially placed there. Peckarsky's original analyses did not correctly discriminate between density-dependent vs. density-independent mechanisms (Sell, 1981), though subsequent evaluation indicated an aggregate dependence on density when all species are pooled together (Peckarsky, 1981). The behavior of individual species with respect to density was much more variable, however, with less than 20% of the taxa exhibiting typical density-dependent patterns of migration. These few cases do not permit a determination of the nature of the biological interaction involved, but another study suggested that some colonization patterns may be influenced by the presence of "potential competitors" (Peckarsky and Dodson, 1980).

In summary, carefully designed field experiments can provide a strong source of evidence that competitive interactions within and between species occur in streams. When combined with direct observations, it is sometimes possible to identify the actual competitive mechanism, and occasionally such studies provide valuable insights regarding how such interactions lead to particular patterns of distribution and abundance.

Laboratory Experiments

Experimental studies of competition have been carried out more frequently in the laboratory, probably due to the greater

ease with which the process can be evaluated. Several studies
have identified interference competition as a process regulating
patterns of spacing and dispersal, including work with isopods
(Culver, 1971; Culver and Ehlinger, 1980), caddisfly larvae
(Glass and Bovbjerg, 1969) and salmonids (Chapman, 1962; Dill et
al., 1981). Hildrew and Townsend (1980) noted a size-dependent
agonistic contest for web ownership in predacious caddis larvae,
and commented on its relationship to patterns of patch selection.
Jansson and Vuoristo (1979) documented a similar pattern of
size-dependent interference in hydropsychids. Hart (1979a)
studied territorial behavior in filter-feeding blackfly larvae,
demonstrating that larval defense of attachment sites allows
individuals to enhance their rate of food intake. In a study of
territorial defense by brook charr, McNicol and Noakes (1981)
showed that the level of aggression towards intruders by a
resident depended on both the distance and direction to a given
intruder, such that less energetically expensive aggressive
behaviors were directed towards those intruders posing less of a
threat. Wiley (1981) carried out a laboratory study of
emigration rates of midges as a function of larval density and
substrate composition, and found a clear density-dependent
component to the pattern. He argued that the densities of larvae
in the field often exceeded the point where density-dependent
emigration became apparent in the laboratory, through possible
differences in food availability between the laboratory and field
might complicate this comparison. It is difficult to rigorously
utilize laboratory-derived coefficients to determine the
likelihood that competition operates in the field, but Wiley's
approach is suggestive of its role in nature.

Wiley's is one of many laboratory studies which focus on
factors influencing the probability that animals will leave the
substrate and be transported downstream by the current. This
phenomenon, termed stream drift, has been documented in the field
for more than a half century, and many of these studies have been
reviewed by Hynes (1970). Over the years, increasing attention
has been focused on whether the relationship between the number
of animals drifting and the number of animals on upstream
substrates is density-dependent or density-independent. Müller
(1954) suggested that invertebrate drift resulted from
competition for limiting resources on the stream bottom, and
Waters (1961) argued more explicitly that drift occurred when
population densities exceeded the carrying capacity. If such
competitive interactions do underlie drift activity, then a
density-dependent relationship between drift and benthic
densities would be predicted. Although a few field studies have
been carried out which describe such a pattern (e.g., Dimond,
1967; Gyselman, 1980), far more of these studies have been
conducted in the laboratory. Walton et al. (1977) and Walton
(1980) both found that such factors as species, substrate type,

and current velocity influenced whether drift rates were density-dependent or -independent. It is premature to judge whether the majority of drift activity follows this density-dependent pattern, given the limited number of careful studies conducted to date. But the fact that some of these patterns are clearly density-dependent, in combination with other laboratory experiments showing that drift rates rise as food availability declines (Bohle, 1978), suggests that competitive interactions can be an important factor influencing stream drift.

Other laboratory experiments have focused on the manner in which competitive interactions influence such factors as the growth rates of individuals and populations. For example, Nilsson and Otto (1977) examined intra- and interspecific competitive effects on the larval growth of a caddisfly by manipulating the density of a potential competitor. They noted that growth declined significantly with increasing consumer densities due to both intra- and interspecific competition (cf. Peters and Barbosa, 1977). Colbo and Porter (1979) found that both growth rate and probability of survival declined with increased larval blackfly densities. Sculpin food consumption was shown to decline not only with increasing densities of conspecifics, but also with increases in the density of stonefly nymphs (Davis and Warren, 1965). This result reinforces the notion that competitive interactions need not be restricted to taxonomically similar species, as will be discussed in more detail later. Asymmetric competitive interactions have also been demonstrated in laboratory experiments, in which sculpin had a negative effect on food consumption by trout, whereas trout had little effect on the sculpin (Brocksen et al., 1968). Competitive replacement of one isopod species by another was illustrated in the experiments of Hynes and Williams (1965). Klotz et al. (1976) identified how the outcome of competition between diatoms and green algae varied as a function of nutrient availability, and related these laboratory results to similar patterns of relative abundance in the field. All of these laboratory studies are useful in their own right, and they allow investigators to systematically evaluate the role of particular factors while holding others constant in the experimental design. However, at best they can only demonstrate the potential (rather than acutal) role played by competition in the field setting, given the inherent difficulties in extrapolating laboratory results to the field.

Descriptive Studies

Descriptive studies of competition generally involve the comparison of natural patterns of distribution, abundance, and resource use with those predicted as a result of competitive interactions. Some of the natural patterns that have been evaluated in this regard include the partitioning of resources along environmental gradients by co-occurring species, especially the spatial arrangement of such niches along one or more gradients. A more quantitative version of this kind of approach compares various indices of resource use by species (such as frequencies of species co-occurrence, differences between trophic structures or other morphological features, relative abundance patterns, etc.) with those predicted from statistical null models in which the effect of interspecific competition is explicitly removed (e.g., Caswell, 1976).

Natural "experiments". The term natural "experiments" (sensu Connell, 1975) is actually a misnomer. They differ from a true experimental approach in that they lack one critical element--the control. For example, one might observe a system with two similar species, one of which occupies the upper and the other the lower reaches of a stream, with little or no overlap in altitudinal distribution. This pattern could then be explained by hypothesizing that the species exclude each other due to competition. To test this hypothesis, one could look for a second stream where one of the two species is missing. If competition determines the altitudinal boundary when the two species co-occur, the species present in the second stream should extend beyond that boundary.

This hypothetical pattern is quite similar to the one described for two stoneflies in Great Britain (Hynes, 1970). Diura bicaudata is normally replaced by Perlodes microcephala in streams at lower elevations, but the former species occurs all the way down to sea level on the Isle of Man, where P. microcephala is absent (presumably for historical reasons). Although this pattern fits the predicted relationship that would result from competition, it suffers from the flaw shared by all natural experiments: without the control, there is no certainty that all factors other than the absence of one species remain constant between the locations being compared. Thus, one is left with information providing a correlation between two variables, but no certainty that we have correctly identified the causal mechanism (see Diamond, 1978 for an alternate view).

Lock and Reynoldson (1976) examined how the distribution and abundance of a stream-dwelling triclad responded to natural variations in the abundance of a potentially competing triclad. In a segment of stream in which the abundance of the principal triclad was greatly reduced, the other species increased substantially in density, while remaining relatively unchanged elsewhere. They consider this natural experiment to provide relatively strong evidence indicating the operation of interspecific competition.

Studies of resource partitioning. Yet another form of correlative evidence for competition comes from the study of resource partitioning (Schoener, 1974), which describes differences in the way species in the same guild utilize resources. These studies attempt to analyze the limits interspecific competition places on the number of species that can coexist in a stable community. At the heart of these analyses lies the concept that the niche of a species can be displayed graphically (cf. Hutchinson, 1957), with resource spectra represented along niche axes, and resource utilization functions portraying the frequency with which certain portions of a given resource spectrum are utilized by a particular species. Thus, one might describe how two or more species utilize a series of microhabitats which differ in substrate particle size, and then examine the degree to which these resource utilization functions overlap along this niche dimension. Given such data, what patterns of resource utilization implicate the role of interspecific competition? Schoener (1974) argued that the mere demonstration of differences between utilization functions is not enough, since even if niches were arranged randomly with respect to each other, differences would exist (cf. Strong et al., 1979). Instead, he felt that competition would result in a regular spacing of niches (resource utilization functions) in niche space (along a resource axis). Where niches are uniformly and widely spaced over one or more dimensions, it may be possible to reject the null hypothesis of randomly generated differences.

Schoener's review described three principal niche or resource dimensions which can be partitioned among species: time, habitat (space), and food. Furthermore, he concluded that of these three axes, species are most likely to be separated due to habitat segregation, second most due to food segregation, and that temporal segregation is a relatively uncommon means of achieving resource partitioning (i.e. avoiding or reducing competition) among the taxa surveyed. He added, however, that aquatic animals appeared to segregate by habitat less often than terrestrial taxa.

Much of the study of resource partitioning involves a consideration of the degree to which the resource utilization functions of different species overlap, and the relationship of such niche overlap to competition per se. Colwell and Futuyma (1971) and Kohn (1971) have pointed out a number of considerations to keep in mind when assessing niche overlap, but perhaps the sternest warning they offer is that niche overlap may have little or no relationship to competition. Niche overlap (i.e. the sharing of resources between two or more species) is harmless in a competitive sense if the resources are not acting to limit the consumers. Thus, caution must be used in determining whether overlap can be equated with competition, and

it will often be necessary to carry out supplementary studies of
resource supply and demand before resources can be said to be
limiting.

a. Patterns of temporal segregation. Coleman and Hynes
(1970) studied the life history patterns of "closely allied
species," and found that the temporal patterns of development
differed between the species. They concluded that such
differences probably allowed the species "to coexist without
intense interspecific competition." Oswood (1976) states that
"separation in time is the most likely strategy of ecological
segregation in closely related Hydropsychidae." He theorizes
that these species are constrained to high niche overlap with
respect to food and habitat, and therefore concludes that
"divergence of life history patterns allows ecologically distinct
instars to partition available resources." In a refinement of
this approach, Cather and Gaufin (1976) demonstrate a seasonal
succession in the maximum absolute growth rates of three Zapada
species. Temporal segregation is thought to reduce interspecific
competition as a result of differences in the times at which the
species undergo maximum growth (i.e. periods of high energy
demand are separated). Vannote and Sweeney (1980) present data
for several different guilds of insects, describing the role that
thermal seasonality plays in producing temporally staggered
periods of maximum growth, and the manner by which such temporal
segregation may reduce competition within a habitat.

b. Patterns of habitat segregation. Two species occupying
totally different habitats are unlikely to compete, since it is
difficult for them to deplete each others' resources. However,
the degree to which two species are spatially segregated is
highly dependent upon the spatial scale being considered. For
example, Otto (1976) shows large-scale differences in the
habitats utilized by three related caddis larvae (e.g., one
species occupies a small stream, another inhabits temporary
pools, and the third lives in nearby ponds). He quantifies some
of the factors leading to the differential success of each
species in its particular habitat, and suggests that
interspecific competitive interactions during the evolutionary
history of these species may have contributed to the patterns of
segregation. The distributions of seven species of congeneric,
pool-dwelling darters are almost entirely non-overlapping within
a single drainage system (Page and Schemske, 1978), and the
authors propose that this medium-scale pattern of partitioning
results from competition for habitat space. Microspatial
separation of species has been demonstrated by Ulfstrand (1967),
who described various benthic invertebrates that are segregated
on sand vs. gravel vs. cobble substrates at the same site along a
stream. On an even finer scale of spatial resolution, there is
some evidence that species may be specialized to inhabit separate

microhabitats on the same stone, as indicated by Linduska's
(1942) description of the different habitat preferences of
several species of mayfly nymphs on a single, 60 cm boulder (cf.
Hart, 1978; Trush, 1979). To the extent that resource
partitioning occurs at such small spatial scales, it may well go
undetected in conventional studies. Fortunately, new sampling
methodologies are being developed which make it possible to
quantify within-stone microspatial separation, and to carry out
experimental studies evaluating the role played by competition in
generating these patterns (e.g., Alstad, 1981; McAuliffe, this
volume; Hart, 1982a).

Bovbjerg (1970) carried out an elegant study of spatial
resource partitioning which began with the observation that two
species of crayfish are ordinarily ecologically isolated, with
Orconectes immunis occupying ponds and O. virilis occupying
streams. However, he found one river in which the two species
co-occurred, with the species overlapping where pools alternated
with riffles. By conducting a series of laboratory experiments,
he determined that O. immunis is better adapted than O. virilis
to live in relatively stagnant habitats, where low oxygen
concentrations and summer drying are prevalent. In the absence
of their congeners, O. virilis is poorly adapted to these harsh
conditions, whereas O. immunis is not only able to live in both
pools and ponds, but also the riffle habitats ordinarily occupied
by O. virilis. Bovbjerg observed behavioral interactions between
the two species and was able to determine the factor that
prevents O. immunis from occupying riffles. When the two species
interact, especially while seeking crevices for shelter, O.
virilis is much more aggressive, and is thus better able to
acquire and defend such apparently limiting shelter sites in its
preferred habitat. This study provides an excellent demonstration of the included niche phenomenon (Miller, 1967), where
coexistence between two competitors results from the fact that
the inferior competitor has a refuge outside the fundamental
niche of the dominant competitor.

Tuskes (1975) supplemented descriptive studies with some
experimental work to evaluate the effect of competition on the
distribution and abundance of two congeneric pyralid moth larvae.
He studied Parargyractis jaliscalis and P. confusalis in both
allopatry and sympatry. P. confusalis has a lower tolerance to
reduced oxygen concentration than does P. jaliscalis, and thus
the former species is found primarily in higher elevation
streams. Because P. jaliscalis is more aggressive than P.
confusalis, and can actively displace the latter species from its
shelters, this also acts to set a lower elevation limit to the
range of P. confusalis. On the other hand, the upper elevation
limit for P. jaliscalis may in part be determined by increased
current velocities at higher elevations, which causes their

dislodgement from stones during winter storms. Thus, the upper limit to the distribution of P. jaliscalis seems to be determined especially by physical factors, whereas the lower limit to P. confusalis apparently results from both physiological tolerances and competitive interactions. It is interesting to compare this pattern to commonly observed patterns of vertical zonation in the rocky intertidal. There, high intertidal species tend to be poor competitors but tolerant to widely ranging abiotic conditions, whereas low intertidal species are often less tolerant physiologically but better competitors (Connell, 1972).

c. Patterns of food segregation. In his review of the trophic relations of aquatic insects, Cummins (1973) states that these animals are usually food generalists. To the extent that he is correct, this either suggests that species avoid competition by being segregated along other niche axes (i.e., space and time), or that food resources are not limiting, in which case broad, overlapping diets do not lead to competition. Although Cummins prefers the former explanation, I am aware of very few experimental studies that are sufficiently detailed to reject the latter hypothesis. For example, Fuller and Stewart's (1977) study of food utilization patterns in stoneflies demonstrate a high degree of dietary overlap in temporally coexisting species. They provide some evidence that competition is not occurring in this system due to the high productivity of prey groups (i.e. food resources may not be limiting). Devonport and Winterbourn (1976) explain the high degree of food overlap between two invertebrate predators in a similar fashion, indicating that prey taxa are abundant year round. The absence of a shift in the diet of one of the predators, where it occurs in the absence of the other, supports the hypothesis that extensive niche overlap with respect to food is not an indication of competition between the two species. Although observations of feeding specialization and partitioning of food resources are less frequent, they nonetheless exist. Certain leptocerid caddis larvae are specialized to feed upon freshwater sponges (Lehmkuhl, 1970), and some hydroptilid caddis larvae feed primarily by piercing the cells of filamentous algae and extracting the contents (Nielsen, 1948). Partitioning of food resources by similar species has been suggested to occur both on the basis of particle size for filter feeders (e.g., Wallace, 1975; Malas and Wallace, 1977), and by differences in the types of prey taxa taken (e.g., Thut, 1969).

d. Patterns of partitioning along several resource dimensions. Schoener's (1974) review provides qualitative predictions regarding the effect of competition on patterns of resource partitioning along more than one resource dimension (or niche axis). In particular, he discusses the separation of species along complementary dimensions, suggesting that

similarity of two species' resource utilization functions on one
axis must be offset by dissimilarity of their functions along
another. For example, he considers cases where species that
overlap in habitat eat different foods, where species consume the
same food resources at different times, etc. The consideration
of multiple axes has already been introduced in this paper, but
will now be analyzed in more detail.

Cummins (1964) performed an early and detailed study of
partitioning along several resource dimensions by limnephilid
caddis larvae. He found that Pycnopsyche lepida and P. guttifer
showed distinct microhabitat segregation, in addition to
differences in temporal overlap of growth periods. Mackay (1972)
and Mackay and Kalff (1973) also analyzed patterns of resource
partitioning in Pycnopsyche. They detected little temporal
segregation of the dominant species, but showed some segregation
with respect to food, and even more complete separation in
habitat preference. Although they interpret much of this pattern
in light of avoidance of competition, a supplementary laboratory
experiment indicated that patterns of habitat selection differ
very little in the presence and absence of congeneric
competitors, which tends to suggest that ongoing competition is
not playing an important role in generating the observed patterns
of resource partitioning. On the other hand, this segregation
may result from prior competition during the evolutionary history
of these species. Unfortunately, this explanation is a potential
alternate hypothesis for many experimental studies of resource
partitioning in which species do not alter their patterns of
resource use in the presence vs. absence of competitors (cf.
Schoener, 1974; Diamond, 1978). Evaluating the credibility of
this hypothesis will not be an easy task (cf. Connell, 1980).

Sheldon (1972) chose a well-suited study system, where four
perlodid stoneflies were known to overlap temporally.
Recognizing that temporal segregation was minimal, he then
explored patterns of resource partitioning with respect to food
and habitat. The study indicated that major differences in food
habits were present, but that habitat preferences did not differ
greatly. Further, those species pairs occupying the most similar
habitats showed the greatest difference in diets. This study
fits well with the predictions derived from the theory of
resource partitioning. In a more recent study, however, Sheldon
(1980a) found a minimal degree of ecological segregation between
two perlid stoneflies along niche dimensions related to habitat,
food, and (in part) season. Though certain potentially important
spatial and temporal scales went unanalyzed, he implies that
these two species are likely to be competing for limiting
resources given their extensive niche overlap. Magdych (1979)
found a high degree of niche overlap with respect to food, space,
and time between herbivorous mayfly nymphs living in macrophyte

beds. Though he notes that non-equilibrium processes may lead to such patterns in streams, he argued that this particular consumer-resource system is likely to be more stable than many. Instead, he concluded that resources were not in short supply, though it is unclear what mechanisms prevent these populations from increasing in size, and thus reducing resource availability.

Allan's (1975) research on longitudinal zonation simultaneously analyzed niche overlap along several spatial scales. Congeneric species that exhibited sharp mutual truncation in longitudinal distribution had very similar microhabitat preferences, whereas other congeners that showed more overlap differed in their microhabitat preferences. This result suggests that competition may influence the patterns of resource utilization by the species. Hildrew and Edington (1979) examined patterns of longitudinal distribution in net-spinning caddis larvae. Though some species had disjunct distributions related to differences in thermal preferenda, others co-occurred at the same sites. They suggested that some within-site patterns of coexistence might be maintained by non-equilibrium processes, whereas others involved microspatial segregation. They believe that space is more likely to be a limiting resource than food for these filter-feeders, which agrees with theoretical arguments derived from considerations of resource depressibility (Charnov et al., 1976; Hart, 1979a).

Townsend and Hildrew (1979) conjecture that diet specialization is more effective than habitat specialization in reducing competition between two insect predators, due to the low degree of prey specificity for particular habitat types. They observed that the degree of dietary overlap between the two species varied seasonally, with high overlap during seasons of resource abundance and much lower overlap when resources were scarce. Zaret and Rand (1971) documented a similar pattern of minimal food overlap during periods of resource scarcity between tropical stream fishes. In a study of resource partitioning by stream cyprinids, Baker and Ross (1981) noted segregation on several dimensions. Some species occupied different drainages, and others which co-occurred within a given reach partitioned space according to vertical position in the water column, and degree of association with vegetation. One pair of species that showed very similar patterns of spatial resource utilization partitioned time, with one species largely nocturnal and the other diurnal. Mendelson (1975) found that minnows in a Wisconsin stream segregated primarily on the basis of habitat rather than prey type. Instead of invoking ongoing competition as the factor producing this segregation, however, he suggests that the pattern of resource partitioning is more likely to result from morphological preadaptations.

As a last example, one may consider the widely cited work of
Grant and Mackay (1969). They argue that closely related species
must be ecologically segregated in order to avoid competing.
Having made this assumption, they go on to quantitatively
estimate the relative importance of temporal and spatial scales
of separation. They develop an index of separation that is
questionably related to resource overlap and competition (as they
admit), and their analysis of data is weak in that no replicate
samples were taken. They conclude that whereas some degree of
segregation is accounted for by their measures, a number of
superficially similar species must coexist by utilizing different
resource exploitation patterns (e.g., microhabitat or dietary
differences) that went unmeasured.

What conclusions can be reached regarding the extent to
which these and other studies of resource partitioning elucidate
the role of competitive interactions in stream communities?
Basically, since differences in the resource utilization patterns
of species may often be unrelated to competition (either in the
present or the past), such patterns will ordinarily be difficult
to interpret. Studies evaluating both the degree to which
species overlap in their use of resources, and the availability
of these resources relative to consumer demand, are more likely
to identify whether niche overlap and competition are necessarily
related (cf. Birch, 1979). Even in ecological settings in which
resources are known to be limiting, however, few studies examine
in a quantitative fashion whether the patterns of spacing between
resource utilization functions differ from those expected by
chance. Despite all of these weaknesses, such studies of
resource partitioning represent the most common approach towards
the examination of potential interspecific competitive
interactions. As such, they provide much raw material from which
to generate and test more rigorous hypotheses regarding the
degree to which competition actually influences resource
utilization patterns.

General Null Models

The previous studies of resource partitioning represent one
qualitative version of a general approach to the evaluation of
competitive interactions within communities. The basic objective
of such studies is to describe the observed patterns of
distribution, abundance, morphology, etc., of different species,
and compare these with a null model in which the effect of
species interactions is absent. The value of this approach is
that it explicitly provides a null hypothesis which can be
statistically compared with the natural pattern. The approach is
weakened, however, because the patterns predicted by such models
are often quite sensitive to relatively small changes in
assumptions, with no clear indication regarding which is the

"appropriate" null model. Thus, there is frequently more than
one null model that can be tested against a given data set.
Furthermore, even when the null model can be confidently
rejected, it is difficult to ensure that competitive interactions
have therefore produced the observed patterns.

One of the simplest types of these studies is the analysis
of interspecific association patterns. Basically, this approach
involves collecting samples of organisms in a community, and
determining whether any given species is distributed across these
samples independently of other species. One way to test for
independence in distributions is to take a pair of species,
identify for every sample whether each species is present or
absent, and construct a 2 x 2 contingency table which lists the
number of cases where: 1) species 1 occurs alone; 2) species 2
occurs alone; 3) the two species are jointly present; or 4)
jointly absent. It is a simple matter to compare statistically
these frequencies with those resulting from a null model of
independent distributions. Similar analyses using numerical
abundance rather than simple presence-absence data can be
evaluated by correlation statistics. In either case, the
presence of competitive interactions might be suggested by
patterns of negative association or inverse correlation between a
species pair.

Unfortunately, this descriptive approach has several
short-comings which prevent it from providing a definite
conclusion regarding the importance of competitive interactions
(cf. Hart, 1979b). First, the degree to which a pair of species
is positively or negatively associated can be highly dependent on
the size of sampler used in collecting the raw data, and such
sampler sizes are often selected arbitrarily (see chapters by
Allan and Shiozawa, this volume). Second, many different indices
of association are available, and the use of two different
measures with the same raw data often can produce different
conclusions regarding patterns of association. Even more
fundamental than artifacts resulting from non-robust indices and
arbitrary sampler sizes is the fact that negative associations
can result either from active segregation of species as a result
of competitive interactions or from differences in habitat
preference that are unrelated to any ongoing competition. In
sum, it will be difficult to determine how frequently such
measures of association over- or underestimate the presence of
segregation resulting from competition. As a result, studies of
species association should be regarded as a tool that can
potentially suggest the importance of competitive interactions,
rather than as a method for documenting the presence of such
interactions. Reice (1981) provides similar cautions regarding
the use of this approach, though he ultimately concluded that the
low frequency of negative associations detected in his study
reflects the minimal importance of competitive interactions.

Another type of study which explicitly compares a null model with an observed pattern is the study of morphometry, especially as it relates to the use of resources by consumers. Such studies begin with the premise that particular aspects of an organism's morphology determine what kinds of resources it can use. Thus, by quantifying morphometric patterns, it is possible to identify patterns of morphological overlap and compare these with a null model in which morphometric features are distributed among species at random. Gatz (1979) followed this approach in describing the morphological niches of co-occurring stream fishes. He attributed the non-random spacing of these niches, and the relatively constant degree of morphological overlap, to the role of competition in constraining species packing. Sheldon (1980b) takes a different approach to the problem, by first quantifying the morphometric similarity of various perlid stoneflies living in the western U.S. He used these patterns of similarity to make predictions regarding which species should be capable of coexisting, demonstrating that greater morphometric similarity implies decreased likelihood of coexistence.

Studies of community-level morphometric patterns are only as good as the validity of their assumptions. When they examine both resource utilization and morphology, it becomes possible to document how functional morphology constrains foraging efficiency, and thus to predict and experimentally test how differences in competitive ability between species result in particular patterns of resource partitioning (e.g., Werner, 1977). Unfortunately, a number of studies leave unevaluated the precise relationship between resource use and morphology, greatly weakening the approach as a result. This failure to examine the validity of such a premise is especially serious when resources are known to be shared by many taxa of widely differing morphology. For example, the periphyton resource occurring on the surface of stones is frequently utilized by many taxa, including caddisflies, mayflies, true flies, beetles, tadpoles, and fish. These taxa clearly differ in morphology, and a morphometric assessment of niche overlap would probably conclude that the greatest competitive effect occurs between the most closely-related taxa, since they are most likely to be similar morphologically. Yet morphologically and phylogenetically distinct taxa often overlap markedly in the kinds of resources they utilize. Furthermore, these measures fail to account for the differing rates of resource consumption between taxa, in that a tadpole may consume periphyton at a rate several orders of magnitude greater than (say) a mayfly nymph, leading to strongly asymmetrical competive effects (cf. Lawton and Hassell, 1981). Thus, inferences drawn from a study of morphologically similar consumers may produce a very unrealistic picture of the ecological and evolutionary pressures which affect patterns of resource utilization within a guild. Without firmly grounding studies of community-level morphometry in a knowledge of the constraints placed on resource utilization by morphology, such studies seem destined to cloud rather than clarify our understanding of the role of competition in species packing.

One final set of natural patterns that have received a great deal of attention in terms of their ability to implicate the importance of competition are relative abundance distributions. Any natural community contains species which have varying degrees of commonness or rarity. By censusing the abundance of species in a community, a probability density function can be produced which indicates the relative numbers of species which occur in these different abundance categories. Three distributions that have received much attention are the broken-stick, the lognormal, and the geometric, which represent points along a continuum from relatively even to highly skewed distributions. A number of authors consider the lognormal to represent a null model for relative abundance patterns in which species occupy niche space independently of one another (but see Sugihara, 1980). Thus, workers often conclude that deviations from the lognormal infer that competition is an important factor shaping the community. Such a conclusion is plagued with difficulty, since a multiplicity of factors undoubtedly influence observed patterns of relative abundance. As a result, similar patterns can result from very different processes, once again pointing up the short-comings of testing the conclusions of community-level deductive models without examining the validity of their premises. That more than one set of assumptions can lead to similar or identical predictions regarding relative abundance distributions is widely recognized (e.g., Cohen, 1968; Pielou, 1969; May, 1975; Caswell, 1976), and one of the early proponents of this inferential approach to examine the role of competition later concluded that the approach did not warrant further study (MacArthur, 1966).

This review of approaches involving the comparisons of observed community patterns with null models is admittedly brief, but it serves to point at a major short-coming of the general method. Because of the complex array of factors which interact to determine such community-level patterns, it is extremely difficult to infer the operation of any particular process (e.g., interspecific competition) merely by describing a natural pattern. The recent flurry of arguments regarding the idea that many purported competitively-structured systems might really represent random assemblages of species documents one aspect of this problem (e.g., Strong et al., 1979). However, even when a null model predicts the observed community pattern as well as a model incorporating competitive effects, this does not rule out the possibility that competitive interactions are responsible for such patterns. It merely indicates that more than one set of

premises lead to the same prediction. One answer to this dilemma
is to build null models which incorporate more realistic
biological assumptions, and to test these models with more
powerful and unbiased statistical methods (e.g., Hendrickson,
1981; Diamond and Gilpin, 1982; Colwell and Winkler, 1982).
However, a more direct approach would be to test both the
assumptions and predictions of community organization models,
incorporating documented mechanisms into their structure in a way
that achieves more realism.

Towards a Balance of Approaches

I have documented the wide variety of approaches used by
investigators to better understand the role of competitive
interactions in streams. Although I have argued that some
approaches provide more direct evidence regarding the operation
of competition than others, this should not be construed to mean
that a single method of study can be expected to answer all of
the relevant questions concerning competitive interactions. Upon
considering the criteria necessary to demonstrate the occurrence
of competition in nature (e.g., Reynoldson and Bellamy, 1971;
Menge, 1979), it becomes clear that descriptive approaches need
to be supplemented with well-designed experimental field
manipulations (e.g., Connell, 1975; Birch, 1979). Paine's (1977)
view that research in the marine intertidal zone has contributed
more to ecology due to its emphasis on a field experimental
approach than to any single theoretical contribution underscores
the critical importance of this methodology in population and
community studies. Pianka (1976) also makes a strong case for
the role that field experiments should play in studies of
competition, in spite of the fact that his own studies in this
area have largely been descriptive. But field experiments per se
are not a panacea. In some cases, they are impossible to carry
out for logistical or ethical reasons. Even when they are
performed, many artifacts can be introduced due to the
manipulation itself, such as the various side effects resulting
from caging (e.g., Hulberg and Oliver, 1980; Underwood, 1982).
The frequent statement that laboratory experiments are limited
due to their lack of realism (e.g., Warren and Davis, 1971) is
but one version of a broader truth that any experiment,
laboratory or field, involves the introduction of artificiality
(cf. Dayton and Oliver, 1980). To avoid being misled by such
problems, it will often be necessary to compare and contrast the
results and insights gained from a variety of approaches,
including both description and experimentation (cf. Virnstein,
1980). Clearly, descriptive and experimental studies must work
hand-in-hand in order to provide a clearer understanding of
competitive interactions. Description provides the raw material
necessary to generate hypotheses regarding the factors which
produce natural patterns. Although these hypotheses can

occasionally be tested unambiguously without experimentation, the covariation of many potentially important factors often prevents us from reaching a definite conclusion regarding exactly which factors are responsible. This uncontrolled variation is remedied by the use of experiments, and when these can be conducted in the field, an especially powerful blend of experimental control and in situ realism frequently can be achieved.

Not only has stream ecology tended to under-utilize field experimental approaches, it has also relied heavily on but one of several possible descriptive approaches. Most descriptive research has involved the sampling of populations and communities using a variety of nets, scoops, and grabs. By the time the organisms collected by such techniques have been preserved and counted, it is often too late to learn many critical details regarding their natural history, especially those aspects related to behavioral ecology. Greater emphasis on direct, observational natural history can provide critical insights into the relationships of foraging behavior, activity patterns, aggressive encounters, habitat preferences, etc, to niche relations. To put this remark in perspective, one can only guess how much less avian ecologists would know about the organization of bird communities if they had been unable to carry out direct, observational studies of their subjects. And given the fact that knowledge regarding the biology of coral reefs has increased dramatically since the advent of SCUBA (Ehrlich, 1975), it seems likely that a greater incorporation of underwater observations into studies of stream populations and communities will also have a beneficial effect on our understanding.

Overall then, I believe that a combination of approaches will best serve our interests in learning how competitive interactions (and other biotic interactions, for that matter) operate in streams. Descriptive studies, including both sampling and in situ observation, make it possible to document natural patterns, which are the source of most ecological hypotheses. To identify the processes responsible for these patterns, experimental methods which systematically control for natural variation will be useful. Where possible, manipulative field experiments can often provide an especially powerful means of identifying the processes that generate such natural patterns.

ALTERNATIVE MODELS OF COMMUNITY ORGANIZATION

Although only a limited amount of evidence currently exists demonstrating the operation of competitive interactions in stream populations and communities, I suspect that this result is partially attributable to a lack of appropriately designed studies, rather than any inherent characteristic of stream ecosystems which minimizes circumstances of resource limitation. But there is an important distinction to be made here. Even though species may be competing for limiting resources, this need not result in community patterns that necessarily conform to idealized models of resource partitioning. In the final sections of this paper, I wish to consider some of the factors that might prevent the development of stream communities characterized by evenly dispersed niches (sensu Schoener, 1974), in spite of the fact that competition may occur relatively frequently.

If interspecific competition is an important force, then the operation of the competitive exclusion principle should lead to narrow niches and minimal niche overlap. Yet many stream consumers are characterized as generalists, and community analyses frequently demonstrate high degrees of niche overlap. How is it possible to reconcile the paradox that species can be resource-limited, yet communities apparently structured little by competitive interactions?

One possibility is that these results suggesting a "loose" organization of stream communities are really illusions, resulting from artifacts in which community patterns are sampled inappropriately. Remember that resource partitioning may occur at a variety of spatial scales, and that our ability to detect such partitioning depends on the degree to which we can systematically sample at the relevant scales. Traditionally, stream ecologists have sampled at rather large scales, with the ≈ 0.1 m^2 Surber sampler being one of the smallest scales at which patterns of resource partitioning are typically described. There is increasing evidence, however, that competitive interactions are played out much more locally than this (Hart, 1982b; McAuliffe, this volume; Wiley and Kohler, 1981), and that community patterns are a direct function of environmental heterogeneity (and possibly habitat specialization) at these microspatial scales (Hart, 1978; Trush, 1979). These kinds of results suggest that more attention should be focused on the description of patterns at local scales, and the development of experimental methods to determine the processes which lead to local patterns of species abundance.

But I do not mean to suggest that all stream community patterns will fall into place when we examine these smaller scales. Though I believe that patterns will often be more apparent locally, there are sound reasons to believe that stream communities will frequently appear unstructured when in fact competition is operating. Much of the development of competition theory relies on the assumption that communities are composed of homogeneously distributed populations which are at or near equilibrium. By making these assumptions, mathematical analyses

can be made more tractable. But assumptions such as these should
not be made merely for convenience; it is critical that we know
how often competitive interactions reach such equilibria, and
whether the outcome of these interactions is influenced by
environmental heterogeneity. There is a growing body of
theoretical and empirical results which suggest that the dynamics
of ecological systems are quite sensitive to environmental
patchiness, and frequently involve a variety of non-equilibrium
phenomena (cf. Wiens, 1977; Levin, 1976, 1981). Below, I
consider how a few of these might operate in streams.

     When models of competition incorporate temporal and spatial
heterogeneity, the probability that species can coexist often is
increased greatly. One model of community organization posits
that environments are divided into a series of patches, and that
the composition of species within patches is determined by the
rate at which species colonize and go extinct in each patch
(Levins and Culver, 1971; Slatkin, 1974). Species that are poor
competitors and would otherwise go extinct in a homogeneous
environment may persist if natural disturbance provides vacant
patches to which they may immigrate. By virtue of their superior
dispersal ability, such fugitive species (cf. Hutchinson, 1951)
are able to persist by remaining one step ahead of their
slowly-dispersing competitors. Stout and Vandermeer (1975)
discuss how streams might vary latitudinally with regard to the
relative importance of immigration and extinction in contributing
to patterns of coexistence, though further data would be required
to document if and how this process operates in stream
communities.

     Caswell (1978) has modeled a process in which extinction
within patches is caused by predators. As long as there are many
patches in the system, and their dynamics are relatively
independent, it is possible for inferior competitors to coexist
almost indefinitely. Such predator-mediated coexistence has been
observed in space-limited marine benthic systems by Paine (1966),
in which an invertebrate predator preferentially feeds upon (and
thus removes) the dominant competitor, allowing the subordinate
species to flourish. In addition to predation, physical
disturbance can open patches in a manner that bears on the
persistence of competitors. Connell (1978) and Huston (1979)
discussed the manner by which varying intensities and frequencies
of natural disturbance might affect the local diversity of
competing species. They believe that the relationship between
diversity and disturbance will often be hyperbolic, such that
intermediate levels of disturbance promote maximum levels of
species richness. Sousa (1979) has shown how disturbance rate
varies across substrate size in a rocky intertidal boulder field,
and that intermediate sized substrates (which are disturbed at
intermediate frequencies) hold the greatest diversity of sessile

taxa. It seems likely that a similar process may operate in
stream benthic communities (cf. McAuliffe, this volume).
Disturbance in streams is often more extreme than the
intermediate level which maximizes diversity, however, as
indicated by studies which have focused on the devastating
effects of flooding on the benthic community (e.g., Siegfried and
Knight, 1977; Anderson and Lehmkuhl, 1968).

Sale (1977) proposed a model that describes how high local
diversity might be maintained in a space-limited system such as a
coral reef. The ability to colonize available space is
determined entirely by chance, such that individuals are drawn as
in a lottery from a pool of unspecialized competitors. The
details of the recruitment process are poorly known for both
coral reefs and streams, making it difficult to determine whether
the assumptions of this model obtain in nature. Nonetheless,
this process may play a role in the maintenance of high local
diversity. Whether such a mechanism can maintain high diversity
on larger scales is less clear, however (cf. Chesson and Warner,
1981).

All natural environments undergo change, and theorists have
addressed the consequences of different forms of temporal
heterogeneity for competitive processes. Hutchinson (1961) has
suggested that when an environment changes at rates comparable to
those at which competitive exclusion occurs, competitive
advantages may shift from species to species in ways that prevent
the elimination of any single one. Wiens (1977) argues from a
similar perspective, indicating that communities may go through
periods of brief, intense competition followed by longer
intervals when resources are not limiting in abundance. This
variability in competitive intensity also can permit coexistence
of competitors. In streams, we know little about seasonal or
between-year changes in resource spectra, or about temporal
variations in other environmental parameters relevant to
competitive interactions. Nonetheless, it is apparent that
various characteristics of stream consumers reflect this
variability (e.g., Illies, 1979), and workers are aware that such
temporal heterogeneity may have important implications for both
population (Wright et al., 1981) and community-level (Horwitz,
1978; Townsend and Hildrew, 1979) aspects of competitive
interactions. Clearly, long term studies of stream population
and community dynamics are needed before it will be possible to
evaluate the role played by these longer scales of temporal
heterogeneity in competitive processes.

One other factor which can influence the likelihood that
competitors will coexist is the nature of the competitive
interaction itself. Much of competition theory rests on the
assumption that interactions can be represented by exploitation

alone (e.g., Levins, 1968; MacArthur, 1972). Yet there are strong theoretical reasons for believing that such exploitative interactions will evolve toward interference (e.g., Case and Gilpin, 1974; Gill, 1974), and interference competition is documented in the field much more frequently than exploitation (Menge, 1979). Incorporating interference into theoretical models of interspecific competition can result in the prediction of stable equilibra (e.g., Case and Gilpin, 1974), and where such interference involves competition for space, multiple stable points can result (e.g., Schoener, 1976; cf. Yodzis, 1978). The degree to which such interactions contribute to the high local diversity of stream communities is largely unknown. However, the variety of circumstances in which interference has been observed in stream consumers suggests that further attention to the community-level consequences of this interference is warranted.

INCORPORATING COMPETITIVE EFFECTS INTO PREDICTIVE MODELS OF RESOURCE USE

The majority of this review has emphasized the quantity and quality of different kinds of evidence regarding the importance of competitive interactions in streams. Given that such interactions do sometimes occur, a separate concern regards the kinds of models that are needed to predict the effect of competition on patterns of resource use. Earlier, I suggested that such models are likely to require quantitative determinations of the individual-level effects of interspecific competitive interactions. If we take the simple case of exploitative competition, one valuable approach (e.g., Pulliam, 1980; Werner, 1982) involves a determination of the factors influencing the energy (or other currency, sensu Schoener, 1971) intake rates of individual consumers, assuming, of course, that these rates are related to probabilities of survival and reproductive success. Thus, simple foraging experiments can be performed to determine how these rates vary with such factors as prey abundance, prey size, prey type, habitat structure, and mortality risk. Given this information, it is then possible to make predictions regarding which habitats and prey items are most profitable to a given consumer, and thus, how and where it might be expected to feed. Because this approach explicitly determines the efficiency of resource use by different consumers in various resource settings, it then becomes possible to predict how a given individual will be affected by the addition of an interspecific competitor due to its effect on resource characteristics (since, by definition, exploitative competitors interact only through their resources). Optimal foraging theory has proven to be a useful tool for making such micro-ecological predictions regarding how patterns of resource utilization should change in the presence and absence of competitors (e.g., Werner,

1982). The power of this kind of approach is that it allows the ecologist to incorporate explicit mechanisms of consumer-resource interactions into a theory of competition which can be tested empirically. Whether it will be adopted more widely by ecologists, and whether it can be extended to accommodate competitive effects resulting from interference, remains to be seen.

CONCLUSION

More than two decades ago, Macan (1962) observed that stream ecologists tended to emphasize the importance of abiotic factors while largely ignoring the importance of biotic ones. In a speculative and thoughtful essay, he attempted to counter balance the prevailing view, by suggesting some natural circumstances in which the operation of competition appeared to be important. Some twenty years later, I conclude that prevailing views have a great deal of inertia, and that we still have a long way to go if we are to understand the operation of biological interactions such as competition at a level equivalent to that known for abiotic factors. I do not interpret the substantial evidence documenting the effects of abiotic factors to mean that competitive interactions are of little importance in stream populations and communities. On the contrary, a small but growing body of research demonstrates that competition can play a critical role in determining patterns of foraging behavior, habitat selection, growth, survivorship, and, ultimately, Darwinian fitness in stream consumers. It remains to be seen whether future workers will overcome the historical bias against the pursuit of such questions, and the epistemological difficulties involved in relating patterns with processes, in order to reach a clearer understanding of the frequency and importance of competitive interactions within stream populations and communities. From my perspective, there is considerable basis for optimism in this regard.

ACKNOWLEDGMENTS

During the development of this paper, I was generously supported by National Science Foundation grants DEB-7912604 and DEB-8111305. I would like to thank J. Dixon, J. McAuliffe, S. Reice, and S. Schroeter for their critical reviews of an earlier version of this manuscript. This is Contribution number 470 from the W. K. Kellogg Biological Station.

REFERENCES

Allan, J. D. 1975. The distributional ecology and diversity of benthic insects in Cement Creek, Colorado. Ecology 56:1040-1053.

Alstad, D. N. 1981. Nearest-neighbor analysis of microhabitat partitioning in stream insect communities. Hydrobiologia 79:137-140.

Anderson, N. H. and D. M. Lehmkuhl. 1968. Catastropnic drift in a woodland stream. Ecology 49:198-206.

Baker, J. A. and S. T. Ross. 1981. Spatial and temporal resource utilization by southeastern cyprinids. Copeia 1981:178-189.

Birch, L. C. 1957. The meanings of competition. Am. Nat. 91:5-18.

Birch, L. C. 1979. The effect of species of animals which share common resources on one another's distribution and abundance. Fortschr. Zool. 25:197-221.

Bohle, H. W. 1978. Beziehungen zwischen dem Nahrungsangebot, der Drift und der raumlichen Verteilung bei Larven von Baetis rhodani (Pictet) (Ephemeroptera: Baetidae). Arch. Hydrobiol. 84:500-525.

Bovbjerg, R. V. 1970. Ecological isolation and competitive exclusion in two crayfish (Orconectes virilis and Orconectes immunis). Ecology 51:255-236.

Brocksen, R. W., G. E. Davis, and C. E. Warren. 1968. Competition, food consumption, and production of sculpins and trout in laboratory stream communities. J. Wildl. Mgmt. 32:51-75.

Brown, J. C. 1969. Territorial behavior and population regulation in birds: a review and re-evaluation. Wilson Bull. 81:293-329.

Calow, P. 1981. Adaptational aspects of growth and reproduction in Lymnaea peregra (Gastropoda: Pulmonata) from exposed and sheltered aquatic habitats. Malacologia 21:5-13.

Case, T. E. and M. E. Gilpin. 1974. Interference competition and niche theory. Proc. Natl. Acad. Sci. 71:3073-3077.

Caswell, H. 1976. Community structure; a neutral model analysis. Ecol. Monogr. 46:327-54.

Caswell, H. 1978. Predator-mediated coexistence: a nonequilibrium model. Am. Nat. 112:127-154.

Cather, M. R. and A. R. Gaufin. 1976. Comparative ecology of three Zapada species of Mill Creek, Utah. Am. Midl. Nat. 95:464-471.

Chapman, D. W. 1962. Aggressive behavior in juvenile coho salmon as a cause of emigration. J. Fish. Res. Board Can. 19:1047-1080.

Charnov, E. L., G. H. Orians and K. Hyatt. 1976. Ecological implications of resource depression. Am. Nat. 110:247-259.

Chesson, P. L. and R. R. Warner. 1981. Environmental variability promotes coexistence in lottery competitive systems. Am. Nat. 117:923-943.

Cohen, J. E. 1968. Alternate derivations of a species-abundance relation. Am. Nat. 102:165-172.

Colbo, M. H. and G. N. Porter. 1979. Effects of food supply on the life history of Simuliidae (Diptera). Can. J. Zool. 57:301-306.

Coleman, M. J. and H. B. N. Hynes. 1970. The life histories of some Plecoptera and Ephemeroptera in an Ontario stream. Can. J. Zool. 48:1133-1139.

Colwell, R. K. and D. J. Futuyma. 1971. On the measurement of niche breadth and overlap. Ecology 52:567-576.

Colwell, R. K. and D. Winkler. 1982. A null model for null models in biogeography In: D. R. Strong, Jr., D. S. Simberloff, and L. G. Abele (eds.), Ecological communities: conceptual issues and the evidence. Princeton Univ. Press, New Jersey (in press).

Connell, J. H. 1972. Community interactions in marine rocky intertidal shores. Ann. Rev. Ecol. Syst. 3:169-192.

Connell, J. H. 1975. Some mechanisms producing structure in natural communities: a model and evidence from field experiments. pp. 460-90. In: M. L. Cody and J. M. Diamond (eds.), Ecology and evolution of communities. Belknap, Harvard.

Connell, J. H. 1978. Diversity in tropical rain forests and coral reefs. Science 199:1302-1310.

Connell, J. H. 1980. Diversity and the coevolution of competitors, or the ghost of competition past. Oikos 35:131-138.

Culver, D. C. 1971. Analysis of simple cave communities. III. Control of abundance. Amer. Midl. Nat. 85:173-187.

Culver, D. C. and T. J. Ehlinger. 1980. Effect of microhabitat size and competitor size on two cave isopods. Brimleyana 4:103-113.

Cummins, K. W. 1964. Factors limiting the microdistribution of larvae of the caddisflies Pycnopsyche lepida (Hagen) and Pycnopsyche guttifer (Walker) in a Michigan stream. Ecol. Monogr. 34:271-295.

Cummins, K. W. 1973. Trophic relations of aquatic insects. Ann. Rev. Ent. 18:183-206.

Davis, G. E. and C. E. Warren. 1965. Trophic relations of a sculpin in laboratory stream communities. J. Wildl. Mngmt. 29:846-71.

Dayton, P. K. 1973. Two cases of resource partitioning in an intertidal community: making the right prediction for the wrong reason. Am. Nat. 107:662-670.

Dayton, P. K. and J. S. Oliver. 1980. An evaluation of
    experimental analyses of population and community patterns
    in benthic marine communities. pp. 93-120. In: K. R.
    Tenore and B. C. Coull (eds.), Marine benthic dynamics.
    Univ. South Carolina Press, Columbia, South Carolina.
Devonport, B. F. and M. J. Winterbourn. 1976. The feeding
    relationships of two invertebrate predators in a New Zealand
    river. Freshwat. Biol. 6:167-176.
Diamond, J. M. 1978. Niche shifts and the rediscovery of
    interspecific competition. Amer. Sci. 66:322-331.
Diamond, J. M. and M. E. Gilpin. 1982. Examination of the
    "null" model of Connor and Simberloff for species
    co-occurrences on islands. Oecologia 52:64-74.
Dill, L. M., R. C. Ydenberg, and A. H. G. Fraser. 1981. Food
    abundance and territory size in juvenile coho salmon. Can.
    J. Zool. 59:1801-1809.
Dimond, J. B. 1967. Evidence that drift of stream benthos is
    density related. Ecology 48:855-857.
Ehrlich, P. R. 1975. The population biology of coral reef
    fishes. Ann. Rev. Ecol. Syst. 6:211-247.
Fausch, K. D. and R. J. White. 1981. Competition between brook
    trout (Salvelinus fontinalis) and brown trout (Salmo trutta)
    for positions in a Michigan stream. Can. J. Fish. Aquat.
    Sci. 38:1220-1227.
Fuller, R. L. and Stewart, K. W. 1977. The food habits of
    stoneflies in the Upper Gunnison River, Colorado. Env.
    Entomol. 6:293-302.
Gatz, A. J., Jr. 1979. Community organization in fishes as
    indicated by morphological features. Ecology 60:711-718.
Gill, D. E. 1974. Intrinsic rate of increase, saturation
    density, and competitive ability. II. The evolution of
    competitive ability. Amer. Nat. 108:103-116.
Glass, L. W. and R. V. Bovbjerg. 1969. Density and dispersion
    of laboratory populations of caddisfly larvae
    (Cheumatopsyche). Ecology 50:1082-1084.
Grant, P. R. and R. J. Mackay. 1969. Ecological segregation of
    systematically related stream insects. Can. J. Zool.
    47:691-94.
Gray, L. J. 1981. Species composition and life histories of
    aquatic insects in a lowland Sonoran Desert stream. Am.
    Midl. Nat. 106:229-243.
Gregory, S. V. 1978. Phosphorous dynamics on organic and
    inorganic substrates in streams. Verh. Internat. Limnol.
    20:1340-1346.
Gregory, S. V. 1980. Effects of light, nutrients, and grazing
    on periphyton communities in stream. Ph.D. thesis, Oregon
    State University, Corvallis. 151 pp.

Gyselman, E. C. 1980. The mechanisms that maintain population
    stability of selected species of Ephemeroptera in a
    temperate stream. pp. 309–319. In: J. F. Flannagan and K.
    E. Marshall (eds.), Advances in Ephemeroptera biology.
    Plenum Press, New York.

Hart, D. D. 1978. Diversity in stream insects: regulation by
    rock size and microspatial complexity. Verh. Internat.
    Verein. Limnol. 20:1376–1381.

Hart, D. D. 1979a. Patchiness and ecological organization:
    experimental studies within stream benthic communities.
    Ph.D. Dissertation, Univ. California, Davis.

Hart, D. D. 1979b. Association analysis, species interactions,
    and the structure of benthic communities. Can. Spec. Publ.
    Fish. Aquat. Sci. 43:63–71.

Hart, D. D. 1981. Foraging and resource patchiness: field
    experiments within a grazing stream insects. Oikos
    37:46–52.

Hart, D. D. 1982a. Asymmetric interference competition and
    patterns of habitat selection in tunnel–dwelling caddis
    larvae. Unpublished manuscript.

Hart, D. D. 1982b. Causes and consequences of territoriality in
    a grazing stream insect. Unpublished manuscript.

Hart, D. D. 1982c. Food depletion and food limitation: field
    experiments examining exploitative competition in grazing
    larval caddisflies. Unpublished manuscript.

Hendrickson, J. A., Jr. 1981. Community–wide character
    displacement reexamined. Evolution 35:794–810.

Hildrew, A. G. and J. M. Edington. 1979. Factor facilitating
    the coexistence of hydropsychid caddis larvae (Trichoptera)
    in the same river system. J. Anim. Ecol. 48:557–576.

Hildrew, A. G. and C. R. Townsend. 1980. Aggregation,
    interference and foraging by larvae of Plectrocnemia
    conspersa (Trichoptera:Polycentropidae). Anim. Behav.
    28:553–560.

Horwitz, R. J. 1978. Temporal variability patterns and the
    distributional patterns of stream fishes. Ecol. Monogr.
    48:307–321.

Hulberg, L. W. and J. S. Oliver. 1980. Caging manipulations in
    marine soft–bottom communities: importance of animal
    interactions or sedimentary habitat modifications. Can. J.
    Fish. Aquat. Sci. 37:1130–1139.

Huston, M. 1979. A general hypothesis of species diversity.
    Am. Nat. 113:81–101.

Hutchinson, G. E. 1951. Copepodology for the ornithologist.
    Ecology 32:571–77.

Hutchinson, G. E. 1957. Concluding remarks. Cold Spring Harbor
    Symp. Quant. Biol. 22:415–427.

Hutchinson, G. E. 1961. The paradox of the plankton. Am. Nat.
    95:137–146.

Hynes, H. B. N. 1970. The ecology of running waters. Univ. Toronto Press, Toronto.

Hynes, H. B. N. and W. D. Williams. 1965. Experiments on competition between two Asellus species (Isopoda, Crustacea). Hydrobiologia 26:203-210.

Illies, J. 1979. Annual and seasonal variation of individual weights of adult water insects. Aquatic Insects 1:153-163.

Jansson, A. and T. Vuoristo. 1979. Significance of stridulation in larval Hydropsychidae (Trichoptera). Behaviour 171:167-186.

Klotz, R. L., J. R. Cain, and F. R. Trainor. 1976. Algal competition in an epilithic flora. J. Phycol. 12:363-368.

Kohn, A. J. 1971. Diversity, utilization of resources, and adaptive radiation in shallow-water marine invertebrates of tropical oceanic islands. Limnol. Oceanogr. 16:332-348.

Lawton, J. H. and M. P. Hassell. 1981. Asymmetrical competition in insects. Nature 289:793-795.

Lehmkuhl, D. M. 1970. A North American trichopteran larva which feeds on freshwater sponges (Trichoptera: Leptoceridae; Porifera: Spongillidae). Am. Midl. Nat. 84:278-80.

Levin, S. A. 1976. Population dynamic models in heterogeneous environments. Ann. Rev. Ecol. Syst. 7:287-310.

Levin, S. A. 1981. Mechanisms for the generation and maintenance of diversity in ecological communities. pp. 173-194. In: R. W. Hiorns and D. Cooke (eds.), The mathematical theory of the dynamics of biological populations. II. Academic Press, New York.

Levins, R. 1968. Evolution in changing environments. Princeton University Press, New Jersey.

Levins, R. and D. Culver. 1971. Regional coexistence of species and competition between rare species. Proc. Natl. Acad. Sci. 68:1246-1248.

Linduska, J. P. 1942. Bottom type as a factor influencing the local distribution of mayfly nymphs. Can. Ent. 74:26-30.

Lock, M. A. and T. B. Reynoldson. 1976. The role of interspecific competition in the distribution of two stream dwelling triclads, Crenobia alpina (Dana) and Polycelis felina (Dalyell), in North Wales. J. Anim. Ecol. 45:581-592.

Macan, T. T. 1962. Biotic factors in running water. Schweiz. Zeitschr. Hydrol. 24:386-407.

MacArthur, R. H. 1965. Patterns of species diversity. Biol. Rev. 40:510-533.

MacArthur, R. H. 1966. Note on Mrs. Pielou's comments. Ecology 47:1074.

MacArthur, R. H. 1972. Geographical ecology. Harper and Row, New York.

Mackay, R. J. 1972. Temporal patterns in life history and flight behavior of Pycnopsyche gentilis, P. luclenta, and P. scabripennis (Trichoptera: Limnephilidae). Can. Ent. 104:1819–1835.

Mackay, R. J. and J. Kalff. 1973. Ecology of two related species of caddis fly larvae in the organic substrates of a woodland stream. Ecology 54:499–511.

Magdych, W. P. 1979. The microdistribution of mayflies (Ephemeroptera) in Myriophyllum beds in Pennington Creek, Johnston County, Oklahoma. Hydrobiologia 66:161–175.

Malas, D. and J. B. Wallace. 1977. Strategies for coexistence in three species of net spinning caddisflies (Trichoptera) in second-order southern Appalachian streams. Can. J. Zool. 55:1829–1840.

May, R. M. 1975. Patterns of species abundance and diversity. pp. 81–120. In: M. L. Cody and J. M. Diamond (eds.), Ecology and evolution of communities. Belknap, Harvard.

McNicol, R. E. and D. L. G. Noakes. 1981. Territories and territorial defense in juvenile brook charr (Salvelinus fontinalis) (Pisces: Salmonidae). Can. J. Zool. 59:22–28.

Mendelson, J. 1975. Feeding relationships among species of Notropis (Pisces: Cyprinidae) in a Wisconsin stream. Ecol. Monogr. 45:199–230.

Menge, B. A. 1979. Coexistence between the seastars Asterias vulgaris and A. forbesi in a heterogeneous environment: a non-equilibrium explanation. Oecologia 41:245–272.

Miller, R. S. 1967. Pattern and process in competition. Adv. Ecol. Res. 4:1–74.

Minshall, G. W. and J. N. Minshall. 1977. Microdistribution of benthic invertebrates in a Rocky Mountain Stream. Hydrobiologia 55:231–249.

Müller, K. 1954. Investigations on the organic drift in North Swedish streams. Rep. Inst. Freshwater Res. Drottningholm 35:133–148.

Nielsen, A. 1948. Postembryonic development and biology of the Hydroptilidae. A contribution to the phylogeny of caddisflies and to the origin of case building. K. Dan. Vidensk. Selsk. Skr. 5:1–200.

Nilsson, L. M. and C. Otto. 1977. Effects of population density and of presence of Gammarus pulex L. (Amphipoda) on the growth in larvae of Potamophylax cingulatus Steph. (Trichoptera). Hydrobiologia 54:109–112.

Oswood, M. W. 1976. Comparative life histories of the Hydropsychidae in a Montana lake outlet. Am. Midl. Nat. 96:493–97.

Otto, C. 1976. Habitat relationships in the larvae of three Trichoptera species. Arch. Hydrobiol. 77:505–17.

Otto, C. and B. W. Svensson. 1976. Consequences of removal of pupae for a population of Potamophylax cingulatus (Trichoptera) in a south Swedish stream. Oikos 27:40–43.

Page, L. M. and D. W. Schemske. 1978. The effect of interspecific competition on the distribution and size of darters of the subgenus Catonotus (Percidae: Etheostoma). Copeia 1978:406-412.

Paine, R. T. 1966. Food web complexity and species diversity. Am. Nat. 100:65-75.

Paine, R. T. 1977. Controlled manipulations in the marine intertidal zone, and their contributions to ecological theory. pp. 245-270. In: C. E. Goulden (ed.), Changing scenes in natural sciences, 1776-1976. Academy of Natural Sciences, Philadelphia, Pennsylvania.

Parry, G. D. 1981. The meanings of r- and K-selection. Oecologia 48:260-264.

Patrick, R. 1972. Benthic communities in streams. Trans. Conn. Acad. Arts Sci. 44:272-84.

Patrick, R. 1975. Stream communities. pp. 445-59. In: M. L. Cody and J. M. Diamond (eds.), Ecology and evolution of communities. Belknap, Harvard.

Patrick, R. 1978. Effect of trace metals in the aquatic ecosystem. Amer. Sci. 66:185-191.

Peckarsky, B. L. 1979. Biological interactions as determinants of distributions of benthic invertebrates within the substrate of stony streams. Limnol. Oceanogr. 24:59-68.

Peckarsky, B. L. 1981. Reply to comment by Sell. Limnol. Oceanogr. 26:982-987.

Peckarsky, B. L. and S. I. Dodson. 1980. An experimental analysis of biological factors contributing to stream community structure. Ecology 61:1283-1290.

Peters, T. M. and P. Barbosa. 1977. Influence of population density on size, fecundity, and developmental rate of insects in culture. Ann. Rev. Entomol. 22:431-450.

Pianka, E. R. 1976. Competition and niche theory. pp. 114-141. In: R. M. May (ed.), Theoretical ecology, principles and applications. Saunders, Philadelphia.

Pielou, E. C. 1969. An introduction to mathematical ecology. Wiley-Interscience, New York.

Pulliam, H. R. 1980. On digesting a theory. Auk 97:418-420.

Rabeni, C. F. and G. W. Minshall. 1977. Factors affecting microdistribution of stream insects. Oikos 29:33-43.

Reice, S. R. 1980. The role of substratum in benthic macroinvertebrate microdistribution and litter decomposition in a woodland stream. Ecology 61:580-590.

Reice, S. R. 1981. Interspecific associations in a woodland stream. Can. J. Fish. Aq. Sci. 38:1271-1280.

Reynoldson, T. B. and L. S. Bellamy. 1971. The establishment of interspecific competition in field populations, with an example of competition in action between Polycelis nigra (Mull.) and P. tenuis (Ijima) (Turbellaria, Tricladida). Proc. Adv. Stud. Inst. Dyn. Numbers Pop. 1970:282-297.

Sale, P. F. 1977. Maintenance of high diversity in coral reef fish communities. Am. Nat. 111:337-359.

Schoener, T. W. 1971. Theory of feeding strategies. Ann. Rev. Ecol. Syst. 2:369-404.

Schoener, T. W. 1974. Resource partitioning in ecological communities. Science 185:27-39.

Schoener, T. W. 1976. Alternatives to Lotka-Volterra competition: models of intermediate complexity. Theor. Pop. Biol. 10:309-333.

Sell, D. W. 1981. Comment on "Biological interactions as determinants of distributions of benthic invertebrates within the substrate of stony streams" (Peckarsky). Limnol. Oceanogr. 26:981-982.

Sheldon, A. L. 1972. Comparative ecology of Arcynopteryx and Diura in a California stream. Arch. Hydrobiol. 69:521-546.

Sheldon, A. L. 1980a. Resource division by perlid stoneflies (Plecoptera) in a lake outlet ecosystem. Hydrobiologia. 71:155-161.

Sheldon, A. L. 1980b. Coexistence of perlid stoneflies (Plecoptera): predictions from multivariate morphometrics. Hydrobiologia 71:99-105.

Siegfried, C. A. and A. W. Knight. 1977. The effects of washout in a Sierra foothill stream. Amer. Midl. Nat. 98:200-207.

Slatkin, M. 1974. Competition and regional coexistence. Ecology 55:128-134.

Sousa, W. P. 1979. Disturbance in marine intertidal boulder fields: the non-equilibrium maintenance of species diversity. Ecology 60:1225-39.

Southwood, T. R. E. 1981. Bionomic strategies and population parameters. pp. 30-52. In: R. M. May (ed.), Theoretical ecology (Second edition). Blackwell Scientific Publications, Oxford.

Stearns, S. C. 1976. Life-history tactics: a review of the ideas. Quart. Rev. Biol. 51:3-47.

Stockner, J. G. and K. R. S. Shortreed. 1978. Enhancement of autotrophic production by nutrient addition in a coastal rainforest stream on Vancouver Island. J. Fish. Res. Board Can. 35:28-34.

Stout, J. and J. Vandermeer. 1975. Comparison of species richness for stream inhabiting insects in tropical and mid-latitude streams. Am. Nat. 109:263-280.

Strong, D. R., Jr., L. A. Szyska, and D. S. Simberloff. 1979. Tests of community-wide character displacement against null hypotheses. Evolution 33:897-913.

Sugihara, G. 1980. Minimal community structure: an explanation of species abundance patterns. Am. Nat. 116:770-787.

Thut, R. N. 1969. Feeding habits of larvae of seven Rhyacophila (Trichoptera: Rhyacophilidae) species with notes on other life-history features. Ann. Ent. Soc. Amer. 62:894-898.

Townsend, C. R. and A. G. Hildrew. 1979. Resource partitioning by two freshwater invertebrate predators with contrasting foraging strategies. J. Anim. Ecol. 48:909-920.

Trush, W. J., Jr. 1979. The effects of area and surface complexity on the structure and formation of stream benthic communities. Va. Polytech. Univ., Blackburg. Unpublished M.S. thesis. 149 pp.

Tuskes, P. M. 1975. Biological observations, and larval competition in two species of aquatic pyralid moths of the genus *Parargyractis*. Unpublished Ph.D. dissertation, Univ. California, Davis.

Ulfstrand, S. 1967. Microdistribution of benthic species (Ephemeroptera, Plecoptera, Trichoptera, Diptera: Simuliidae) in Lapland streams. Oikos 18:293-310.

Underwood, A. J. 1982. Prevailing paradigms of intertidal communities. In: D. R. Strong, Jr., D. S. Simberloff, and L. G. Abele (eds.), Ecological communities: conceptual issues and the evidence. Princeton Univ. Press, New Jersey (in press).

Vannote, R. L. and B. W. Sweeney. 1980. Geographical analysis of thermal equilibria: a conceptual model for evaluating the effect of natural and modified thermal regimes on aquatic insect communities. Am. Nat. 115:667-695.

Virnstein, R. W. 1980. Measuring effects of predation on benthic communities in soft sediments. pp. 281-290. In: V. S. Kennedy (ed.), Estuarine perspectives. Academic Press, New York.

Wallace, J. B. 1975. Food partitioning in net-spinning Trichoptera larvae: *Hydropsyche venularis*, *Cheumatopsyche etrona*, and *Macronema zebratum* (Hydropsychidae). Ann. Ent. Soc. Amer. 68:463-472.

Walton, O. E., Jr. 1980. Invertebrate drift from predator-prey associations. Ecology 61:1486-1497.

Walton, O. E., Jr., S. R. Reice, and R. W. Andrews. 1977. The effects of density, sediment particle size, and velocity on drift of *Acroneuria abnormis* (Plecoptera). Oikos 28:291-298.

Warren, C. E. and G. E. Davis. 1971. Laboratory stream research: objectives, possibilities, and constraints. Ann. Rev. Ecol. Syst. 2:111-144.

Warren, C. E., J. H. Wales, G. E. Davis, and P. Doudoroff. 1964. Trout production in an experimental stream enriched with sucrose. J. Wildl. Manag. 28:617-660.

Waters, T. F. 1961. Standing crop and drift of stream bottom organisms. Ecology 42:532-537.

Werner, E. E. 1977. Species packing and niche complementarity in three sunfishes. Am. Nat. 111:553-578.

Werner, E. E. 1982. The mechanisms of species interactions and community organization in fish. In: D. R. Strong, Jr. and D. Simberloff (eds.), Conceptual issues in community ecology. Princeton Univ. Press, New Jersey (in press).

Westlake, D. F. 1975. Macrophytes. pp. 106-128. In: B. A.
    Whitton (ed.), River Ecology. University of California
    Press, Berkeley.
Whitton, B. A. 1975. Algae. pp. 81-105. In: B. A. Whitton
    (ed.), River Ecology. University of California Press,
    Berkeley.
Wiens, J. A. 1977. On competition and variable environments.
    Amer. Sci. 65:590-597.
Wilbur, H. M., D. W. Tinkle and J. P. Collins. 1974.
    Environmental certainty, trophic level, and resource
    availability in life history evolution. Amer. Nat.
    108:805-517.
Wiley, M. J. 1981. Interacting influences of density and
    preference on the emigration rates on some lotic chironomid
    larvae. Ecology 62:426-438.
Wiley, M. J. and Kohler, S. L. 1981. An assessment of
    biological interactions in an epilithic stream community
    using time-lapse cinematography. Hydrobiologia 78:183-188.
Wilson, D. S., M. Leighton, and D. R. Leighton. 1978.
    Interference competition in a tropical ripple bug
    (Hemiptera: Veliidae). Biotropica 10:302-306.
Wright, J. F., P. H. Hiley, and A. D. Berrie. 1981. A 9-year
    study of the life cycle of Ephemera danica Mull.
    (Ephemeridae: Ephemeroptera) in the River Lambourn, England.
    Ecol. Entomol. 6:321-31.
Yodzis, P. 1978. Competition for space and the structure of
    ecological communities. Lecture Notes in Biomathematics,
    vol. 25. Springer-Verlag.
Zaret, T. M. and A. S. Rand. 1971. Competition in tropical
    stream fishes: support for the competitive exclusion
    principle. Ecology 52:336-342.

COMPETITION, COLONIZATION PATTERNS, AND DISTURBANCE

IN STREAM BENTHIC COMMUNITIES

Joseph R. McAuliffe

Department of Zoology
University of Montana
Missoula, Montana   59812

INTRODUCTION

A major goal of research on the ecology of communities is to
build an understanding of the mechanisms responsible for the
distribution, abundance, and coexistence of the member species.
The richness and complexity of benthic communities in streams
has long been recognized but very few factors that influence
these patterns are understood. A hierarchy of mechanisms
responsible for the distribution and abundance of organisms
exists within any community (Brown, 1981; Reynoldson, 1981).
Ultimately, organisms in streams are limited to specific ranges
of physical environments by their physiological tolerances to
ranges of temperature, water chemistry, etc. (Fiance, 1978;
Vannote and Sweeney, 1980). Within a suitable range of these
physical limitations, organisms may be restricted to particular
ranges of current velocities, substrate types, or food
availability within which they can successfully procure food
resources (Harrod, 1965; Edington, 1968; Carlsson et al., 1977;
Malas and Wallace, 1977). In addition, predation, interspecific
competition, and many kinds of physical disturbances may further
restrict the distribution and abundance of individuals within
otherwise suitable microhabitats.

The vast majority of studies on benthic stream communities
have either descriptively or experimentally addressed the
effects of physical microhabitat variables on the distribution
of benthic invertebrates (Edington, 1968; Cummins and Lauff,
1969; Higler, 1975; Haddock, 1977; Minshall and Minshall, 1977;
Williams and Mundie, 1978; Williams, 1980). Competition often
has been assumed to play a minor role in structuring these

137

communities (Minshall and Minshall, 1977). In this paper, I discuss the need for experimental approaches in studies of competition in stream communities. My goal is to present an overview of some different kinds of competitive interactions that may occur between benthic organisms and discuss how I have approached their investigation through experimentation. I also present observations and discuss the ways that different types of disturbances characteristic to streams may influence the effects of competition.

Approaches to the study of interspecific competition in stream benthic communities have been largely descriptive and have dealt with mechanisms such as temporal or spatial segregation of related taxa that may result in the partitioning of resources (Grant and Mackay, 1969; Sheldon, 1972; Williams and Hynes, 1973; Malas and Wallace, 1977; Hildrew and Edington, 1979). This approach may help to explain the faunal richness of stream communities in relation to the diversity of food resources or microhabitats but it does not address the problem of the effects of ongoing competitive interactions. Another approach is to compare distributions or relative abundances of species that are suspected to interact. In these studies, negative associations or correlations of abundances of individuals are inferred to be the result of interspecific interactions (Allan, 1975; Hynes and Williams, 1962; Chutter, 1968; Pavlichenco, 1977; Townsend and Hildrew, 1977; Reice, 1981). Hart (this volume) gives a more thorough review of these and other approaches. These descriptive approaches may generate hypotheses about the possible evolutionary or ecological consequences of competition but are not capable of testing the same hypotheses.

Detailed field experiments are needed to investigate the importance of biological interactions in stream communities. The impacts of these interactions on community structure are unlikely to be understood until such experiments are employed. Field experiments have been used widely in marine rocky intertidal and subtidal communities (Connell, 1961a, b; Paine, 1966, 1977, 1979; Dayton, 1971, 1975; Menge, 1976, 1979; Lubchenco, 1978; Lubchenco and Menge, 1978; Peterson, 1979; Sousa, 1979). The success and clarity of their work is due largely to their use of experiments that separately analyze the effects of individual biological and physical factors. Investigators of terrestrial plant communities (Harper, 1977; Fowler, 1981) and freshwater zooplankton communities (Dodson, 1976; Lynch, 1979) also have demonstrated the value of experiments. The mechanics of this experimental approach have been outlined by Connell (1975), Menge (1979) and Reynoldson and Bellamy (1971).

The lack of experimental studies on effects of competition, predation, and physical disturbance in stream communities may be

due to the difficulty of applying this approach. Many types of
manipulative experiments that have been effectively used in other
systems are impractical to employ in streams. Fencing, caging,
and additions of animals to stream substrates are often
difficult or inappropriate due to habitat change (current
velocity reduction, sedimentation effects, etc.) and the small
size and mobility of many stream invertebrates. Despite these
difficulties, many possibilities exist for the design of
experiments to test hypotheses about effects of biological
interactions. In the next section, I discuss some experimental
approaches that I have begun to use in studies of competition in
stream benthic communities.

FIELD OBSERVATIONS AND EXPERIMENTS

     Unless otherwise stated, the work described here has been
conducted in Owl Creek, Missoula County, Montana. Owl Creek is a
fourth-order outlet stream of 449 ha mesotrophic Placid Lake.
The stream has an average gradient of about 1.5% and the
substrate is composed of glacially polished boulders, cobbles,
and metamorphosed, bedded sedimentary rock. In Owl Creek, I have
dealt with competitive interactions among (1) sessile species,
(2) sessile and mobile species, and (3) mobile periphyton
grazers.

Interactions Among Different Sessile Foragers

     Insect larvae that do not travel over substrate surfaces to
procure food but instead reside within permanently affixed cases
or retreats are abundant in Owl Creek. This sessile fauna
includes the following taxa:

Trichoptera: Hydropsychidae (Hydropsyche cockerelli, H.
              occidentalis, H. oslari, Cheumatopsyche gracilis
              - all filter feeders)

Trichoptera: Hydroptilidae (Leucotrichia pictipes - grazer)

Lepidoptera: Pyralidae (Parargyractis confusalis - grazer)

   Diptera: Chironomidae (Rheotanytarsus sp. - filter feeder,
              Eukiefferiella sp. - grazer)

     In riffles, the Hydropsychidae are abundant on moss-covered
or rough-surfaced substrates (Oswood, 1979). Leucotrichia,
Parargyractis, Rheotanytarsus, and Eukiefferiella are prevalent
on smooth surfaces of stone substrates. Of this smooth-surface
fauna, Leucotrichia and Parargyractis are the most conspicuous
members. Final instar Leucotrichia larvae reside within
flattened, elliptical silk cases that are attached to the

substrate. A larva extends the anterior portion of its body (the head, thorax, and first few abdominal segments) from openings at either end of the case to graze periphyton (McAuliffe, in press). Parargyractis larvae construct fixed, irregular silken retreats from which they extend to graze surface periphyton (Lange, 1956). Both insects have a single generation per year in Owl Creek. Adults of each species emerge during June and July, new larvae colonize rock surfaces in mid and late summer, both overwinter in cases or webs as fully developed larvae and pupate in late spring. These two species as well as Rheotanytarsus and Eukiefferiella all coexist on the flat upper surfaces of stones. This observation suggested that the preferred microhabitat of each species in terms of current velocity, exposure to current, and surface complexity overlaps considerably. Although all species were commonly found together on the same surfaces, space often would be completely monopolized by Leucotrichia to the near exclusion of the other sessile species. When Leucotrichia and Parargyractis occurred together, the minimum distance between the case of Leucotrichia and the web of Parargyractis was approximately equal to the distance that Leucotrichia could extend from its case to forage. Leucotrichia larvae seemed capable of preventing encroachment by Parargyractis and overgrowth of Leucotrichia by Parargyractis was never observed. Instead, the presence and shape of Parargyractis webs conformed to the presence and location of individual Leucotrichia.

The distributional patterns of Leucotrichia and Parargyractis suggested that they competed for space on stone surfaces and I performed a preliminary experiment in August 1980 to test this hypothesis. The experiment was designed to test the prediction that colonization by Parargyractis would be inversely related to Leucotrichia density. Thirty stones with dense accumulations of Leucotrichia were collected from the stream and one of three manipulations of Leucotrichia densities was randomly assigned to each stone surface. All Leucotrichia larvae and cases were removed with forceps from 10 stone surfaces, approximately 40 to 60% were removed from another 10, and the high Leucotrichia densities were unmanipulated on the last set of 10 stones. All stones were gently brushed with a soft-bristled toothbrush to remove insects other than Leucotrichia. The light brushing was not sufficient to dislodge the firmly attached cases of Leucotrichia. The 30 stones were placed in triplets (2 manipulated density and 1 high-density control treatment) in a single riffle and 15 days were allowed for colonization. After this period, the stones were withdrawn, all insects were collected from the upper surface of each, and the outline of the upper surface area of each stone was traced on an acetate sheet. There was a significant inverse relationship between the number

of <u>Parargyractis</u> that colonized the stone surfaces and the
density of <u>Leucotrichia</u> (F = 4.35, p < 0.05) (Fig. 1).

Interactions Between Sessile and Mobile Grazers

Although the sessile grazing insects of Owl Creek are the
most conspicuous part of the epilithic fauna, non-sessile grazers
are also abundant. These mobile grazers include the mayfly
genera <u>Baetis</u> and <u>Epeorus</u>, and the caddis larvae <u>Glossosoma</u> and
<u>Oligoplectrum</u>. The sessile grazers <u>Leucotrichia</u> and
<u>Parargyractis</u> may compete with mobile grazers for periphyton.
The mechanism of such a competitive interaction may be either
exploitation of periphyton and/or interference interactions.
Given a sufficient renewal rate of the periphyton resource,
sessile or territorial foragers are predicted to be able to
interfere with mobile exploitation foragers with low cost and
high effect (Case and Gilpin, 1974; Charnov et al., 1976; Davies,
1979). If sessile species compete with mobile grazers either
through exploitation or interference, I expected that
colonization of substrate surfaces by mobile grazers should be
inversely related to the density of already-present sessile
grazers.

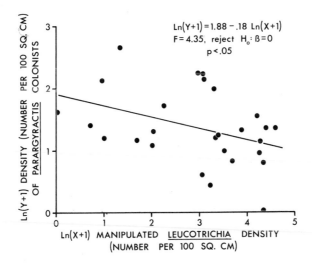

Fig. 1.   Colonization by <u>Parargyractis</u> in response to manipulated
<u>Leucotrichia</u> densities.

I tested this hypothesis in an experiment similar to the one previously discussed. Twenty stones with high Leucotrichia densities were withdrawn from a riffle and all Leucotrichia were removed with forceps from 10 randomly assigned stone surfaces. On the other set of 10 stones that served as controls, all Leucotrichia were left intact. Both control and manipulated stone surfaces were brushed with a toothbrush to remove all invertebrates other than Leucotrichia. The 10 pairs of manipulated and control substrates were placed in a riffle for a 7-day colonization period to examine the effect of Leucotrichia on colonization by the mobile grazer Baetis. After 7 days, the stones were lifed from the stream, all Baetis present on each upper stone surface were collected, and the stone surface area was measured. Fewer Baetis colonized stones with high Leucotrichia densities than those with Leucotrichia absent (Mann-Whitney U-test, P < .003) (Fig. 2). Hart (this volume) and I have observed aggressive responses of Leucotrichia to Baetis nymphs that venture into their foraging territories. Thus, interference plays a large role in this competitive interaction.

Interactions Among Mobile Grazers

Competition among animals that utilize renewable resources such as periphyton may differ greatly from competition for space.

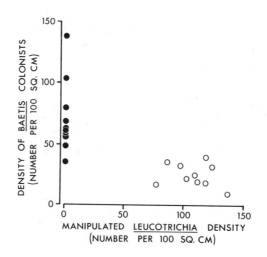

Fig. 2. Effect of experimental removal of Leucotrichia on colonization by Baetis. Open and solid circles indicate the two density manipulation treatments discussed in text.

Competition for a renewable resource need not involve interference. Instead, it may occur when individuals depress the
supply of a resource that would otherwise be available to others.
Even though the bulk of competition and niche theory rests upon
the assumption that the effects of exploitation determine the
distribution and abundance of individuals, competitive effects of
resource exploitation have seldom been shown conclusively (Case
and Gilpin, 1974). Assemblages of invertebrates that graze
periphyton on the upper surfaces of stones in streams offer many
possibilities to study potential effects of exploitation. Hart
(1981) provided data on rates of periphyton consumption by
Dicosmoecus gilvipes as a function of periphyton renewal rates
that suggested individuals were competing through exploitation.
The following experiments suggest that competition through
exploitation may be important in assemblages of stream grazing
insects.

In order for exploitation to occur, the activities of
organisms must depress the abundance of a renewable resource.
There is limited evidence to suggest that grazing invertebrates
in streams are able to depress the abundance of periphyton food
on hard substrates (Douglas, 1958). In Cottonwood Creek,
Missoula County, Montana, I tested the hypothesis that naturally
occurring densities of caddis larvae of the genus Glossosoma were
able to depress the abundance of periphyton on the upper surfaces
of substrata. I manipulated the densities of the caddis larvae
and recorded the subsequent abundances of periphyton. I
attempted to prevent Glossosoma from immigrating to the upper
surfaces of 10 X 10 X 2 cm brick substrates by applying a thin
(approximately 0.5 mm thick) petroleum jelly fence that covered
the four 10 X 2 cm vertical sides of the bricks. Glossosoma
larvae pass through five instars; at each molt a larva unsually
leaves the top of a stone surface, discards part of the old,
undersized case during ecdysis, reconstructs a larger case, and
then crawls back to the top of stone surfaces to forage. This
petroleum jelly fence should be effective in impeding immigration
to upper surfaces of bricks.

Forty of the petroleum-jelly rimmed treatment bricks were
paired with 40 untreated control bricks. Ten rows of 4 brick
pairs were placed in one riffle zone. The rows were spaced
approximately 1.5 m apart along the length of the riffle. Ten
treated bricks and 10 control bricks were randomly chosen and
withdrawn at 7, 14, 28, and 56 days after the introduction date
(28 July 1980). All insects were brushed from the upper surface
of each withdrawn brick into a collection pan and were preserved
for later identification and enumeration. On day 56,
quantitative periphyton samples were collected from the upper
surfaces of 10 control and 10 treatment bricks. Periphyton was
removed with a stiff-bristled toothbrush and preserved in a known

volume of formalin. Periphyton densities (algal cells per unit
area of substrate) were later determined in a counting cell at
450X magnification.

The vaseline "fence" treatment was effective in impeding the
colonization of brick substrates by <u>Glossosoma</u>. Colonization of
<u>Glossosoma</u>, as indicated by the slopes of the lines that describe
accumulation of individuals over time (Fig. 3), was signficantly
lower on the vaseline-rimmed treatment bricks than on control
bricks (t-test for homogeneity of slopes, t = 24.85, p < 0.001).
On day 56, the mean density of algal cells (primarily diatoms)
was significantly higher on the vaseline-rimmed bricks where
<u>Glossosoma</u> densities were low (t = 9.20, p < 0.001) (Fig. 4).   I
concluded from these results that grazing insects can depress
periphyton resources.

In a complementary experiment, I tested the hypothesis that
periphyton densities can affect colonization by mobile grazers.
Thirty-six 10 X 10 X 2 cm bricks were placed in an Owl Creek
riffle and were colonized by both periphyton and invertebrates.
After 28 days, the bricks were withdrawn and all invertebrates
were removed with forceps and gentle washings so that the
attached periphyton would be minimally disturbed. These
periphyton-covered bricks were paired with 36 additional bricks

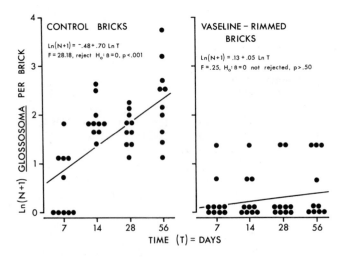

Fig. 3.   Reduction of <u>Glossosoma</u> densities on upper surfaces of
vaseline-rimmed bricks.

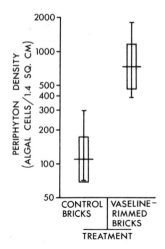

Fig. 4. Day 56 periphyton densities (log scale) on control
bricks with high Glossosoma densities and
vaseline-rimmed bricks with fewer Glossosoma. Verticle
lines are ranges, long horizontal lines are means of log
transformed densities, the bars represent 1 S.D. (of
transformed data) added and subtracted from mean. Log-
transformed data was used in t-test discussed in text.
Sample size is 10 bricks in each treatment on reintroduced bricks with high initial periphyton densities.

that had no periphyton cover and were returned to the stream.
Twelve of the experimental pairs were randomly assigned to be
withdrawn on each of three sampling dates: 3, 7, and 14 days
after initiation of the experiment. On retrieval of the
substrates, all invertebrates and periphyton were collected as
described for the Cottonwood Creek experiment. Periphyton
densities on reintroduced bricks with high initial periphyton
densities (Treatment A) and newly introduced bricks with low
initial periphyton densities (Treatment B) are shown in Figure 5.
The slope of the line describing change in periphyton density
over time for Treatment A was not significantly different from
zero (F = .55, .50 < p < 0.75) (Fig. 5a). Thus I concluded that
in Treatment A, periphyton densities did not change over the 14
day recolonization period and that the initial removal of insects
with forceps and water washings had little impact on periphyton
densities. The density of periphyton on the newly introduced
(Treatment B) bricks increased over time as indicated by the
significant positive slope of the line describing change in
periphyton density over time (F = 142.84, p < 0.001) (Fig. 5b).
The mean density of algal cells on reintroduced bricks (A) was
significantly greater than mean algal cell density on newly
introduced bricks (B) on day 3 (t = 18.49, p < 0.001) and day 7
(t = 13.48, p < 0.001) but not on day 14 (t = .58, p < 0.25).

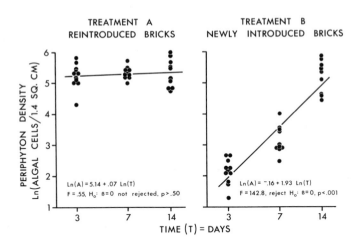

Fig. 5.   Periphyton densities on (A) reintroduced substrates and
(B) newly introduced substrates.

        Colonization of the reintroduced substrates by predominant
grazers (Baetis, Glossosoma, Oligoplectrum, and Epeorus) was
rapid and reached peak levels by the first sampling date (day 3)
(Fig. 6a).   The density of grazers on these substrates with high
initial periphyton densities showed no significant linear
increase or decrease after the third day of colonization
($F = .05$, $p < 0.50$).   Colonization of Treatment B substrates with
low initial periphyton densities by the same set of grazers was
much slower.   The mean densities of grazers on newly introduced
bricks on days 3 and 7 were significantly less than the mean
densities on reintroduced bricks (t's = 5.95, 5.77; p's < 0.001),
but not on day 14 ($t = .28$, $p < 0.50$) (Fig. 6).   I concluded that
these grazers colonized substrates with higher periphyton
densities more rapidly.   During the 14-day colonization period,
the number of grazers on Treatment A and B bricks was positively
correlated with periphyton densities on individual bricks
($r = .76$, $p < 0.01$) (Fig. 7).   This experiment provided evidence
that grazers track the abundance of the periphyton resource.   The
first experiment in Cottonwood Creek demonstrated the ability of
grazing insects to depress periphyton abundances.   Thus, whether
depression of the food supply is produced by experimental
manipulation or by grazing, the result should be the same: a
lowering of both the colonization rates and equilibrium densities
of other grazers.   This hypothesis will be tested through
experimental manipulations of densities of individual grazing
species.

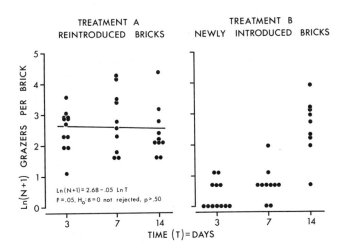

Fig. 6.   Colonization by grazers (<u>Glossosoma</u>, <u>Baetis</u>, <u>Oligoplectrum</u>, and <u>Epeorus</u>) in response to periphyton densities in two treatments shown in Figure 5.

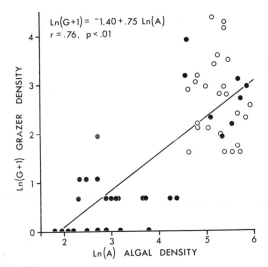

Fig. 7.   Correlation between algal and grazer densities on individual brick substrates. Solid circles represent Treatment B bricks over the 14-day period; open circles represent Treatment A bricks.

Competition, Disturbance, and Coexistence of Species

The coexistence of several sessile and mobile species on planar substrate surfaces in Owl Creek and the responses of certain species to Leucotrichia removal indicate that microhabitat preferences of these species overlap. The factors that allow such coexistence in stream communities are poorly understood. Diversity of trophic specializations in benthic invertebrates and microhabitat diversity may influence the diversity of species that coexist in streams (Hynes, 1970; Hart, 1978; Trush, 1979) but such a fine-scale level of resource or habitat partitioning is probably not the only factor contributing to the coexistence of species on rock surfaces in Owl Creek. In this section, I discuss possible ways that different disturbances may influence the coexistence of species.

Patchy distributions of the four sessile species (Leucotrichia, Parargyractis, Rheotanytarsus, and Eukiefferiella) on qualitatively similar substrates suggests that some dynamic phenomenon is acting to produce such patchiness. This patchiness is evident on both a fine scale (distributions of sessile species on single stone surfaces) as well as on a large scale (differences in sessile fauna composition from stone to stone). In many types of communities, disturbances may act to create such patchiness (Dayton, 1971; Levin and Paine, 1974; Paine, 1977). In marine intertidal communities, disturbances that eliminate dominant space-monopolizing species open up space for colonization by other species. In these systems, diversity is enhanced by intermediate levels of disturbance (Lubchenco, 1978).

Many kinds of physical disturbances occur in streams, including spates and seasonally high discharges that scour and overturn rocks, anchor ice formation on stone surfaces, and scour due to drift ice during spring ice breakup. Predation may also act as a disturbance by freeing space or renewable resources that are utilized by dominant competitors. Little is known about the ecological effects of such natural disturbances in streams.

The following observations demonstrate the possible ecological effects of a disturbance that results in overturning stone substrates in Owl Creek. A moss (Hygrohypnum sp.) is commonly found growing completely submerged in Owl Creek riffles. In August 1980 I examined the relationship between stone size and extent of moss cover on stones in a single riffle. Stone length X width was used as an index of stone size (the approximate area of the upper surface) and the percentage area covered by moss (0 to 100%) was estimated visually in incrememts of 10 percentage points. This examination revealed that small stones are usually devoid of any moss cover, but moss cover increased as a function of stone size (Fig. 8). Hygrohypnum appears to be a very

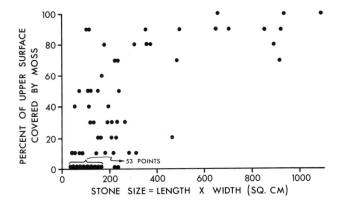

Fig. 8.   Relationship between rock size and surface moss cover in
an Owl Creek riffle.

slow-growing moss. Moss on surfaces of 20 stones exhibited no
noticeable growth between August and November 1980. Several
years may be required for Hygrohypnum to completely cover the
surface of even a small stone. Small stones would have a higher
probability of being overturned by spring flood waters or lifted
and displaced by winter anchor ice than larger stones. The
slow-growing moss may never be able to successfully establish on
very small rocks or predominate for long periods of time, even on
intermediate rock sizes. Only very large, stable stones and
boulders were completely covered with the moss. Sousa (1979)
found a similar relationship between rock size, levels of
disturbance, and marine macroalgae cover and composition in a
California coastal boulder field. The increase in moss cover on
a stone surface results in great changes in the associated
invertebrate fauna. Although Leucotrichia is unable to use
moss-covered surfaces, net spinning caddisflies (Hydropsyche
cockerelli and H. occidentalis) use the moss to support their
food capture nets. As moss cover on stone substrates increases,
the invertebrate assemblage shifts from one that is dominated by
Leucotrichia to one dominated by hydropsychids. If the extent of
physical disturbance limits the distribution and abundance of
Hygrohypnum, it is also ultimately responsible for great
difference in the composition of the invertebrate community. In
addition to the effect of rock size on disturbance and moss
cover, the overturn of rocks may free space formerly occupied by
univoltine sessile insects such as Leucotrichia or Parargyractis.
Such freeing of space may contribute to the global persistence of
multivoltine sessile species or mobile grazers.

"A stone that is rolling can gather no moss
For master and servant, oft changing is loss."
Thomas Tusser (1524-1580)

Other types of disturbances. Frazil and anchor ice
formation may be important agents of disturbance in high latitude
streams. Hynes (1970) reviewed the known effects of these
factors and commented that frazil ice crystals can severely scour
the surfaces of bottom substrata. Solid build-ups of anchor ice
form on substrates during extremely cold nights and usually
detach from the bottom and wash downstream during the day. This
drift of ice can produce daily scouring of the stream bed and can
also move or lift small stones. Frazil and anchor ice may also
subject invertebrates on stone surfaces to lethal freezing
temperatures. The daily detachment and downstream drift of
frazil and anchor ice from the stream bottom may eventually
result in a solid surface cover of ice over a stream. Compressed
snow can add to the thickness of these solid ice covers.
Extensive scouring of the stream bottom may result during the
early spring thaw when solid ice bridges collapse and drift
downstream. In larger rivers, solid ice jams that extend to the
river bottom (Ashton, 1979) may also have a great impact.
Relatively little is known about the magnitudes of ice-caused
disturbances in streams.

Predation can be considered a special kind of disturbance
that may serve to free space or resources monopolized or
exploited by a dominant competitor. Studies of sessile
invertebrate and macroalgae communities of the marine intertidal
zone have shown that either selective predation on competitive
dominants or non-selective predation may enhance species
diversity within an area (Paine, 1966; Dayton, 1971; Lubchenco,
1978). Predation may be important in facilitating local
coexistence of species in Owl Creek. The most conspicuous
predators in Owl Creek riffles are the stonefly nymphs
Hesperoperla pacifica and Calineuria californica. Sheldon (1980)
showed that large Parargyractis larvae were common in the diet of
Hesperoperla pacifica in Owl Creek. If Parargyractis or any of
the other sessile species are capable of maintaining an exclusive
hold on space through competition, predation upon them would make
space available for colonization by other species. Predation may
also serve to reduce densities of abundant mobile grazers,
consequently reducing the resource-depressing effect of these
grazers and allowing greater abundances of other species.

The importance of predation in influencing the distribution
and abundance of prey species in streams is poorly understood
(see Allan this volume). Peckarsky (1980) and Peckarsky and
Dodson (1980a, b) experimentally showed that stonefly predators
can significantly influence the distributions and abundances of
certain prey species. Wiley and Kohler (1981) also demonstrated
that the predaceous caddis larva Rhyacophila was responsible for
changes of several different mobile prey species on surfaces of
natural substrates. The ultimate effect of predator-prey

interactions on community structure in streams is little known. It is possible that invertebrate predators can have a significant impact in mitigating the effects of competition for either space or renewable resources in stream benthic communities. Predation and its possible effects remain relatively unexplored in these communities, however.

EXPERIMENTAL STUDIES OF STREAM COMMUNITY STRUCTURE: AN OUTLOOK

Experimental analyses of biological interactions are relatively new to the study of stream communities. Understanding the nature and extent of biological interactions should not be considered ends in themselves, but rather, they can contribute to more encompassing questions of how these interactions and various other phenomena such as physical disturbance interplay to produce characteristic patterns in the distribution and abundance of stream organisms.

For example, the patchy (contagious) distributions of stream invertebrates has long been recognized. This patchiness has often been considered a stumbling block in sampling methodology (Needham and Usinger, 1965; Chutter and Noble, 1966; Chutter, 1972). Such patchiness, however, is more than simply a source of problems for sampling design; it presents intriguing ecological questions that await answers from experimental studies. Studies of marine intertidal communities (Paine, 1977; Sousa, 1979; Paine and Levin, 1981) have demonstrated the effect of unpredictable physical disturbance in opening up a patchwork of free space that allows the wide-ranging (global) persistence of high biological diversity. A close examination of patchiness and other distinctive patterns in stream communities may reveal that competition, predation, and physical disturbance operate in ways that may be similar to their operation in widely different terrestrial and aquatic communities. In addition, many ecological problems such as the ecological effects of resource exploitation that have been less thoroughly studied in other communities are well suited for experimental study in stream communities. Such experimentation will yield valuable information on ways in which stream communities are structured as well as contribute to a general understanding of ecological processes.

ACKNOWLEDGMENTS

I thank John Dixon, Rebecca Everett, David Hart, Richard Hutto, Barbara Peckarsky, Stephen Schroeter, Andy Sheldon, J. Bruce Wallace and an anonymous reviewer for their helpful criticisms of earlier versions of this manuscript. Nancy

Williams assisted in many aspects of my work. The idea for the "petroleum-jelly fence" experiment resulted from a lively discussion with Bill Trush in September 1979. Portions of this work were funded by the University of Montana, Department of Zoology and a Sigma Xi Grant-in-Aid of Research.

REFERENCES

Allan, J. David. 1975. The distributional ecology and diversity of benthic insects in Cement Creek, Colorado. Ecology 56:1040-1053.

Ashton, G. D. 1979. River ice. Amer. Sci. 67:38-45.

Brown, J. H. 1981. Two decades of homage to Santa Rosalia: toward a general theory of diversity. Amer. Zool. 21:877-888.

Carlsson, M., L. M. Nilsson, Bj. Svensson, and S. Ulfstrand. 1977. Lacustrine seston and other factors influencing the blackflies (Diptera:Simuliidae) inhabiting lake outlets in Swedish Lapland. Oikos 29:229-238.

Case, T. J. And M. E. Gilpin. 1974. Interference competition and niche theory. Proc. Nat. Acad. Sci. U.S.A. 71:3073-3077.

Charnov, E. L., G. H. Orians and K. Hyatt. 1976. Ecological implications of resource depression. Am. Nat. 100:247-259.

Chutter, F. M. 1968. On the ecology of the fauna of stones in the current in a South African river supporting a very large Simulium (Diptera) population. J. Appl. Ecol. 5:531-561.

Chutter, F. M. 1972. A reappraisal of Needham and Usinger's data on the variability of a stream fauna when sampled with a Surber sampler. Limnol. Oceanogr. 17:139-141.

Chutter, F. M. and R. G. Noble. 1966. The reliability of a method of sampling stream invertebrates. Arch. Hydrobiol. 62:95-103.

Connell, J. H. 1961a. Effects of competition, predation by Thais lapillus, and other factors on natural populations of the barnacle Balanus balanoides. Ecol. Monogr. 31:61-104.

Connell, J. H. 1961b. The influence of interspecific competition and other factors on the distribution of the barnacle Chthamalus stellatus. Ecology 42:710-723.

Connell, J. H. 1975. Some mechanisms producing structure in natural communities: A model and evidence from field experiments. In: M. L. Cody and J. M. Diamond (eds.), Ecology and evolution of communities. Belknap, Harvard.

Cummins, D. W. and G. H. Lauff. 1969. The influence of substrate particle size on the microdistribution of stream macrobenthos. Hydrobiologia 34:145-181.

Davies, N.B. 1978. Ecological questions about territorial behavior. In: J. R. Krebs and N. B Davies (eds.), Behavioral ecology: an evolutionary approach. Blackwell, Oxford.

Dayton, P. K. 1971. Competition, disturbance and community organization: the provision and subsequent utilization of space in a rocky intertidal community. Ecol. Monogr. 41:351-389.

Dayton, P. K. 1975. Experimental evaluation of ecological dominance in a rocky intertidal algal community. Ecol. Monogr. 45:137-159.

Dodson, S. I. 1974. Zooplankton competition and predation: an experimental test of the size-efficiency hypothesis. Ecology 55:605-613.

Douglas, B. 1958. The ecology of the attached diatoms and other algae in a small stony stream. J. Ecol. 46:295-322.

Edington, J. M. 1968. Habitat preferences in net-spinning caddis larvae with special reference to the influences of water velocity. J. Anim. Ecol. 37:675-692.

Fiance, S. B. 1978. Effects of pH on the biology and distribution of Ephemerella funeralis (Ephemeroptera). Oikos 31:332-339.

Fowler, N. 1981. Competition and coexistence in a North Carolina Grassland. II. The effects of the experimental removal of species. J. Ecol. 69:843-854.

Grant, P. R. and R. J. Mackay. 1969. Ecological segregation of systematically related stream insects. Can. J. Zool. 47:691-694.

Haddock, J. D. 1977. The effect of stream current velocity on the habitat preference of a net-spinning caddis fly larva, Hydropsyche oslari Banks. Pan-Pac. Entomol. 53:169-174.

Harper, J. T. 1977. Population biology of plants. Academic Press, London.

Harrod, J. J. 1965. Effect of current speed on the cephalic fans of the larvae of Simulium ornatum var. nitidifrons Edwards (Diptera:Simuliidae). Hydrobiologia 26:8-12.

Hart, D. D. 1978. Diversity in stream insects: regulation by rock size and microspatial complexity. Verh. Internat. Verein. Limnol. 20:1376-1381.

Hart, D. D. 1981. Foraging and resource patchiness: field experiments with a grazing stream insect. Oikos 37:46-52.

Higler, L. W. G. 1975. Reactions of some caddis larvae (Trichoptera) to different types of substrate in an experimental stream. Freshwat. Biol. 5:151-158.

Hildrew, A. G. and J. M. Edington. 1979. Factors facilitating the coexistence of hydropsychid caddis larvae (Trichoptera) in the same river system. J. Anim. Ecol. 48:557-576.

Hynes, H. B. N. 1970. The ecology of running waters. Univ. Toronto Press, Toronto.

Hynes, H. B. N. and T. R. Williams. 1962. The effect of DDT on the fauna of a central African stream. Ann. Trop. Med. Parasit. 56:78-91.

Lange, W. H. 1956. Aquatic Lepidoptera, pp. 271-288. In: R.
    L. Usinger (ed.), Aquatic insects of California. Univ.
    Calif. Press, Berkeley.
Levin, S. A. and R. T. Paine. 1974. Disturbance, patch
    formation and community structure. Proc. Nat. Acad. Sci.
    U.S.A. 68:1246-1248.
Lubchenco, J. 1978. Plant species diversity in a marine
    intertidal community: importance of herbivore food
    preferences and algal competitive abilities. Am. Nat.
    112:23-39.
Luchenco, J. and B. A. Menge. 1978. Community development and
    persistence in a low rocky intertidal zone. Ecol. Monogr.
    59:67-94.
Lynch, M. 1979. Predation, competition, and zooplankton
    community structure: an experimental study. Limnol.
    Oceanogr. 24:253-272.
Malas, D. and J. B. Wallace. 1977. Strategies for coexistence
    in three species of net-spinning caddisflies (Trichoptera)
    in second-order Appalachian streams. Can. J. Zool.
    55:1829-1840.
McAuliffe, J. R. (In Press). Behavior and life history of
    Leucotrichia pictipes (Banks) (Trichoptera:Hydroptilidae)
    with special emphasis on case reoccupancy. Can. J. Zool.
Menge, B. A. 1976. Organization of the New England rocky
    intertidal community: role of predation, competition and
    environmental heterogeneity. Ecol. Monogr. 46:355-369.
Menge, B. A. 1979. Coexistence between seastars Asterias
    vulgaris and A. forbesi in a heterogeneous environment: A
    non-equilibrium explanation. Oecologia 41:245-272.
Minshall, G. W. and J. N. Minshall. 1977. Microdistribution of
    benthic invertebrates in a Rocky Mountain (U.S.A.) stream.
    Hydrobiologia 55:231-249.
Needham, P. R. and R. L. Usinger. 1956. Variability in the
    macrofauna of a single riffle in Prosser Creek, California,
    as indicated by the Surber sampler. Hilgardia 24:383-409.
Oswood, M. W. 1979. Abundance patterns of filter-feeding
    caddisflies (Trichoptera:Hydropsychidae) and seston in a
    Montana (U.S.A.) lake outlet. Hydrobiolgia 63:177-183.
Paine, R. T. 1966. Food web complexity and species diversity.
    Am. Nat. 100:65-75.
Paine, R. T. 1977. Controlled manipulations in the marine
    intertidal zone, and their contributions to ecological
    theory. In: C. E. Goulden (ed.), Changing scenes in
    natural sciences 1776-1976. Special Publ. No. 12. Acad.
    Nat. Sci. Phila., Penn.
Paine, R. T. 1979. Disaster, catastrophe, and local persistence
    of the sea palm Postelsia palmaeformis. Science
    205:586-687.

Paine, R. T. and S. A. Levin. 1981. Intertidal landscapes: disturbance and the dynamics of pattern. Ecol. Monogr. 51:145-178.

Pavlichenko, V. I. 1977. Role of larvae of Hydropsyche angustipennis Curt (Trichoptera:Hydropsychidae) in the destruction of blackfly larvae in running water of Zaporozke Oblast. Soviet J. Ecol. 8:84-85.

Peckarsky, B. L. 1980. Predator-prey interactions between stoneflies and mayflies: behavioral observations. Ecology 61:932-943.

Peckarsky, B. L. and S. I. Dodson. 1980a. Do stonefly predators influence benthic distributions in streams? Ecology 61:1275-1282.

Peckarsky, B. L. and S. I. Dodson. 1980b. An experimental analysis of biological factors contributing to stream community structure. Ecology 61:1283-1290.

Peterson, C. H. 1979. The importance of predation and competition in organizing the intertidal epifaunal communities of Barnegat Inlet, New Jersey. Oecologia 39:1-24.

Reice, S. R. 1981. Interspecific associations in a woodland stream. Can. J. Fish. Aq. Sci. 38:1271-1280.

Reynoldson, T. B. 1981. The ecology of the Turbellaria with special reference to the freshwater triclads. Hydrobiologia 84:87-90.

Reynoldson, T. B. and L. S. Bellamy. 1971. The establishment of interspecific competition in field populations, with an example of competition in action between Polycelis nigra (Mull.) and P. tenuis (Ijima) (Turbellaria, Tricladida). Proc. Adv. Stud. Inst. Dyn. Numbers Pop. 1970:282-297.

Sheldon, A. L. 1972. Comparative ecology of Arcynopteryx and Diura (Plecoptera) in a California stream. Arch. Hydrobiol. 69:521-546.

Sheldon, A. L. 1980. Resource division by perlid stoneflies (Plecoptera) in a lake outlet ecosystem. Hydrobiologia 71:155-161.

Sousa, W. P. 1979. Disturbance in marine intertidal boulder fields: the nonequilibrium maintenance of species diversity. Ecology 60:1225-1239.

Townsend, C. R. and A. G. Hildrew. 1976. Field experiments on the drifting, colonization, and continuous redistribution of stream benthos. J. Anim. Ecol. 45:759-772.

Trush, W. J. 1979. The effects of area and surface complexity on the structure and formation of stream benthic communities. Va. Polytech. Univ. Blacksburg, Va. Unpublished M.S. thesis.

Vannote, R. L. and B. W. Sweeney. 1980. Geographic analysis of thermal equilibria: a conceptual model for evaluating the effect of natural and modified thermal regimes on aquatic insect communities. Am. Nat. 115:667-695.

Wiley, M. J. and S. L. Kohler. 1981. An assessment of biological interactions in an epilithic stream community using time-lapse cinematography. Hydrobiologia 78:183-188.

Williams, D. D. 1980. Some relationships between stream benthos and substrate heterogeneity. Limnol. Oceanogr. 25:166-172.

Williams, D. D. and J. H. Mundie. 1978. Substrate size selection by stream invertebrates and the influence of sand. Limnol. Oceanogr. 23:1030-1033.

Williams, N. E. and H. B. N. Hynes. 1973. Microdistribution and feeding of the net-spinning caddisflies (Trichoptera) of a Canadian stream. Oikos 24:73-84.

PLANT-HERBIVORE INTERACTIONS IN STREAM SYSTEMS

Stanley V. Gregory

Department of Fisheries and Wildlife
Oregon State University
Corvallis, Oregon 97331

INTRODUCTION

Ecological theories of herbivory have been derived primarily from terrestrial research because man has long depended on domesticated grazers and has suffered the detrimental consequences of outbreaks of phytophagous insects. Only in recent decades has the amount of information on herbivory in aquatic ecosystems grown sufficiently to stimulate development of general theories of herbivory for these systems.

Herbivory in streams has had little impact on the development of general ecological theory or the conceptual framework of stream ecology. Stream researchers are becoming more aware of differences in both qualities and quantities of sources of organic matter for consumers in streams and that herbivory is a critical process in maintaining the organization of stream ecosystems. However, no one has addressed the implications of plant-herbivore relationships in flowing water environments for general ecological theory (but see McAuliffe this volume). Stream ecosystems appear to have more dynamic and rigorous environmental conditions than terrestrial, marine, or lentic ecosystems. These conditions should impose unique constraints that determine evolutionary patterns of stream organisms and shape the general organization and processes of lotic ecosystems. As a result, plant and herbivore communities in streams should exhibit distinct sets of characteristics not found in terrestrial, marine, or lake systems.

Consumption of plants by animals forms the foundation of trophic structure in all ecosystems and the term herbivory is

often broadly applied to this general process. Unfortunately,
this definition obscures major distinctions in the processes
through which plant matter enters animal food chains. In a
strict sense, herbivory is the consumption of living plants or
their parts and detritivory is the consumption of dead organic
matter, which includes dead plant material. This is a critical
conceptual distinction though in practice it may be difficult to
separate living and dead components of plant matter. This paper
addresses herbivory in streams, the interaction between living
plants and their consumers; however, the reader should realize
that consumption of dead aquatic plants may be a major route of
entry of plant matter into animal food chains in many instances.

     The study of plant-herbivore relationships in streams is
critical for rigorous development of the conceptual framework of
stream ecology and the general ecological theory of herbivory.
The volume of literature on plant-herbivore relationships is
immense for terrestrial ecosystems and quite substantial for
marine and freshwater planktonic systems, marine intertidal
zones, and lake benthos; however, the body of literature on
herbivory in lotic systems is diminuitive in comparison. The
following section of this paper is a review of available
literature on plant-herbivore relationships in stream ecosystems.
This synthesis is intended to describe patterns of plant-
herbivore interactions in stream ecosystems, identify deficien-
cies and problems in our conceptual framework of stream ecology,
and demonstrate the value of research on herbivory in streams to
general ecological theory of plant-herbivore relationships.

Herbivory in Stream Ecosystems

     Herbivory, like many other ecological processes, involves
the interaction of living organisms, capable of adapting to
changing conditions. As a result, plant-animal interactions in
streams may exhibit a wide array of responses. Unfortunately,
ecologists usually study herbivory from the perspective of either
the plant or the herbivore, describing the dynamics of one
component in detail and greatly simplifying the other.
Realistically, this problem is often unavoidable; but unless
stream ecologists accurately characterize the taxonomic,
physiological, and ecological properties of both plant and
herbivore communities, we will undoubtedly amass a body of
seemingly contradictory information on herbivory in lotic
ecosystems.

Characteristics of Plants and Herbivores in Streams

     Development of ecological theory requires that the array of
interactions within ecosystems must be simplified and

generalized, but a delicate balance must be maintained between complexity and generality. The benefits of broad generalizations are quickly lost if critical distinctions are obscured by well-intended efforts to reduce complex systems to common levels. As a result, important information is lost and apparent contradictions arise.

Though all major lotic plant forms usually will be found in any river drainage, the distributions of mosses, liverworts, and vascular macrophytes are relatively patchy whereas benthic algae are generally cosmopolitan. Responses and susceptibility of aquatic plants to grazing are determined by morphology, cell wall structure, mode of attachment, reproduction, and chemical defenses. Recovery of plants from grazing depends on mode of reproduction, rate of reproduction and colonization, and growth rate of the plants. Aquatic plants exhibit a wide range of these characteristics; therefore, plants respond in many different ways to herbivory.

Responses of grazers in plant-herbivore interactions in streams are determined by methods of feeding, motility, patterns of reproduction, rates of ingestion and assimilation, and temporal patterns of life history. Morphological specializations for feeding on plants in streams are not uncommon in aquatic herbivores, for example, the blade-like mandibles of glossosomatid caddisflies, the rasping radula of snails, the brush-like mandibles of several mayflies, the chewing mouthparts of baetid mayflies, the piercing mandibles of aquatic mites and hydroptilid caddisflies, and molariform pharyngeal teeth of carp. Mode of feeding has been used as the major criterion for a functional group classification of aquatic insects (Cummins, 1973) and includes several types of grazing specializations, particularly scrapers. A scraper gathers its food by scraping across a surface and most commonly is a herbivore; however, herbivores obtain their food by scraping, collecting, shredding, or piercing. The terms "scraper" and "grazer" are not synonymous; a scraper is a specific type of grazer. The misuse of the term grazer has commonly resulted in underestimation of the importance of herbivores in stream ecosystems because scrapers are incorrectly assumed to be the only form of grazer. The concept of functional feeding groups plays a central role in the theory of stream ecology and stream ecologists must demand precise use of the concept or we will develop our science on a foundation of misconception and obfuscuity.

Effects of Herbivores on Plants in Streams

Benthic algae are the principal food source for aquatic herbivores in stream ecosystems (Chapman and Demory, 1963; Hynes, 1970; Cummins, 1974; Mann, 1975). Coffman et al. (1971)

examined gut contents of 75 common macroinvertebrates in a
woodland stream during two autumn seasons; algal material made up
50% or more of gut content in 50 of the 75 taxa examined and
algal material was not found in the guts of only one species.
These authors estimated that 17-21% of the standing crop of
macroinvertebrate tissue in Linesville Creek was supported by
benthic algae. However, many herbivores in streams are
generalists and their gut contents reflect the relative
availability of their food (Brown, 1961; Chapman, 1966; Hynes,
1970; Mecom, 1972; Moore, 1977). For example, Cummins (1973)
compared gut content of Glossosoma nigrior, a caddisfly scraper,
from Augusta Creek to gut content of the same species in
Linesville Creek from the study of Coffman et al. (1971). In
Augusta Creek, a stream with less primary production, algal
material in guts of all instars averaged only 25.9% as compared
to 98.4% in Linesville Creek.

Ecologists frequently assume that herbivores in streams
rarely consume filamentous algae; however, many grazers readily
consume and often prefer these plants. Brown (1961) observed
that the mayfly Chloeon dipterum fed on filamentous algae and
suggested that the slow gut passage time of filamentous algae
resulted in a greater assimilation efficiency. Caddisflies,
mayflies, isopods, and amphipods all have been observed to feed
on filamentous diatoms and green algae in stream systems
throughout the world (Jones, 1949; Cummins, 1964; Mecom, 1972;
Moore, 1975). Certain filamentous algae may not be palatable,
however. In English rivers, the isopod Asellus aquaticus and the
amphipod Gammarus pulex consumed the filamentous diatoms,
Melosira varians and Diatoma vulgare; but the filamentous green
alga, Cladophora, and filamentous blue-green alga, Phormidium
foveolarum, were not found in the animals' guts though both
species of algae were present (Moore, 1975). Gray and Ward
(1978) observed that significant quantities of filamentous algae
were consumed by aquatic insects but only as "new growths or as
decomposing fragments." Though particular grazers may not
utilize filamentous algae and certain species of filamentous
algae may be unpalatable, filamentous algae can be a significant
source of food for many aquatic herbivores.

Ecologists often assume that blue-green algae are unsuitable
food sources for aquatic herbivores because several blue-green
algae produce toxins or chemical deterrants. However, this
inference is not valid for all grazers and all species of
blue-green algae. Gajevskaja (1958) did suggest that blue-green
algae are usually avoided by consumers. Also, Sumner (1980)
observed that growth of the snail, Juga plicifera (reported as
Oxytrema silicula by Sumner) in laboratory streams was inversely
correlated to standing crop of blue-green algae (primarily
Schizothrix calcicola) and suggested that snail growth was

inhibited by blue-green algae. However, the mayfly collector
Tricorythodes minutus consumed two genera of blue-green algae,
Anabaena and Lyngbya, in laboratory stream channels; rates of
assimilation were high for Lyngbya and moderate for Anabaena but
assimilation efficiencies were equal for the two algae
(McCullough et al., 1979). Izvekova (1971) also found that
chironomid larvae readily assimilated blue-green algae.

Bryophytes in streams usually are not consumed but there are
a number of exceptions. In laboratory studies, amphipods failed
to grow or survive when fed the aquatic liverwort, Nardia
(Willoughby and Sutcliffe, 1976). Many of the invertebrates that
inhabit mosses in streams use the plant as habitat and feed on
organic matter trapped among the branches and other invertebrates
living in the moss. However, a number of aquatic insects feed
predominantly on moss, in particular the caddisfly genus
Micrasema (Decamps and Lafont, 1974; Wiggins, 1977). Stoneflies
and caddisflies may consume large amounts of mosses (Hynes, 1941;
Chapman and Demory, 1963). Jones (1949) found that the mayfly
Ephemerella notata consumed Fontinalis when filamentous algae
were unavailable; he also noted that moss was consumed by a
species of black-fly, Simulium tuberosum, and another mayfly,
Baetis rhodani. Dance flies in the subfamily Hemerodomiinae feed
primarily on aquatic mosses (Brindle, 1964). Aquatic mites
pierce the cells of mosses and suck out the cytoplasm
(Richardson, 1981). The mayflies, Caudetella cascadia, C.
hystrix, C. edmundsi, Serretella teresa, and S. levis, consume
significant quantities of moss in mountain streams of the Pacific
Northwest (Charles Hawkins, personal communication). Although
these macroinvertebrates consume aquatic mosses, they appear to
be exceptions rather than the rule. Inspection of the large
standing crops of mosses in headwater streams reveals little
apparent damage to leaves to stems by consumers. As in the case
of vascular macrophytes, the detrital food chain is the most
probable route of entry of mosses into primary consumers. This
process may be slower for mosses than vascular plants because of
inhibitory compounds, particularly sphagnol, that retard
decomposition of mosses (Rosswall et al., 1975). Bryophytes are
dominant forms of plants in many headwater streams and may be a
major source of organic matter for consumers; however, knowledge
of moss-herbivore relationships is insufficient to accurately
describe this process.

Living aquatic vascular macrophytes generally are not a
major component of the diet of herbivores in lotic ecosystems.
Tough cell walls, lignified structures, and low nitrogen content
decrease the ability to ingest and digest these plants. Aquatic
macrophytes primarily enter the consumer food chain as detritus
(Whitehead, 1935; Hynes, 1970; Mann, 1975; Westlake, 1975). Live

aquatic macrophytes in the Fort River, Massachusetts, were not
grazed and entered the food chain as detritus (Fisher and
Carpenter, 1976). Studies have found that diets of stream
macroinvertebrates did not include living aquatic macrophytes,
though aquatic macrophytes were present (Cummins, 1964; Koslucher
and Minshall, 1973; Gray and Ward, 1979). Distributions of
aquatic gastropods have been correlated with distributions of
particular species of aquatic macrophytes but this correlation
does not prove that there was direct nutritional dependency (Pip,
1978). It is estimated that animals require food with C:N ratios
of 17 or less (Russell-Hunter, 1970). The low C:N ratio of algae
(4:1 to 8:1) makes periphyton a more nutritional food source than
aquatic vascular macrophytes (13:1 to 69:1, based on Boyd, 1970).
One of the few examples of a grazer that consumes living vascular
macrophytes in streams is a caddisfly, Limnephilus lunatus, that
feeds on watercress, an aquatic macrophyte, and is grown
commercially (Gower, 1967). Although vascular herbs are among
the most palatable and digestable plants available to terrestrial
herbivores, living aquatic vascular macrophytes are one of the
least preferred foods of aquatic grazers.

Abundance of Plants

A primary effect of herbivores on plant communities is
removal of vegetation, which potentially reduces numbers or
biomass of primary producers. This cropping effect has been
observed both in laboratory streams that contain only primary
producers and grazers and natural streams with a full complement
of benthic invertebrates and plants.

Studies of plants and animals in artificial streams have
enabled researchers to focus on the process of herbivory and the
majority of these studies have shown that grazing can reduce the
abundance of plants in streams. The mayfly Baetis rhodani
reduced standing crops of benthic algae in laboratory streams and
their drift rate increased when biomass of algae decreased
(Bohle, 1978). Orthoclad chironomid larvae decreased standing
crops of algae in outdoor channels but standing crops of algae
increased in channels treated with insecticides; after
insecticide addition was stopped, biomass of benthic algae
declined abruptly as chironomids recolonized the channels
(Eichenberger and Schlatter, 1978). The snail Juga plicifera
reduced standing crops of periphyton by as much as ten-fold as
compared to ungrazed channels (McIntire, 1975; Sumner, 1980). A
study of the effects of grazing intensity of Juga on benthic
algae found that heavy and moderate grazing intensity reduced
biomass of algae but light grazing caused no observable decrease
in standing crop in comparison to an ungrazed algal community
(Gregory, 1980). Lack of reduction of algae by light grazing

intensity may reflect either compensation for cropping by enhanced primary productivity or the inability to detect small changes in standing crop. Grazing by another aquatic snail, Physa gyrina, in laboratory streams caused no apparent reduction of periphyton biomass (Kehde and Wilhm, 1972). Thus herbivores seem to be capable of altering the abundance of aquatic plants but grazing intensity is not always sufficient to decrease the biomass of plants.

The cropping effect of herbivores in artificial streams is not surprising because these are simple systems that usually contain a plant community and a single type of grazer; however, in natural streams, herbivory is only one of many factors that regulate plants and grazers. Even so, herbivory is a dominant controlling factor and many studies have demonstrated that grazers reduce standing crops of algae in natural streams. Cell numbers of the diatom Achnanthes were negatively correlated with densities of a grazing caddisfly, Agapetus fuscipes, in an English stream (Douglas, 1958). Estimates of total grazing rates of macroinvertebrates in a woodland stream equalled estimates of periphyton production rates and the authors concluded that standing crops of periphyton were controlled by grazers (Elwood and Nelson, 1972). Periphyton standing crops were significantly reduced at higher densities of the caddisfly Glossosoma in a Montana stream (McAuliffe, this volume). Studies also have indicated that elimination of herbivores in natural streams released plant communities from grazing pressure and resulted in increased abundance of benthic algae. After an application of DDT eliminated all aquatic insects in a Canadian stream, standing crops of benthic algae increased (Ide, 1967). In Africa, insecticide from cattle dipping tanks was released to a stream and destroyed the entire fauna (Chutter, 1970); standing crops of benthic algae increased immediately after the insecticide release; but as the fauna recovered, standing crops of benthic algae decreased to prepoisoning levels. On the other hand, some field studies of macroinvertebrate communities in streams have concluded that grazers exerted little effect on abundance of benthic algae (Cummins et al., 1973; Stockner, 1971; Moore, 1975, 1978; Collins et al., 1976; Stockner and Shortreed, 1976). Two of these examples, Stockner (1971) and Collins et al. (1976), are studies of thermal springs, streams with unique environmental conditions and restricted faunas. The other studies that reported no effects of grazers on algae are based on either estimates of ingestion by macroinvertebrate communities or conjecture. Though estimates of ingestion may provide an excellent basis for examination of this question, accurate assessment of abundance of small aquatic insects and protozoans is extremely difficult and the role of these organisms in grazing in streams is unknown.

Primary Production

Availability of plant material for herbivores is determined
by both the biomass of plants and the rate of primary production;
however, this fundamental ecological concept is neglected
constantly in studies of herbivory in lotic ecosystems.
Measurement of rates of primary production in streams is
certainly more complicated than the simple measurement of plant
biomass or amounts of plant pigments, but this difficulty does
not alter the basic fact that both biomass and rate of primary
production must be measured to determine food availability for
herbivores. The conceptual framework of herbivory in stream
ecosystems will continue to be incomplete until stream ecologists
incorporate assessment of rates of aquatic primary production in
studies of plant-herbivore interactions.

A simulation model of periphyton dynamics in streams
indicated that grazers should be able to reduce standing crops of
benthic algae and maintain them at low levels (McIntire, 1973).
Low biomasses of algae can support much higher standing crops of
aquatic consumers because of the rapid turnover times of algae.
McIntire's (1973) periphyton model predicted that periphyton
could support biomasses of grazers 20 times greater than their
own. Subsequent studies of grazers in artificial streams by
McIntire (1975) and Gregory (1980) demonstrated that algal
communities supported herbivore biomasses 13 and 20 times greater
than their own, respectively. Therefore, herbivory depends not
only on the standing crop of plants but the rate of plant
production as well.

Evaluation of primary productivity within the context of
herbivory is important from the perspective of both plant and
herbivore communities. Stimulation of primary production is
frequently assessed by measuring changes in standing crops of
benthic algae; however, increased rates of primary production may
be reflected in greater abundance of herbivores rather than
plants because of the cropping effect described previously. A
woodland stream in Tennessee was enriched with phosphorus and
standing crops of periphyton increased initially but then
decreased slightly; however, biomass of grazers increased greatly
in the enriched section (Elwood et al., 1981). Grazing pressure
regulated standing crops of periphyton and prevented a sustained
increase in the biomass of benthic algae. Manuel and Minshall
(1980) enriched artificial stream channels with nitrogen and
phosphorus and observed increased standing crops of periphyton;
however, nutrient additions to the adjacent natural stream
resulted in no significant increase in periphyton abundance. The
authors concluded that grazing in the natural stream masked the
increase in primary production that resulted from nutrient

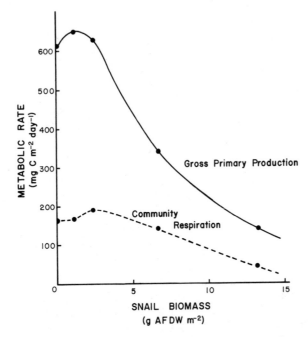

Fig. 1.   Rates of gross primary production and community
          respiration in artificial stream channels with
          different standing crops of snails after 35 days
          (data from manuscript submitted by Gregory,
          McCullough, and Busch).

enrichment. These studies demonstrate that neither plant nor
herbivore dynamics in stream ecosystems can be understood without
evaluation of plant-herbivore interactions.

     Reduction of algal biomass by grazers potentially reduces
rates of primary production but only two studies have examined
this response (Gregory, 1980; Sumner, 1980). Rates of gross
primary production, net community primary production, and
community respiration were measured in laboratory streams stocked
with different densities of the snail Juga plicifera (Gregory,
McCullough, and Busch, manuscript submitted; Gregory, 1980).
Heavy and moderate grazing intensities significantly reduced
rates of gross primary production and net community primary
production; community respiration was significantly lower only at
the higher grazer density (Fig. 1). Turnover times (mean biomass
of periphyton/net community primary production) ranged from a low

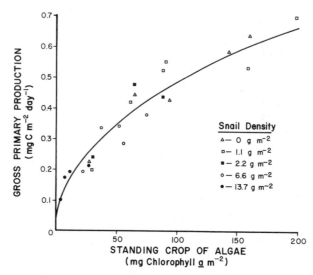

Fig. 2   Rates of gross primary production in relation to
          standing crops of chlorophyll a in periphyton
          communities exposed to different grazing pressures
          in laboratory streams during the 35 day study.

of 1.5 days at high snail density to 21.8 days in the ungrazed
channel. Therefore, it required less time to replace the
standing crop of periphyton at high to moderate densities of
grazers than at low grazer densities or no grazing. Patterns of
primary production in the presence of different densities of
grazers indicated that the potential for production at different
standing crops of periphyton was not a linear function of
biomass of primary producers. Gross primary production increased
as standing crops of primary producers increased (Fig. 2).
However, primary production rates at low standing crops were not
as depressed as might be expected because younger plant
communities are capable of higher rates of production per unit
weight and self-shading is less than for older plant communities.
This response was reflected in the relationship between
assimilation number (rate of gross primary production/amount of
plant pigment) and standing crop of chlorophyll a (Fig. 3).
Greater assimilation number at low standing crops provides a
mechanism for minimizing the impacts of grazing on aquatic
primary producers. As grazers depress standing crop of primary
producers, the productive capacity of the primary producers is
enhanced.

      Sumner (1980), in a subsequent study with Juga, observed no
effect of grazing at low to moderate densities of snails (1.3 and

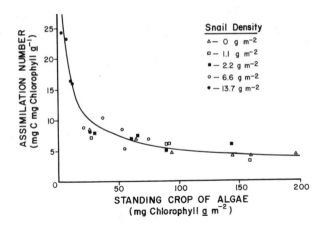

Fig. 3.   Assimilation number in relation to standing crops of chlorophyll a in periphyton communities exposed to different grazing pressures in laboratory streams during a 35 day study.

5.3 g m^{-2}) although he did observe reduced standing crops of algae at these snail densities. Though the different responses found in the studies by Gregory (1980) and Sumner (1980) cannot be fully explained, greater relative productivity of the younger, grazed communities may have resulted in rates of primary production in grazed communities that equalled those of ungrazed communities in Sumner's study. Also, high temperatures in Sumner's study (26°C), different algal communities, and different substrates may have been factors in the different responses. It has been demonstrated that grazers can alter rates of primary production of plant communities in streams; but the extent of that effect is determined by physical, chemical, and biological factors that regulate rates of primary production and the ability of the herbivore to crop the plant community. Therefore, a complex array of responses of aquatic primary production to grazing are possible and additional research is critically needed in this fundamental component of plant-herbivore interactions in streams.

Evidence for stimulation of total quantity of primary production by grazing has been found in aquatic systems, but only in lentic ecosystems (Hargrave, 1970; Cooper, 1973; Flint and Goldman, 1975). All three studies found that primary production could be stimulated by grazing, but the stimulatory effect could be overridden by excessive consumption at high densities of herbivores. Nutrient regeneration or excretion by grazers and cropping to juvenile communities are possible mechanisms for stimulation of primary production by grazing. The only study that examined this phenomenon in streams was the study of the snail _Juga_ in laboratory streams (Gregory, 1980) and direct stimulation of instantaneous rates of primary production was not observed. Total amount of primary production (plant biomass accumulation + export + consumption) was equal at intermediate and high densities of snails even though the grazing pressure was measured to be 75% greater at the high density of snails, but total primary production was lower at both of these densities than under ungrazed conditions (Table 1). However,

Table 1.   Effects of grazing by snails on total primary production in laboratory streams (data from manuscript submitted by Gregory, McCullough, and Busch) (all values in g AFDW m^{-2}).

	Grazing Pressure				
	High	Moderate	Low	Low	None
Biomass Accumulation	-1.3	1.6	18.1	23.4	18.5
Export	3.6	4.1	2.9	0.9	0.9
Consumption	11.7	6.8	2.9	1.4	0
TOTAL	14.0	12.4	23.9	25.8	19.4

total primary production was slightly greater at low densities of snails than without snails. This suggests that there was a stimulatory effect; although it could not be supported statistically. Measures of biomass, rates of primary production, and calculations of total primary production demonstrated that compensatory mechanisms that allow primary producers to support higher grazing pressures must exist because primary production was not reduced as much as would be expected from the consumptive demands of the grazers. At high grazer density, most of the

total primary production went into consumption by the herbivore;
but a low grazing pressure, biomass accumulation accounted for
almost all of the total plant production. Grazing pressure
maintained young, monolayer communities dominated by diatoms in
the channels with high and intermediate densities of snails.
Communities of primary producers in the channels that had low
densities of snails or were ungrazed developed older, thicker
growth forms, dominated by filamentous algae and epiphytic
diatoms. The younger, monolayer communities were more productive
per unit weight than the older, thicker communities. The greater
productivity of periphyton exposed to high grazing pressure
provided partial compensation for the greater grazing demand.

Community Structure

In addition to influencing plant abundance and rate of
production, lotic grazers may also alter the taxonomic structure
of plant communities. Two mechanisms may result in alteration of
plant community structures: 1) herbivores may actively select
for or against a particular plant taxon and 2) plants may differ
in their vulnerability or tolerance to grazing. I restrict the
term "selection" to cases in which the herbivore actively chooses
for or against a particular form or taxon of plant. Selection,
as defined here, does not include differential consumption that
results from passive attributes of the plant which result in
differences in vulnerability to specific grazing behavior (e.g.,
stalked diatoms well above the surface versus diatoms firmly
attached to the surface) or tolerance to grazing pressure (e.g.,
rapidly reproducing algae versus slower growing algae). These
mechanisms may occur simultaneously; however, the mechanism
involving plant vulnerability or tolerance to grazing is most
frequently observed in streams. Regardless of the mechanism,
differential consumption potentially results in alteration of
plant community structure by herbivores.

Differential consumption by herbivores most often is
estimated by comparison of composition of food items in the gut
of the herbivore with the abundance of plant taxa in the
environment. Several researchers have found that diets of
particular herbivores in streams differed from the relative
availability of the food (Hynes, 1941, 1961; Patrick, 1970;
Calow, 1973a; Moore, 1975, 1977a; Gray and Ward, 1979). Calow
(1973a) noted that differential consumption was greatest when
animals were satiated and that starvation eliminated this
behavior. In several cases, differential consumption was
referred to as food selection (Brown, 1961; Calow, 1973a; Moore,
1975); however, it is rarely possible to distinguish true food
selection from differential vulnerability or grazing tolerance of
plants by comparison of diet with relative abundance of plant
taxa. Another problem with this method is the difficulty of

ascertaining the true availability of the food. If community
structure of plants is determined over an area different from the
actual habitat of the herbivore, incorrect conclusions will be
drawn.   Other researchers have found no evidence of differential
consumption by herbivores in streams (Nagagawa, 1952; Nishimura
and Ohgushi, 1958; Scott, 1958; Chapman and Demory, 1963; Mecom
and Cummins, 1964; Chapman, 1966; Mecom, 1972; Cummins, 1973;
Koslucher and Minshall, 1973).

Active selection of particular forms or taxa of plants by
herbivores has been observed in streams. Certain ciliates are
known to feed only on diatoms (Gizella and Gellert, 1958). One
of the few known examples of selection for a specific plant host
in lotic ecosystems involves the blue-green alga Nostoc
parmeloides and the chironomid larvae Cricotopus nostocola
(Wirth, 1957; Brock, 1960). In the absence of the midge, Nostoc
parmeloides is a gelatinous, globular colony; however, in the
presence of the midge, the colony takes on a ear-like morphology.
Cricotopus spends its entire larval stage in the Nostoc colony
and apparently feeds on the alga.

Selection of plants by herbivores in streams most commonly
involves the rejection of filamentous algae or blue-green algae;
although apparently available, filamentous algae frequently are
not consumed by grazers (Cummins, 1964; Moore, 1975; Gray and
Ward, 1979). These cases of apparent selection against
filamentous algae may be due to difficulty in consuming or
digesting certain types of algae. Many grazers feed by scraping
or brushing across the surface of the substrate, and long,
tangled filaments may simply be difficult to ingest. As a
result, many invertebrates that consume filamentous algae may
feed more like shredders than grazers or collectors. Moore
(1975) suggested that grazers may not be able to rupture or
digest the cellulose cell wall of filamentous green algae. He
observed that isopods and amphipods frequently ingested
filamentous diatoms but rarely consumed blue-green filamentous
algae, and Calow (1973a) found that the limpet Ancylus
fluviatilis least preferred blue-green algae of all diets
offered.   Diatoms are a frequent food source for aquatic grazers;
however, some diatom taxa, such as Cocconeis placentula, are
sometimes less numerous in herbivore diets than in the
environment. Mayflies (Moore, 1977), caddisflies (Mecom and
Cummins, 1964), amphipods and isopods (Moore, 1975), and snails
(Patrick, 1970) have been observed to have lower quantities of
Cocconeis in their guts than were available in the environment.
It has been suggested that herbivores are less able to harvest
Cocconeis because of the firm, close attachment of the diatom to
the substrate (Moore, 1975). However, small nymphs of the mayfly
Chloeon dipterum, the oligochaete Stylaria lacustris, and the
caddisfly Lepidostoma have been found to consume large amounts of

Cocconeis (Brown, 1961; Dickman and Gochnauer, 1978). Some diatoms may, in fact, be more easily grazed because of their growth form. The stalked diatom Gomphonema grows above the surface of the substrate and may be more vulnerable to cropping by certain kinds of grazers (Calow, 1973a; Moore, 1975). The characteristics of a particular herbivore or plant must be considered before generalizations can be made.

Alteration of community, structure by grazers in streams has been directly investigated in only a few instances. However, grazing by orthoclad midges prevented the development of filamentous blue-green algal communities (Phormidium and Oscillatoria) in outdoor stream channels (Eichenberger and Schlatter, 1978). Different degrees of alteration of algal community structure were observed at different densities of the snail Juga (Gregory, McCullough, and Busch, manuscript submitted).

Heavy to moderate grazing resulted in a benthic algal community solely dominated by diatoms; heavy grazing led to the dominance of a smaller species of diatom, Achnanthes minutissima, and moderate grazing resulted in dominance of a larger diatom, Achnanthes lanceolata. Low grazing pressure resulted in communities that closely resembled ungrazed periphyton communities. However, the distribution of filamentous algae in channels with the two lower grazer densities appeared to be more patchy than in the control and we assumed that the snails at low densities were keeping small patches clear of filamentous algae and feeding largely on that non-filamentous, monolayer community and that patch size was larger at the slightly greater snail density. Sumner (1980) also observed that snails caused a decrease in the abundance of the filamentous diatom Melosira varians and four species of pennate diatoms, Synedra ulna, Nitzschia linearis, Surirella angustata, and Nitzschia palea. No significant difference in plant community structure was observed on grazed and ungrazed substrates in an English stream (Calow, 1973b). Stones with and without snails were collected in the field and assumed to represent grazed and ungrazed situations, an extremely weak assumption that was acknowledged by Calow. In a laboratory study of Physa, Kehde and Wilhm (1972) also failed to detect a significant difference in species diversity between grazed and ungrazed communities of benthic algae. There are great differences in the abilities of particular herbivores to crop benthic algae and the vulnerability of specific taxa of algae to grazing; therefore, the effect of grazers on community structure of aquatic primary producers depends on grazing intensity and the characteristics of the plants and animals involved.

EFFECTS OF PLANTS ON HERBIVORES IN STREAMS

Abundance of Herbivores

    If herbivores in streams are limited by food resources at
any time, densities or biomasses of grazers could be determined
by the availability of plants. Many studies have found that
numbers of aquatic herbivores are regulated by the amount of
benthic algae available (Beier, 1949; Hughes, 1966; Kownacki and
Kownacki, 1971; Sheldon, 1972; Wiegert and Mitchell, 1973;
Brennan et al., 1978; Durrant, 1977; Marsh, 1980, Towns, 1981.
In a reach of Walker Branch, Tennessee in which primary
production had been enhanced by phosphorus addition, densities of
the herbivorous snail Goniobasis clavaeformis were significantly
greater than snail densities in unenriched sections (Elwood et
al., 1981). Similarly densities of the stream limpet, Ferrissia
rivularis, were greater in a eutrophic stream with high rates of
primary production than in a less productive stream in upper New
York State (Burkey, 1971). McAuliffe (this volume) observed that
immigration of grazers (Baetis, Glossosoma, Oligoplectrum,
chironomids) was greater on substrates with greater abundance of
benthic algae. Distribution of the mayfly Baetis rhodani was
examined in a Danish springbrook that passed in and out of
forested reaches; densities were high in all unshaded reaches but
dropped dramatically in each of the six stream sections that
flowed through forests (Thorup, 1964). Similar patterns were
observed for other herbivores: three caddisflies - Agapetus
fuscipes, Eccliopteryx guttulata, Silo pallipes; a beetle -
Helodes minuta; a limpet - Ancylus fluviatilis. Therefore, plant
communities may be a major determinant of herbivore densities in
lotic ecosystems.

    Standing crops of macroinvertebrate communities in streams
are not always greater at higher standing crops of periphyton.
Towns (1981) artificially shaded a streambed with a 44 m^2 canopy
suspended 1.5 m above the water surface and found no significant
difference in invertebrate numbers between the area under and
outside of the canopy. Pennak (1977) compared 20 small, mountain
streams in Colorado and found no significant relationship between
periphyton standing crops and biomass of macroinvertebrates. It
is quite possible that such conclusions are correct, but
nonsignificant correlations between biomass of plants and biomass
of consumers do not prove that primary producers are not a major
factor in the regulation of macroinvertebrate communities. As
discussed earlier, standing crop of primary producers can be a
poor index of primary production because of the differences in
productivity between young, mature, or senescent communities or
between taxa of plants. Also, amounts of chlorophyll are often

used as an estimate of biomass of algae, but chlorophyll content
of algal cells varies with light intensity, nutrient
concentrations, and age of the cell; therefore, chlorophyll alone
is a poor measure of differences in biomass of algae,
particularly in comparisons of different stream sites unless
site-specific chlorophyll:biomass ratios are known. Invertebrate
production may also increase without significant changes in
numbers or biomass. Many herbivores, particularly midges,
mayflies, and caddisflies, can shorten their generation time
under more favorable growing conditions (Gose, 1970); therefore,
abundance of these taxa may not change while total production
increases. Caution must be exercised in making conclusions based
on abundance alone because of the tremendous plasticity of the
metabolic and growth responses of both plants and herbivores.

    Availability of primary producers not only regulate numbers
of grazers but biomass as well. Densities of the caddisfly
Glossosoma nigrior were 50 percent higher in an open, third-order
site than in a shaded, first-order site of Augusta Creek,
Michigan (Cummins, 1975). Glossosoma in the open site primarily
consumed benthic algae but consumed more detritus than algae at
the shaded site. Individual weights of Glossosoma were less in
the shaded section. Weights of Glossosoma prepupae in the open
site were approximately threefold greater than weights of
prepupae in the shaded site and P/R ratios (gross primary
production divided by respiration for a 24-hour period) were also
three times greater in the open, third-order site. Standing
crops of the limpet Laevapex fuscus were greater in a reach of
river with higher standing crops of periphyton than a less
productive reach (McMahon, 1975). Therefore, enhancement of
primary production may also result in increased biomass of
grazers.

    Stream ecosystems are supported by both allochthonous and
autochthonous organic matter. Frequently, benthic algae provide
the highest quality food source for macroinvertebrates though
much lower amounts than allochthonous organic matter. Increased
availability of a high quality food may have greater impact on
consumers than biomass of food alone would indicate. Total
numbers and biomasses of stream invertebrates are often greater
under open, well-lighted conditions than under shaded conditions,
reflecting potential differences in abundance and production of
benthic primary producers (Allen, 1951). In an unshaded reach of
the River Polenz, Germany, total numbers and biomass of
macroinvertebrates were greater than in an adjacent shaded reach
(Albrecht, 1956); however, standing crops of algae were lower in
the open reaches with higher densities of invertebrates than in
the shaded section of the river that had lower invertebrate
numbers (Albrecht and Bursche, 1957). The authors attributed the
lower algal abundance in the open site to greater grazing

pressure by invertebrates as compared to the shaded area. Hynes
(1970) remarked that this explanation was unlikely because it
would be unstable for large numbers of invertebrates to exist on
lower standing crops of algae and the situation could not last
without the invertebrates depriving themselves of food; however,
Hynes' objection does not consider the ability of benthic algae
to tolerate increased grazing pressure because of the higher
productivity and faster turnover time of younger algal
communities or differences in the percentage of primary
production ingested (Gregory, 1980). Emergence of aquatic
insects was fourfold greater in a stream flowing through a
clearcut than in an upstream reach that flowed through a virgin
coniferous forest although standing crops of insects were only
slightly greater in the unshaded section (Anderson and Grafius,
quoted in Lyford and Gregory, 1975). Similarly, Erman et al.
(1977) compared 47 stream sites and observed that numbers and
biomass of invertebrates were greater in open streams in
clearcuts than in forested streams of California; the presence of
wide bufferstrips (30 m) that shaded the streams appeared to
negate this response of the invertebrates. Invertebrate biomass
and production were lower in forested streams than in unshaded
streams of New Zealand (Hopkins, 1976). Minshall (1978)
described energy budgets for three sites in Deep Creek, Idaho
with different amounts of autochthonous plant production and
similar allochthonous inputs; invertebrate metabolism was
fourfold greater at the site with highest autotrophic production
(12,270 kcal m^{-2} yr^{-1}) and equal at the two sites with lower
autotrophic production (3130 and 5915 kcal m^{-2} yr^{-1}).

Not only do plant communities influence abundances of
herbivores in streams, but they also affect local patterns of
grazer distribution. In laboratory experiments, Calow (1973)
compared food selection and consumer aggregation on algae and
detritus for Ancylus fluviatilis and Planorbis contortus. A.
fluviatilis aggregated on the algal material and the degree of
aggregation increased with time. Calow concluded that the limpet
searched for food randomly but recognized algal material by
contact chemoreception. Bohle (1978) placed artificial
substrates with different standing crops of algae in a laboratory
stream channel with the mayfly, Baetis rhodani. With low
standing crops of algae, drift rates of Baetis were high;
however, when substrates with high standing crops of algae were
introduced, drifting behavior ceased and the mayflies aggregated
on the benthic algae. When feeding reduced the algal standing
crops, drift behavior resumed. Hart (1981) removed benthic algae
in bands from artificial tile substrates and then placed the
caddisfly scraper, Dicosmoecus, on the tiles; caddisflies
accumulated on the bands of algae and moved off areas where the
algal material had been removed. The above examples demonstrate

that grazers in streams actively search for benthic algae and aggregate on areas with more abundant plant material, thus supporting the concept that primary producers in streams can influence the distribution and abundance of herbivores.

Growth and Production of Herbivores

Though there are many studies of growth rates of stream macroinvertebrates, there is little information on the effect of plant type, abundance, and production on growth or production of herbivores in streams. In an excellent study by McMahon, Hunter, and Russel-Hunter (1974), growth and fecundity of two gastropod grazers in six streams were compared not only to biomass of periphyton but also to C/N ratios and organic carbon content of periphyton as well. Quality of the periphyton seemed to play a greater role in determining growth of grazers than quantity of periphyton; maximum growth and fecundity of the limpet Laevapex fuscus and the snail Lymnaea palustris were found at sites with a combination of low C/N ratios and high organic carbon content. Annual net production of Laevapex fuscus was also greatest at the site with low C/N ratios (3.9:1), high organic carbon content (17.6%), and intermediate standing crop of periphyton; limpet production was intermediate at the site with highest periphyton standing crop but low organic carbon content (5.8%); and production was least at the site with lowest standing crop but highest C/N ratio (6.0:1) (McMahon, 1975). Growth and fecundity of the snail Lymnea peregra was compared for diets of diatoms, blue-green algae, filamentous green algae, chopped spinach, and animals (Skoog, 1978). Adult snails grew best on a diet of mixed blue-green algae; however, juvenile snails attained their highest growth on diatoms. Though moderate growth was obtained on spinach, maximum fecundity was attained on this food. Low growth and low fecundity were observed with a diet of pure Cladophora, a filamentous green algae; however, growth of juvenile snails on Cladophora covered with epiphytic diatoms was the second highest attained. Diets of animal matter resulted in poor growth of this herbivore. Humpesch (1979) found that field growth rates for the mayflies Baetis rhodani and B. lutheri were lower than growth rates obtained in the laboratory on epilithic algae. Hynes (1941) suggested that one of the possible causes for faster growth of the stonefly Isoperla in rivers than in small streams was the greater amount of algae in rivers. In the study of the effects of the snail Juga plicifera on periphyton communities in laboratory streams, ingestion, assimilation, and growth rates of the snails were determined (Gregory, McCullough, and Busch – manuscript submitted, Gregory, 1980); although ingestion and assimilation were only slightly lower at lower algal abundance, growth rates of the snail were more than an order of magnitude lower at low algal abundance than at the highest periphyton standing crop (Table 2). Sumner (1980) observed no relationship

between periphyton biomass and absolute growth rates of Juga, but
there was a slight negative correlation with abundance of
blue-green algae, indicating that consumption of blue-green algae
may have reduced snail growth. Although it is clear that growth
and assimilation efficiency of grazers feeding on benthic algae
in streams are high as compared to other food, further studies on
the effects of types and amounts of plant material on growth of
herbivores in streams are needed.

Community Structure

If plant communities can affect the abundance of herbivores
as previously described, relative proportions of herbivores in
animal communities in streams could be altered by differences in
plant availability. Unfortunately such analyses have not been
made because of the magnitude of the effort required to
accurately assign taxa into specific trophic levels. A few
studies have described the trophic structures of a single
location. Energy budgets were developed for Linesville Creek,
Pennsylvania for five dates and herbivores accounted for 31% to
72% of the fauna based on ingestion data (Coffman et al., 1971).
Trophic structure of Linesville Creek was analyzed for a single
date and herbivores comprised 31 percent of the biomass (Cummins
et al., 1966). Minshall (1967) described the trophic structure
of a small springfed stream in Kentucky and found that a small
fraction of the fauna was herbivorous. The relative contribution
of herbivores to the trophic structure of streams ecosystems is
determined by the availability of plant matter.

Though it is often difficult to ascertain the trophic
structure of stream ecosystems, stream ecologists have found that
community structure analysis based on mechanisms of gathering
food can often yield valuable information (Cummins, 1973, 1974;
Merritt and Cummins, 1978). The scraper feeding functional group
most commonly consists of herbivores (Cummins and Klug, 1979);
therefore, relative abundances of this functional group on a
seasonal or longitudinal basis might be influenced by factors
that affect primary production in streams. This idea is a major
component of the relationships between stream size and ecosystem
structure and function described in the river continuum concept
(Vannote et al., 1980), in which it was hypothesized that
the relative abundance of scrapers would be greatest in
intermediate reaches along a longitudinal gradient from
first-order, headwater streams to large rivers. Wiggins and
Mackay (1978) proposed that the greater numbers of genera of
Trichoptera found in headwater streams of western North America
as compared to eastern North America might be attributed to
greater growth of periphyton in streams of coniferous forests
than in shaded streams of eastern deciduous forests. Cummins and
Klug (1979) observed that scrapers comprised a greater percent of

total invertebrate biomass in Augusta Creek, Michigan, during
summer and fall, the period of the year with the highest rates of
primary production. In streams of the Cascade Mountains,
scrapers were more abundant in open, intermediate reaches along a
longitudinal continuum (Hawkins and Sedell, 1980); however,
scraper abundances were not significantly correlated with
standing crop of chlorophyll but rather net daily metabolism, a

Table 2.   Ingestion rate, assimilation rate, and growth rate of
           snails at different standing crops of periphyton (data
           from manuscript submitted by Gregory, McCullough, and
           Busch).

Standing Crop of Periphyton (g AFDW m^{-2})	Ingestion Rate (mg g^{-1} day^{-1})	Assimilation Rate (mg g^{-1} day^{-1})	Snail Growth Rate (mg g^{-1} day^{-1})
21.76	91.20	19.17	7.06
18.14	82.56	17.36	5.06
3.61	72.00	15.13	2.66
0.49	63.12	13.27	0.64

measure of primary production. In a comparison of open
and shaded streams of similar size in the Cascade Mountains,
Hawkins, Murphy, and Anderson (1982) observed greater numbers of
scrapers in open sites and a trend toward higher biomasses of
scrapers in open sites with higher standing crops of periphyton.
These authors also observed lower biomass of periphyton at higher
densities of scrapers. Changes in feeding functional groups
after canopy removal were simulated in a computer model of stream
ecosystems (McIntire and Colby, 1978); the proportion of biomass
and consumer production attributed to the grazing process
increased dramatically to a maximum at three years and declined
gradually thereafter. In general there is evidence to support
the concept that distribution and abundance of scraper functional
groups are influenced by patterns of primary production in
streams, but there are several possible modes of interaction
between aquatic plants and herbivores that may modify the degree
of relationship between scrapers and primary production.

Effects of Reset Mechanisms on Plants and Herbivores in Streams

Plant-herbivore interactions in lotic ecosystems are greatly
influenced by catastrophic events that cause dramatic reductions
in populations of plants and animals. These resetting events
occur frequently in streams; the major reset mechanism in lotic
ecosystems is fluctuation in streamflow. Flows of some streams,
such as spring-fed streams, are relatively constant, but flows of
other streams, such as desert streams, can be extremely
irregular. Storm discharges scour plant communities, dislodge
substrates, and blanket depositional areas with sediments.
Although flow regimes vary geographically and temporally, return
intervals of bankfull floods generally range from one to three
years.

Numerous studies of plant communities in streams have
demonstrated that high flow causes significant reduction in
standing crops of benthic algae (Douglas, 1958; King and Ball,
1967, Stockner and Shortreed, 1976; Moore, 1978; Jones, 1978;
Malmqvist et al., 1978; Busch and Fisher, 1981) and vascular
macrophytes (Haslam, 1978). The main regulator of algal biomass
in the Mechums River, Virginia was high flow (Tett et al., 1978).
In streams of the Pacific Northwest, minimum standing crops of
periphyton occur in winter (Gregory, 1980); however, in a
drought year, biomass of algae in streams of the Cascade
Mountains did not decline in winter until the occurrence of a
small storm the following spring (Rounick and Gregory, 1981).
Debris torrents and sediment avalanches occur during floods and
cause almost total removal of plant communities over short
reaches. The time interval between these mass failures is
extremely long under natural conditions; however, watershed
perturbation by man's activities drastically increases the
frequency of these events (Swanston and Swanson, 1976). Storm
discharge is the dominant, most frequent reset mechanism for
plant communities in stream ecosystems.

Herbivores in streams are exposed to the same catastrophic
disturbances as plant communities of grazers as well as plants in
lotic ecosystems. Numerous investigations have observed reduced
abundances of macroinvertebrates after storms (Ellis, 1936;
Moffet, 1936; Motley et al., 1939; Sprules, 1947; Allen, 1951,
1959; Jones, 1951; Pomeisl, 1953; McGraw, 1961; Minckley, 1963;
Maitland, 1964; Lehmkuhl and Anderson, 1972; Siegfried and
Knight, 1977). Moffet (1936) compared streams with different
flood history and found that chironomids and blackflies, taxa
with short life cycles, dominated the fauna in streams that
frequently flooded. Hynes (1970) cited a number of studies
(Nevin, 1936; Jones, 1948; Engelhardt, 1951; Badcock, 1953; Ivlev
and Ivassik, 1961) which demonstrated that streams with high
flood frequency contained less abundant and less diverse faunas.

Though magnitudes, frequencies, and seasonal patterns of high
flows vary among streams, floods are an important, ubiquitous
factor influencing the structure and function of stream
ecosystems.

Catastrophic disturbance is more frequent in streams than in
lentic, marine, or terrestrial ecosystems. Frequent reset
mechanisms impose rigorous constraints on plant and herbivore
communities in streams and are a major determinant in
evolutionary patterns. In unpredictable environments, plant
defense mechanisms are of little advantage because a major reset
event will most likely regulate consumer populations before plant
communities would be destroyed. Host specifity and co-evolution
of plants and animals involve great risk because catastrophic
events may eliminate or greatly reduce the necessary organism.
Therefore, plants in streams have little protection from grazers
and large proportions of plant biomass are consumed. Herbivores
in lotic ecosystems are generalists with respect to the types of
plants that they consume. However, specialized methods of
gathering food are not uncommon in lotic herbivores because the
general habitats and plant growth forms are predictable and not
changed by catastrophic events. Because of these factors,
grazers may temporarily reduce the abundance of plants in streams
and frequently may be food-limited. All ecosystems may be
"perturbation-dependent" to some degree (Vogl, 1980) and lotic
ecosystems represent systems with intense and frequent
disturbance. Therefore, plant-herbivore interactions in streams
reflect not only attributes that play a role in immediate
responses to grazing but also temporal and spatial patterns that
affect evolutionary development of plants and herbivores.

SUMMARY

Plant-herbivore interactions in streams are complex and
reflect the array of characteristics of plant and animal taxa
involved. The morphological, physiological, reproductive, and
ecological attributes of plant systems determine their
vulnerability and tolerance to grazing pressure. In general,
benthic algae provide the most palatable, nutritious food source
for stream grazers; the structural properties of mosses and
aquatic vascular macrophytes make these plants less desirable to
herbivores. Though some types of primary producers are consumed
to a greater degree than others, no plant type totally escapes
grazing. Grazing can result in decreased numbers or biomass of
primary producers. Rates of primary production also may be
reduced by intensive grazing pressure although changes in the age
of the community and community structure of the plants may result
in higher productivity as grazing pressure increases. Also, the
proportion of primary production that is routed through the

Table 3. Attributes of developmental and mature systems in a model of ecological succession that are also attributes of autochthonous and allochthonous-based systems (selected from Odum, 1969). (Gregory, 1980).

Ecosystem Attributes	Developmental Stages	Mature Stages
Gross primary production/community respiration	Greater than 1.0	Approaches 1.0
Gross primary production/biomass	High	Low
Net community production	High	Low
Total organic matter	Small	Large
Inorganic nutrients	Extrabiotic	Intrabiotic
Species diversity	Low	High
Biochemical diversity	Low	High
Spatial Heterogeneity	Poorly organized	Well organized
Niche specialization	Broad	Narrow
Size of organism	Small	Large
Life cycles	Short, simple	Long, complex
Mineral cycles	Open	Closed
Nutrient exchange rate	Rapid	Slow
Role of detritus in nutrient regeneration	Less Important	Important
Growth form	For rapid growth (r-selection)	For feedback control (K-selection)
Stability	Poor	Good
Entropy	High	Low
	Autochthonous Systems	Allochthonous Systems

primary consumer may increase at higher grazing rates. Community
structure of plant systems may be altered by grazing because of
differential vulnerability or tolerance to grazing pressure
between different plant taxa. True selective grazing, though
apparently rare, may greatly influence plant community structure.

Herbivores also may be influenced by plant communities.
Availability of plants may regulate the distribution and
abundance of grazers. Growth rates of herbivores are determined
by the amounts and types of plant food available. Distributions
of herbivores in stream ecosystems are often determined by the
potential for primary production and, therefore, community
structure of stream fauna may be influenced by plant systems.
Both plants and animals are regulated to some degree by
catastrophic events; the major reset mechanism in lotic
ecosystems is high flow with recurrence intervals of several
times per year or more.

The biological and physical characteristics of plant and
herbivores create a dynamic, interactive system that plays a
major role in the maintenance and organization of stream
ecosystems. The attributes of autochthonous-based and
allochthonous-based components provide a framework for
understanding the functioning of lotic ecosystems (Table 3).
Autochthonous systems generally have low amounts of organic
matter but have high rates of primary production as opposed to
allochthonous systems with high standing crops of organic matter
but low rates of primary production. Species diversity is low
and generalists are more common in systems based on aquatic
primary producers; species diversity is greater and specialists
are more common in systems based on terrestrial inputs.
Herbivores tend to be small, have fast turnover rates and short,
simple life cycles; however, detritivores are generally larger
with slower turnover rates and longer, more complex life cycles.
Nutrient turnover is more rapid and detritus is less important in
nutrient regeneration in autochthonous-based systems.
Plant-herbivore systems are more typical of r-selected organisms
but detritivores are more like K-selected organisms. These
characteristics of autochthonous-based and allochthonous-based
systems are also attributes that Odum (1969) used to describe
developmental and mature stages of ecosystems. Plant-herbivore
relationships in lotic ecosystems integrate well within the
general theory of ecosystem organization and function and offer
a unique set of characteristics that promises to broaden the
scope of ecological theory of herbivory.

REFERENCES

Albrecht, M.L., and E.M. Bursche. 1957. Fischereibiologische
    Untersuchungen an Fliessgewassern. I.
    Physiographisch-biologische Studies an der Polenz. Z.
    Fisherei 6:209-240.
Allen, K.R. 1951. The Horokiwi stream. A study of a trout
    population. Fish. Bull N.Z. 10:1-231.
Allen, K.R. 1959. The distribution of stream bottom fauna.
    Proc. N.Z. Ecol. Soc. 6:5-8.
Badcock, R.M. 1953. Comparative studies in the populations of
    streams. Rep. Inst. Freshwat. Res. Drottningholm 35:21-34.
Beier, M. 1949. Zur Kenntnis von Korperbau und Lebenweise der
    Helminen (Col. Dryopidae). Eos, Wein. 24:123-211.
Bohle, V.H.W. 1978. Beziehungen zwischen dem Nahrungsangebot,
    der Drift und der raumlichen Verteilung bei larven von
    Baetis rhodani (Pictet) (Ephemeroptera: Baetidae). Arch
    Hydrobiol. 84:500-525.
Boyd, C.E. 1970. Amino acid, protein and caloric content of
    vascular aquatic macrophytes. Ecology 51:902-906.
Brennan, A., A.J. McLachlan, and R.S. Wotton. 1978. Particulate
    material and midge larvae (Chironomidae: Diptera) in an
    Upland River. Hydrobiologia 59:67-73.
Brindle, A. 1964. Taxonomic notes on the larvae of British
    Diptera. No. 18--The Hemerodrominae (Empididae).
    Entomologist 97:162-165.
Brock, E.M. 1960. Mutualism between the midge Cricotopus and
    the algae Nostoc. Ecology 41:474-483.
Brown, D.S. 1961. The food of the larvae of Chloeon dipterum L.
    and Baetis rhodani (Pictet) (Insecta, Ephemeroptera). J.
    Anim. Ecol. 30:55-75.
Burkey, A.J. 1971. Biomass turnover, respiration, and
    interpopulation variation in the stream limpet Ferrissia
    rivularis (Say). Ecol. Monogr. 41:235-251.
Busch, D.E., and S.G. Fisher. 1980. Metabolism of a desert
    stream. Freshwat. Biol. 11:301-307.
Calow, P. 1973a. Field observations and laboratory experiments
    on the general food requirements of two species of
    freshwater snail, Planorbis contortus (Linn.) and Ancylus
    fluviatilis Mull. Proc. malac. Soc. Lond. 40:483-489.
Calow, P. 1973b. The food of Ancylus fluviatilis (Mull.), a
    littoral stone-dwelling, herbivore. Oecologia 13:113-133.
Chapman, D.W. 1966. The relative contributions of aquatic and
    terrestrial primary producers to the trophic relations of
    stream organisms. Spec. Publs. Pymatuning Lab. Fld. Biol.
    4:116-30.
Chapman, D.W. and R. Demory. 1963. Seasonal changes in the food
    ingested by aquatic insect larvae and nymphs in two Oregon
    streams. Ecology 44:140-146.

Chutter, F.M. 1970. Hydrobiological studies in the catchment of
     Vaal Dam, South Africa. Part 1. River zonation and the
     benthic fauna. Int. Revue ges. Hydrobiol. 55:445-494.
Coffman, W.P., K.W. Cummins, and J.C. Wuycheck. 1971. Energy
     flow in a woodland stream ecosystem: I. Tissue support
     trophic structure of the autumnal community. Arch.
     Hydrobiol. 68:232-276.
Collins, N.C., R. Mitchell, and R.G. Wiegert. 1976. Functional
     analysis of a thermal spring ecosystem, with an evaluation
     of the role of consumers. Ecology 57:1221-1232.
Cooper, D.C. 1973. Enhancement of net primary productivity by
     herbivore grazing in aquatic laboratory microcosms. Limnol.
     Oceanogr. 18:31-37.
Cummins, K.W. 1964. Factors limiting the microdistribution of
     the caddisflies Pycnopsyche lepida (Hagen) and Pycnopsyche
     guttifer (Walker) in a Michigan stream (Trichoptera:
     Limnephilidae). Ecol. Monogr. 34:271-295.
Cummins, K.W. 1973. Trophic relations of aquatic insects. Ann.
     Rev. Ent. 18:183-206.
Cummins, K.W., R.C. Peterson, F.O. Howard, J.C. Wuycheck, and
     V.I. Holt. 1973. The utilization of leaf litter by stream
     detritivores. Ecology 54:336-345.
Cummins, K.W. 1974. Structure and function of stream
     ecosystems. BioScience 24:631-641.
Cummins, K.W. 1975. Macroinvertebrates. In: B.A. Whitton
     (ed.),   River ecology. University of California Press,
     Berkeley, California. 724 pp.
Cummins, K.W., W.P. Coffman, and P.A. Roff. 1966. Trophic
     relationships in a small woodland stream. Verh. Internat.
     Verein. Limnol. 16:627-638.
Cummins, K.W. and M.L. Klug. 1979. Feeding ecology of stream
     invertebrates. Ann. Rev. Ecol. Syst. 10:147-172.
Decamps, H., and M. Lafont. 1974. Cycles vitaux et production
     des Micrasema Pyrenenes dans les mousses d'eau courante.
     Annls. Limnol. 10:1-32.
Dickman, M.D., and M.B. Gochnauer. 1978. A scanning electron
     microscopic study of periphyton colonization in a small
     stream subjected to sodium chloride addition. Verh.
     Internat. Verein. Limnol. 20:1738-1743.
Douglas, B. 1958. The ecology of the attached diatoms and other
     algae in a small stony stream. J. Ecol. 46:295-322.
Durrant, P.M. 1977. Some factors that affect the distribution
     of Ancylus fluviatilus (Muller) in the river systems of
     Great Britain. J. Molluscan Studies 43:67-77.
Eichenberger, E., and A. Schlatter. 1978. Effect of herbivorous
     insects on the production of benthic algal vegetation in
     outdoor channels. Verh. Internat. Verein. Limnol.
     20:1806-1810.

Ellis, M.M. 1936. Erosion silt as a factor in aquatic
    environments. Ecology 17:29-42.
Elwood, J.W. and D.J. Nelson. 1972. Periphyton production and
    grazing rates in a stream measured with a ^{32}P material
    balance method. Oikos 23:295-303.
Elwood, J.W., J.D. Newbold, A.F. Trimble, and R.W. Stark. 1981.
    The limiting role of phosphorus in a woodland stream
    ecosystem: Effects of P enrichment on leaf decomposition
    and primary producers. Ecology 62:146-158.
Engelhardt, W. 1951. Faunistisch - okologische Untersuchungen
    uber Wasserinsekten an den sudlichen Zuflussen des
    Ammersees. Mitt. munch. ent. Ges. 41:1-135.
Erman, D.C., J.D. Newbold, and K.B. Roby. 1977. Evaluation of
    streamside bufferstrips for protecting aquatic organisms.
    Contribution No. 165. California Water Resources Center,
    University of California, Davis. 48 pp.
Fisher, S.G. and S.R. Carpenter. 1976. Ecosystem and macrophyte
    primary production of the Fort River, Massachusetts.
    Hydrobiologia 47:175-187.
Flint, R.W. and C.R. Goldman. 1975. The effects of a benthic
    grazer on the primary productivity of the littoral zone of
    Lake Tahoe. Limnol. Oceanogr. 20:935-944.
Gajevskaja, N.S. 1958. Le role des groups principaux de la
    flore aquatique dans les cycles trophiques des differents
    bassins d'eau douce. Verh. Internat. Verein. Limnol.
    13:350-362.
Gizella, T., and J. Gellert. 1958. Uber Diatomeen and ciliaten
    aus dem Aufwuchs der Ufersteine am Osturger der Halbinsel
    Tihany. Ann. Inst. Biol. Hung. Acad. Sci 25:240-250.
Gose, K. 1970. Life history and instar analysis of Stenophylax
    griseipennis (Trichoptera). Jap. J. Limnol. 31:96-106.
Gower, A.M. 1967. A study of Limnephilus lunatus Curtis
    (Trichoptera: Limnephilidae) with reference to its life
    cycle in watercress beds. Trans. Royal Ent. Soc. Land.
    119:283-302.
Gray, L.J., and J.V. Ward. 1979. Food habits of stream benthos
    at sites of differing food availability. Am. Midl. Nat.
    102:157-167.
Gregory, S.V. 1980. Effects of light, nutrients, and grazing on
    periphyton communities in streams. Unpublished Ph.D.
    Dissertation. Oregon State University, Corvallis, Oregon
    154 pp.
Gregory, S.V., D.A. McCullough, and D. Busch. (Submitted to
    Ecology). Plant-herbivore interactions in laboratory
    streams.
Hargrave, B.T. 1970. The effect of a deposit-feeding amphipod
    on the metabolism of benthic microflora. Limnol. Oceanogr.
    15:21-30.

Hart, D.H. 1981. Foraging and resource patchiness: field
     experiments with a grazing stream insect. Oikos 37:46-52.
Haslam, S.M. 1978. River plants: the macrophytic vegetation of
     watercourses. Cambridge University Press, Cambridge. 396
     pp.
Hawkins, C.P. and M.L. Murphy and N.H. Anderson. 1982. Effects
     of canopy, substrate composition and gradient on the
     structure of macroinvertebrate communities in Cascade
     streams of Oregon. Ecology 63:1840-1856.
Hawkins, C.P. and J.R. Sedell. 1980. Longitudinal and seasonal
     changes in functional organization of macroinvertebrate
     communities in four Oregon streams. Ecology 62:387-397.
Hopkins, C.L. 1976. Estimate of biological production in some
     stream invertebrates. New Zealand J. Mar. and Freshwat.
     Res. 10:629-640.
Hughes, D.A. 1966. Mountain streams of the Barberton area,
     eastern Transvaal. Part II. The effect of vegetational
     shading and direct illumination on the distribution of
     stream fauna. Hydrobiologia 27:439-459.
Humpesch, U.H. 1979. Life cycles and growth rates of _Baetis_
     spp. (Emphemeroptera: Baetidae) in the laboratory and in two
     stony streams in Austria. Freshwat. Biol. 9:467-479.
Hynes, H.B.N. 1941. The taxonomy and ecology of the nymphs of
     British Plecoptera, with notes on the adults and eggs.
     Trans. R. Ent. Soc. Lond. 91:459-557.
Hynes, H.B.N. 1961. The invertebrate fauna of a Welsh mountain
     stream. Arch. Hydrobiol. 57:344-388.
Hynes, H.B.N. 1970. The ecology of running waters. University
     of Toronto Press, Canada. 555 pp.
Ide, F.P. 1967. Effects of forest spraying with DDT on aquatic
     insects of salmon streams in new Brunswick. J. Fish. Res.
     Bd. Can. 24:769-805.
Ivlev, V.S. and V.M. Ivassik. 1961. Materials to the biology of
     mountain rivers in the Soviet Transcarpathians (Russian).
     Trudy vses. gidrobiol. Obshch. 11:171-188.
Izvekova, E.I. 1971. On the feeding habits of chironomid
     larvae. Limnologica 8:201-202.
Jones, J.R.E. 1949. An ecological study of the River Rheidol,
     North Cardiganshire, Wales. J. Anim. Ecol. 18:67-88.
Jones, J.R.E. 1951. An ecological study of the River Towy. J.
     Anim. Ecol. 20:68-86.
Kehde, P.M. and J.L. Wilhm. 1972. The effects of grazing by
     snails on community structure of periphyton in laboratory
     streams. Am. Midl. Nat. 87:8-24.
King, D.L. and R.C. Ball. 1967. Comparative energetics of a
     polluted stream. Limnol. Oceanogr. 12:27-33.
Koslucher, D.G. and G.W. Minshall. 1973. Food habits of some
     benthic invertebrates in a northern cool-desert stream (Deep
     Creek, Curlew Valley, Idaho-Utah). Trans. Am. Microsc. Soc.
     92:441-452.

Kownacki, A. and M. Kownacki. 1971. The significance of chironomidae in the ecological characteristics of streams in the High Tatra. Limnologica 8:53–59.

Lehmkuhl, D.M. and N.H. Anderson. 1972. Microdistribution and density as factors affecting the downstream drift of mayflies. Ecology 53:661–667.

Lyford, J.H. and S.V. Gregory. 1975. The dynamics and structure of periphyton communities in three Cascade Mountain streams. Verh. Internat. Verein. Limnol. 19:1610–1616.

Maitland, P.S. 1964. Quantitative studies on the invertebrate fauna of sandy and stony substrates in the River Endrick, Scotland. Proc. R. Soc. Edinb. B68:277–301.

Malmqvist, B., L.M. Nilsson, and B.S. Svensson. 1978. Dynamics of detritus in a small stream in southern Sweden and its influence on the distribution of the bottom animal communities. Oikos 31:3–16.

Mann, K.H. 1975. Patterns of energy flow. In: B.A. Whitton (ed). River ecology. University of California Press, Berkeley, California. 724 pp.

Manuel, C.Y. and G.W. Minshall. 1980. Limitations on the use of microcosms for predicting algal response to nutrient enrichment in lotic systems. In: J.P. Giesy (ed). Microcosms in ecological research. Conf-781101, U.S. Dept. of Energy Symposium Series, Technical Information Center, U.S. Dept. of Energy, Oak Ridge, Tennessee. 1112 p.

Marsh, P.C. 1980. An occurrence of high behavioral drift for a stream gastropod. The Am. Midl. Nat. 104:410–411.

McCraw, B.M. 1961. Life history and growth of the snail, Lymnaea humilis Say. Trans. Am. Microscop. Soc. 80:16–27.

McCullough, D.A., G.W. Minshall, and C.E. Cushing. 1979. Bioenergetics of a stream "collector" organism Tricorythodes minutus (Insecta: Ephemeroptera). Limnol. Oceanogr. 24:45–58.

McMahon, R.F. 1975. Growth, reproduction and bioenergetic variation in three natural populations of a freshwater limpet Laevapex fuscus (C.B. Adams). Proc. Malac. Soc. Lond. 41:331–351.

McMahon, R.F., R. Douglas-Hunter, and W.D. Russel-Hunter. 1974. Variation in aufwuchs at six freshwater habitats in terms of carbon biomass and of carbon:nitrogen ratio. Hydrobiologia 45:391–404.

McIntire, C.D. 1973. Periphyton dynamics in laboratory streams: a simulation model and its implications. Ecol. Monogr. 43:399–420.

McIntire, C.D. 1975. Periphyton assemblages in laboratory streams. In: B.A. Whitton (ed). River ecology. Blackwell Scientific Publications.

McIntire, C.D. and J.A. Colby. 1978. A hierarchical model of lotic ecosystems. Ecol. Monogr. 48:167–190.

Mecom, J.O. 1972. Feeding habits of Trichoptera in a mountain stream. Oikos 23:401-407.

Mecom, J.O. and K.W. Cummins. 1964. A preliminary study of the trophic relationships of the larvae of Brachycentrus americanus (Banks) (Trichoptera: Brachycentridae). Trans. Am. Microsc. Soc. 83:233-243.

Merritt, R.W. and K.W. Cummings. 1978. An introduction to aquatic insects of North America. Kendall/Hunt Publishing Co., Dubuque.. 441 pp.

Minckley, W.L. 1963. The ecology of a spring stream Doe Run, Meade County, Kentucky. Wildl. Monogr. 11:1-124.

Minshall, G.W. 1967. Role of allochthonous detritus in the trophic structure of a woodland springbrook community. Ecology 48:139-149.

Minshall, G.W. 1978. Autotrophy in stream ecosystems. BioScience 28:767-771.

Moffett, J.W. 1936. A quantitative study of the bottom fauna in some Utah streams variously affected by erosion. Bull. Univ. Utah Biol. Ser. 26:224-225.

Moore, J.W. 1975. The role of algae in the diet of Asellus aquaticus L. and Gammarus pulex L. Ecology 44:719-730.

Moore, J.W. 1977. Relative availability and utilization of algae in two subarctic rivers. Hydrobiologia 54:201-208.

Moore, J.W. 1978. Seasonal succession of algae in rivers. III. Examples from the Wyle, an eutrophic farmland river. Arch. Hydrobiologia 83:367-376.

Mottley, C. Mc.C., H.J. Rayner, and J.H. Rainwater. 1939. The determination of the food grade of streams. Trans. Am. Fish. Soc. 68:336-343.

Nagagawa, A. 1952. Food habits of hydropsychid larvae. Jap. J. Limnol. 16:130-138.

Nevin, F.R. 1936. A study of the larger invertebrate forage organisms in selected areas of the Delaware and Susquehanna watersheds. Rep. N.Y. St. Conserv. Dep. 25(Suppl.):195-204.

Nishimura, N. and R. Ohgushi. 1958. On the feeding activity of two species of net-spinning caddisfly larvae (Stenopsychidae: Trichoptera). Jap. J. Ecol. 8:49-50.

Odum, E.P. 1969. The strategy of ecosystem development. Science 164:262-270.

Patrick, R. 1970. Benthic stream communities. Amer. Sci. 58:546-549.

Pennak, R.W. 1977. Trophic variables in Rocky Mountain trout streams. Arch. Hydrobiol. 80:253-285.

Pip, E. 1978. A survey of the ecology and composition of submerged aquatic snail-plant communities. Can. J. Zool. 56:2263-2279.

Pomeisl, E. 1953. Der Mauerbach. Wett. Leben Sonderh. 2:103-121.

Richardson, D.H.S.   1981.   The biology of mosses.   Halsted Press,
    New York.   220 pp.

Riley, G.A.   1963.   Theory of food-chain relations in the ocean.
    In: M.N. Hill (ed).   The Sea, Volume 2.   Interscience, New
    York, 554 pp.

Rosswall, T., A.K. Veum, and L. Karenlampi.   1975.   Plant litter
    decomposition at Fennoscandian tundra sites.   In: F.E.
    Wielgolaski (ed). Fennoscandian tundra ecosystems.   I.
    Plants and microorganisms.   Springer-Verlag, New York.   p.
    268-278.

Rounick, J.S. and S.V. Gregory.   1981.   Temporal changes in
    periphyton standing crop during an unusually dry winter in
    streams of the Western Cascades, Oregon.   Hyrdobiologia
    83:197-205.

Russell-Hunter, W.D.   1970.   Aquatic productivity: an
    introduction to some basic aspects of biological
    oceanogrpahy and limnology.   The MacMillan Company, London.

Scott, D.   1958.   Ecological studies on the Trichoptera of the
    River Dean, Cheshire.   Arch. Hydrobiol. 54:340-392.

Sheldon, A.L.   1972.   Comparative ecology of Arcynopterix and
    Diura (Plecoptera) in California stream.   Arch. Hydrobiol.
    69:521-546.

Siegfried, C.A. and A.W. Knight.   1977.   The effects of washout
    in a Sierra foothill stream.   Am. Midl. Nat. 98:200-207.

Skoog, G.   1978.   Influence of natural food items on growth and
    egg production in brackish water populations of Lymnea
    peregra and Theodoxus fluviatilis (Mollusca).   Oikos
    31:340-348.

Sprules, W.M.   1947.   An ecological investigation of stream
    insects in Algonquin Park, Ontario.   Univ. Toronto Stud.
    Biol. Ser. 56:1-81.

Stockner, J.G.   1971.   Ecological energetics and natural history
    of Hedriodiscus truquii (Diptera) in two thermal spring
    communities.   J. Fish. Res. Bd. Can. 28:73-94.

Stockner, J.G. and K.R.S. Shortreed.   1976.   Autotrophic
    production in Carnation Creek, a coastal rainforest stream
    on Vancouver Island, British Columbia.   J. Fish. Res. Bd.
    Can. 33:1553-1563.

Sumner, W.T.   1980.   Grazer-periphyton interactions in laboratory
    streams.   Unpublished M.S. Thesis.   Oregon State University,
    Corvallis, Oregon.   99 pp.

Swanston, D.N. and F.J. Swanson.   1976.   Timber harvesting, mass
    erosion, and steepland forest geomorphology in the Pacific
    Northwest.   In: D.R. Coates (ed). Geomorphology and
    engineering.   Dowden, Hutchinson, and Ross, Stroudsburg,
    Pennsylvania.

Tett, P., C. Gallegos, M.G. Kelly, G.M. Hornberger, and B.J.
    Crosby.   1978.   Relationships among substrate, flow, and
    benthic microalgal pigment density in the Mechums River,
    Virginia.   Limnol. and Oceanogr. 23:785-797.

Thorup, J. 1964. Substrate type and its value as a basis for
    the delimitation of bottom fauna communities in running
    waters. In: K.W. Cummins, C.A. Tryon, and R.T. Hartman
    (eds). Organism - substrate relationships in streams.
    Special publication No. 4, Pymatuning Laboratory of Ecology,
    University of Pittsburgh, Pittsburgh, Pennsylvania. 145 pp.
Towns, D.R. 1981. Effects of artificial shading on periphyton
    and invertebrates in a New Zealand stream. New Zealand J.
    Mar. Freshwat. Res. 15:185-192.
Vannote, R.L., G.W. Minshall, K.W. Cummins, J.R. Sedell, and C.E.
    Cushing. 1980. The river continuum concept. Can. J. Fish.
    Aquat. Sci. 37:130-137.
Vogl, R.J. 1980. The ecological factors that produce
    perturbation-dependent ecosystems, p. 63-94. In: J. Cairns,
    Jr. (Ed.). The recovery process in damaged ecosystems. Ann
    Arbor Science 167 p.
Westlake, D.F. 1975. Macrophytes. In: B.A. Whitton (ed).
    River ecology. University of California Press, Berkely,
    California. 724 pp.
Whitehead, H. 1935. An ecological study of the invertebrate
    fauna of a chalk stream near Great Driffield, Yorkshire. J.
    Anim. Ecol. 4:58-78.
Weigert, R.G. and R. Mitchell. 1973. Ecology of Yellowstone
    thermal effluent systems: Intersects of blue-green algae,
    grazing flies (Paracoenia, Ephydridae) and water mites
    (Partnuniella, Hydrachnellae). Hydrobiol. 41:251-271.
Wiggins, G.B. 1977. Larvae of the North American caddisfly
    genera (Trichoptera). University of Toronto Press, Toronto.
    401 pp.
Wiggins, G.B. and R.J. Mackay. 1978. Some relationships between
    systematics and trophic ecology in nearctic aquatic insects,
    with special reference to Trichoptera. Ecology
    59:1211-1220.
Willoughby, L.G. and D.W. Sutcliffe. 1976. Experiments on
    feeding and growth of the amphipod Gammarus pulex (L.)
    related to its distribution in the River Duddon. Freshwat.
    Biol. 7:577-586.
Wirth, W.W. 1957. The species of Cricotopus midges living in
    the blue-green algae Nostoc in California (Diptera:
    Tendipedidae). The Pan-Pacific Entomologist 33:121-126.

PREDATOR-PREY RELATIONSHIPS

IN STREAMS

J. David Allan

Department of Zoology
University of Maryland
College Park, Maryland 20742

INTRODUCTION

Questions concerning the effects of predation may be divided
into two broad categories--those concerned with community-level
events, and those concerned with the evolutionary responses of
predator and prey to one another. This second category includes
adaptations on the part of the predator in choosing prey, and of
the prey in minimizing its risk of being eaten.

Connell (1975) reviewed the literature on the relative
importance of predation, competition, and other factors in
shaping community structure, with particular emphasis on evidence
from experimental field studies. It is noteworthy that no
examples from running waters were included, reflecting the
paucity of data. Evidence from both terrestrial and aquatic
systems provided wide-spread support for the observation that
many communities are quite sensitive to changes in predation.
Connell concluded that, except under certain circumstances where
competition or disturbance appeared to play a critical role,
predation generally was the single most important factor
controlling structure of plant and animal communities. This idea
is especially well-supported by a number of studies of aquatic
systems. Paine (1966, 1971) demonstrated how the macro-benthic
component of the rocky intertidal community suffered a decline in
species richness when predaceous starfish were removed. Lakes
and ponds are known to support very different assemblages of
zooplankton depending on the presence or absence of fish (Brooks
and Dodson, 1965). Typically, larger zooplankton occur in ponds
without fish, while smaller species occur in ponds with fish.
The preference of fish for large prey over small explains part of

191

this pattern. Large zooplankton survive better than smaller species in the absence of fish either because they are competitively superior or, more likely, because large zooplankton are less vulnerable to predaceous invertebrates which also are more abundant in the absence of fish (Zaret, 1980).

Such examples from other kinds of aquatic systems raise the question of whether predators are equally important in running waters, and whether different classes of predators affect particular components of the stream invertebrate fauna. Many studies of individual predators in streams have demonstrated that some form of choice is exhibited, although the mechanisms vary greatly. Where fish are able to use vision to choose prey, a preference for larger prey appears universal. Visual prey choice leads to strong selection for adaptations that lessen prey apparency. Small size, cryptic coloration, nocturnal behavior, ability to detect a predator and take evasive action all are interpretable as evolved responses to lessen the risk of being eaten.

Invertebrate predators may use visual cues, as do many parasitoids and odonates, but chemical, mechanical, and tactile cues appear to be of greater importance in detecting prey. Certain predators in the marine littoral locate their prey by chemical cues (Fishlyn and Phillips, 1980). Tactile information may come in the form of actual encounter, or via mechanical waves transported in water due to the movement of animals (Giguere and Dill, 1979). The same suite of cues presumably is available to prey species as well. Some snails and sea urchins exhibit an avoidance response to water conditioned by the presence of a starfish (Phillips, 1978), and some mayflies chemically detect their stonefly predators (Peckarsky, 1980).

The main objective of this chapter is to review what is known of the importance of predation in streams, considering first the community level of organization, and second the individual adaptations of predators and prey. It reflects the author's bias toward temperate streams in general, and salmonid streams in particular. This in itself points to wide lacunae in our knowledge. I hope that by presenting a personal view of unanswered questions, and factors likely to be of importance, this chapter will stimulate further and much-needed research into the role of predation in running waters.

THE COMMUNITY LEVEL OF ORGANIZATION

We usually think of a community as a collection of interacting organisms having some discernible and characteristic structure, as distinct from the term assemblage, which only

requires that the species be found together but makes no
assumptions about interactions (Ricklefs, 1973). Actually
demonstrating that the term community is appropriate is no easy
task, and the evidence from streams is not strong. For our
purposes, the question is whether a different group of species,
or different abundances of the same species, would occur if the
predation level was varied. With the above caveat, I shall
discuss these responses as indicative of community-level changes.

Experimental Manipulations

The most direct approach to answering this question is to
conduct an experimental field manipulation. That is, one alters
the abundance of predators and compares some measure of prey
abundance and species composition in experimental plots relative
to control plots. Reducing abundance of predators, thus
releasing the prey community from predation, probably is
preferable to increasing the number of predators, as the latter
may pose unnaturally high predation intensity. Furthermore, if
certain species already are excluded by predation, adding more
predators will not reveal this. A second approach is to use
"natural" experiments, which consist of comparisons between areas
that appear to differ only in their abundance or presence-absence
of predators. However, there is ample room for concern that two
such natural areas differ in more factors than just predators.

Straskraba (1965) reported on the distribution of fish and
invertebrates in a stream in the Silesian Beskydy Mountains. An
amphipod found only in the headwater sections changed in
abundance from common to very scarce in a scant 100 m, coincident
with a high weir and the occurrence of large numbers of brown
trout (Salmo trutta). Pentland (1930) remarked that the
distribution of Gammarus in North America appeared to be
determined by the presence of trout. Allan (1975a) suggested
that greater abundance of invertebrates in the upper, fishless
section of a Rocky Mountain stream might be due to release from
predation. Each of the above studies represents a "natural"
experiment which argues for the importance of predation to stream
communities. Jacobi (1979) investigated two Wyoming streams,
each of which had sections that presumably never had contained
fish due to impassable waterfalls. In perhaps the most direct
"natural" experiment yet investigated, Jacobi observed no
significant differences in species composition, density, or
biomass between sections with cutthroat trout (Salmo clarki
lewisi) and those lacking fish.

In contrast to the above "natural" experiments, a number of
investigators have conducted direct experimental manipulations,
using fish or stonefly predators.

Zelinka (1974) studied three sections of a Moravian brown
trout stream. One had normal fish densities, another section had
elevated numbers (186% of normal trout and 135% of normal sculpin
Cottus poecillopus density), and the third had reduced numbers
(7% of trout, 73% of sculpins). Overall, invertebrate changes
were slight. It appeared that body size of mayfly nymphs was
slightly smaller, and numbers somewhat greater, when fish
densities were lowered. Little interpretation was offered
concerning the meaning of these rather minor changes.

Allan (1982a) conducted a similar study in a Colorado stream
where the only fish present were trout. An experimental section
1.2 km in length was selected in a mountain meadow at 3050 m for
reduction in trout density, while upstream and downstream control
sections were left unmanipulated. Fences were erected at each
end of the experimental section, and repeated electro-fishing was
used to estimate initial fish biomass standing crop, and biomass
remaining over time (Fig. 1). The initial biomass of 25 kg (4.9
gm/m^2) is fairly typical for infertile trout streams (Chapman,
1978). During summer months, the trout removal (TR) section
generally contained 10 to 25% of initial biomass. Each fall, the
fence was removed to prevent damage from spring run-off, and
reinstalled by the end of June in each of the four years of the
study. Trout biomass was 25 to 38% of initial standing crop at
the beginning of each season. Thus the trout reduction was
substantial but not total.

The expected result was for numbers to increase, and
possibly for species composition to change, in section TR
compared to upper (UC) and lower control (LC). The food web of
the predominant trout species (Salvelinus fontinalis) was
constructed from stomach analyses, providing some indication as
to which invertebrates might be expected to benefit most from
release from predation. Baetis bicaudatus (Ephemeroptera),
Simuliidae, and Zapada haysi (Plecoptera) were especially common
in trout diet, but other taxa were investigated as well. In
particular, the dominant invertebrate predators, perlodid
stoneflies, were examined as the most probable alternative
predator that might take the place of trout. Benthic densities
were monitored using a Surber sampler (12 replicates per site,
approximately 4 times per year). Drift density was measured as
an additional estimate of invertebrate numbers because an
increase in benthic density might, due to crowding, simply be
exported via increased drift (Waters, 1966).

Over four years, there was no indication of increased
invertebrate numbers, or altered species composition, in section
TR compared to both controls. Some representative data are shown
in Fig. 2. Given the skepticism often directed at negative

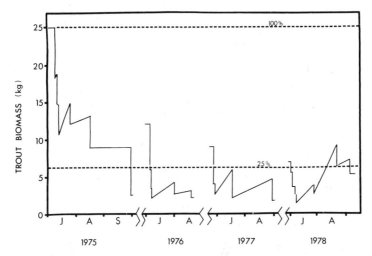

Fig. 1.   Standing crop biomass of trout in an experimental
          section of Cement Creek, Colorado where trout were
          reduced by electro-shocking. A 1.2 km section was
          fenced from July – September each year, and periodically
          electro-shocked to remove trout. From Allan (1982a).
          Copyright© (1982), Duke University Press (Durham, N.C.).

results, the variability of stream data that limits our ability
to detect a difference, and the strong evidence from other
aquatic systems that predator manipulations can affect community
make-up, it seems appropriate to consider possible weaknesses of
this result.

     The manipulation itself does not appear to involve serious
reservations. Initial biomass was estimated with considerable
precision (95% CL ± 18% of the mean), the reduction in trout
biomass, if not total, was substantial, and the initial standing
crop was typical of infertile trout streams in many regions.
However, it must be kept in mind that the study concerned only
one type of stream, and a similar study in a highly fertile trout
stream such as the Horokiwi (Allen, 1951) might lead to different
results. Further, the manipulation concerned only one class of
predators. Elsewhere (Allan, 1982b and below) I have estimated
that predaceous stoneflies consume approximately half of the
amount of prey consumed by trout in this stream. If additional
predaceous invertebrates were considered, it seems likely that
predation due to trout and that due to invertebrates would be of
about the same order of magnitude. A complete removal of all
classes of predators from a section of stream might lead to
different conclusions, if such a manipulation was possible.

     Since the fences were permeable to invertebrates, another
source of concern is the possibility that immigration and

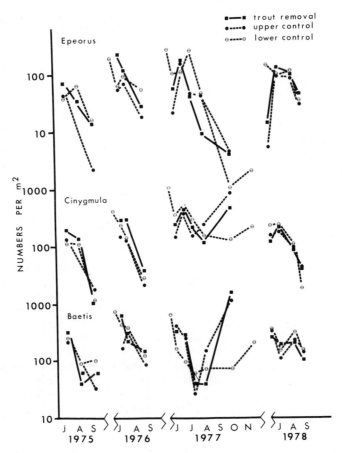

Fig. 2.   Benthic densities of three common mayfly species in
          trout removal and two control sections of Cement Creek,
          based on 12 Surber samples per site. 95% CL omitted for
          clarity, but were approximately ± 50% of the mean. From
          Allan (1982a). Copyright© (1982), Duke University Press
          (Durham, N.C.).

emigration, particularly via drift transport, might mask any
effect of the trout removal. This concern also does not appear
very serious. A relatively large (1.2 km) experimental section
was chosen to minimize this effect, and the census in TR was
conducted near the downstream end, so that roughly 1 km of
experimental section lay above the collecting area. Several taxa
that were rare in the drift, but consumed by trout (e.g.,
Rhyacophila spp.), showed no tendency to increase in section TR.

     A more serious limitation is that of sampling precision, a
problem that affects all field studies. Examination of the

variance of the log-transformed data led to the conclusion that mean density in the section TR would have to nearly double, relative to section UC and LC, to be significant at the .05 level (see also Allan, 1982c). Thus changes of a smaller magnitude could have occurred and gone undetected. While examination of graphed results (e.g., Fig. 2) gave no evidence of this, it is nonetheless true that the variability of counts of benthic invertebrates makes it extremely difficult to detect small changes. In retrospect, it may have been preferable to census benthic invertebrates on fewer dates, perhaps 1 July and 1 September, pay less attention to seasonal phenology, and thus double the sample size. Had this been done, means differing by about 60% (1.6x) could have been distinguished.

To state that influences such as flow or drift may have prevented the detection of an effect due to predators is really no different from saying that trout weren't very important in determining the numbers of benthic invertebrates in this stream, and that appears to be the justifiable conclusion from this study.

Reice (1982) investigated the importance of vertebrate predation in a North Carolina stream, using baskets of substrate placed in a riffle. Some had screen tops to exclude fish and salamanders, and control baskets were topless. Again, no significant differences due to predator manipulation were observed. This particular experiment has one advantage, and one disadvantage, relative to Zelinka (1974) and Allan (1982a). The advantage is that use of a small experimental unit allows for replication of the experimental treatment. A large unit such as both Zelinka and Allan used is too costly for extensive, or often for any, replication of treatments. [Allan (1982c) discusses the design aspects in more detail.] The disadvantage is that small experimental units, which are open to immigration and emigration by the prey community, are much more likely to have the experimental response masked by dispersal.

In contrast to the above studies, Griffith (1981) obtained evidence from a fish introduction in Birtish Columbia, Canada, that indicated a marked effect on insect species composition. Trout were added to a small, fishless headwater within an enclosure, and insect densities compared to a control section. Of nine species examined, three decreased, apparently due to direct predation, and two increased, apparently due to competitive release.

Peckarsky and Dodson (1980a) compared insect colonization of small, substrate-filled cages with and without the presence of predaceous stoneflies. Significantly fewer prey colonized cages with predators, compared to control cages. The authors attributed this result to a combination of direct consumption and prey avoidance behavior. While this does not directly demonstrate a role of predaceous stoneflies in determining community structure, it indicates the potential.

In summary, three studies (Zelinka, 1974; Allan, 1982a; Reice, 1982) involving experimental manipulation of fish in streams have failed to show a significant response of the invertebrate community to that manipulation, although Griffith (1981) did demonstrate a community effect. At present, the weight of evidence indicates that fish do not commonly play a major community structuring role in running waters. Further studies would certainly be of value, as we do not yet have sufficient information to be confident of this generalization. Fishes that consume aquatic invertebrates have a variety of feeding roles (Table 1), which may bear on their influence over the prey community. Total predation may be of a "diffuse" nature, with contributions from such a variety of predators that manipulation of any one fails to elicit a response. Invertebrate predators in particular have been neglected until recently. Before turning to another perspective on this topic, that provided from studies of the food base of streams, it is useful to note that other benthic studies provide supportive evidence for the absence of community response to predator manipulations in fresh waters.

Macan (1966, 1977) conducted a detailed and long-term study of the invertebrate fauna of Hodson's tarn in the English Lake District. Trout (Salmo trutta) were first removed, then stocked, and changes in the invertebrates noted. Certain large species which inhabited the open water (tadpoles, Notonecta, and certain Coleoptera) were eliminated by the addition of trout. Most of the invertebrate fauna appeared unaffected, and was influenced more by changes in the vegetation that occurred independently of the trout manipulations.

Hall et al. (1970) manipulated nutrient input, fish and invertebrate predation in a series of experimental ponds. The fish manipulation affected the species composition (via size composition) of the open water zooplankton, but neither the fish nor the manipulation of predaceous invertebrates had much effect on the benthic invertebrates.

Thorp and Bergey (1981) placed replicate 4 m^2 cages in the soft sediment near-shore area of Par Pond, a reservoir receiving heated effluent water from a nuclear power plant in South Carolina. The density and species richness of benthos in cages open to the likely predators (fish and turtles) did not differ significantly from cages that excluded those predators, except for gastropods. This appeared to be a cage effect, however,

Table 1. Feeding groups of fish that consume aquatic
          invertebrates, from Horwitz (1978). Not included here
          are the piscivores, detritivores, herbivores and
          planktivores.

Feeding role	Description	Representative fishes
Bottom feeders	feed on benthic invertebrates, especially aquatic insect larvae	sculpins (Cottus), most darters (incl. Percina, Etheostoma, Amnocrypta), some minnows, sturgeon (Scaphirhyncus) Noturus, Omiscomaycus
Surface and water column feeders	terrestrial invertebrates, drift (aquatic invertebrates), zooplankton	some minnows (esp. Notropis), killifish (Fundulus), Gambusia, Labidesthes, Hiodon
Generalized invertebrate feeders	includes both of above categories	trout (Salmonidae), many Notropis, some Fundulus, Umbra, Aphredoderus, many sunfish (Lepomis)
Omnivores	consume a wide range of plant and animal food, including terrestrial and aquatic insects, fish, detritus, and mollusks	Some minnows (Notropis, Pimephales, Semotilus), carp and goldfish (Cyprinus and Carassius), bullheads and channel catfish (Ictalurus)
Snail eaters		redear sunfish (Lepomis microlophus)

associated with the use of cage walls by snails, and densities were highest in the cages open to predators.

Fishery biologists have conducted numerous manipulations aimed at enhancing the production of preferred game fish by reducing or eliminating presumed competitor species of little economic or sport value. Moyle (1977) reviewed the evidence that sculpins (Cottus) negatively affect salmonids in streams, and concluded that under natural conditions sculpins were unlikely to limit seriously the food available to salmonids. Little is known about the effect of sculpins on their own food supply, however. The observation that sculpin production may substantially exceed that of salmonids (LeCren, 1969) suggests that this topic bears further study. Holey et al. (1979) reviewed similar evidence concerning the possible adverse effect of suckers (Catostomidae) on food supply of sportfish. In a few instances evidence suggested a benefit to preferred species from reduction of suckers, but strong evidence was lacking. Further, removal of competing species may benefit remaining species by increasing available space and reducing predation on young fish, hence does not demonstrate that food competition was involved (Moyle, 1977).

A possible explanation for the importance of predators in some situations and not others is associated with the availability of refuges. This point will be returned to in a later section.

Comparisons of Food Available to Food Consumed

One of the first, and still among the most extensive efforts to compare food consumption by a predator (brown trout) to prey available, was K.R. Allen's (1951) classic study of the Horokiwi stream in New Zealand. His calculation that trout consumed between 40 and 150 times the standing crop of invertebrates prompted Hynes (1970) to term this "Allen's paradox," as subsequent estimates of invertebrate turnover rates (Waters, 1977) strongly suggest that secondary production cannot support such high prey consumption. Similar studies by Horton (1961) on brown trout in a Dartmoor stream and by Mann (1965) of the entire fish community of the Thames River fit this "paradox" as well. In a review of fish feeding studies, Mann (1978) concluded that very heavy grazing of invertebrate populations was the general rule. An excellent study of dragonfly prey consumption in small ponds (Benke, 1976) provides evidence that invertebrate predators also appear capable of consuming more than production, based on most current production estimates. These studies seem to support two suggestions. First, predators crop a very substantial amount of the prey population. Second, at some point or points, our estimate of prey consumption and availability are in error.

There are some contrary lines of evidence to the argument
that fish severely limit the supply of invertebrates in streams.
Mundie (1974) has pointed out that the amount of drift in streams
of western North America appears adequate to support more young
salmonids than are present. However, Mundie acknowledged that
without better knowledge of the rate at which drift can be
renewed, this is speculation. Hunt (1969, 1971) and White (1975)
report on the effects of habitat alteration, in the form of
creating additional pools and undercut banks, on the standing
crop of trout in a Wisconsin stream. Such habitat alteration can
substantially augment standing crop and production of trout,
implying that food was sufficient to maintain a greater
population than originally existed. Since the habitat alteration
effectively lengthened the stream and increased total stream
area, it is possible that invertebrate abundance also was
augmented. Thus this line of evidence suggesting a surplus of
invertebrate prey also is equivocal.

Allen's (1951) study of the Horokiwi Stream is worthy of
re-examination because of the extensive original data published
in that volume, which may be reconsidered in light of subsequent
information. There has been no substantial criticism of his
estimates of brown trout standing crop biomass. Possibly they
are erroneously high, as the Horokiwi estimates are far greater
than most, but not all, estimates of salmonid production (Waters,
1977; Chapman, 1978). Certainly Allen's estimates of prey
consumption were too high, as were Horton's (1961), due to the
lack of detailed information on the feeding rates of brown trout
available at that time. Gerking (1962) suggested the
over-estimate was about two-fold. With the publication of
Elliott's (1975, 1977) thorough study of brown trout feeding, it
is possible to recalculate total prey consumption. For the four
study sections of the Horokiwi combined, using Elliott's estimate
for the maximum ration consumed by a brown trout, Allan (1982b)
determined that total consumption was only about 20 x the average
standing crop biomass of prey. Macan (1958) pointed out that K.
R. Allen used a relatively coarse mesh net which probably
under-estimated prey standing crop, although the actual extent of
under-estimation is unknown.

Waters' (1977) review of the secondary production literature
indicates that annual production for much of the invertebrate
fauna of temperate streams ranges between 3 and perhaps 10 times
the average standing crop biomass. Some higher values have been
observed, generally for dipterans with many generations per year
or in more warm water habitats (Waters, 1979). Obviously
consumption cannot exceed production, and we might presume that
wherever the former is a large fraction of the latter, predators
are heavily cropping their prey. Thus, while the revised
estimates of prey consumption in the Horokiwi are closer to what

appears possible, a discrepancy, due to consumption apparently
exceeding production, still remains.

In the same mountain stream where Allan (1982a) conducted a
trout-removal experiment, Allan (1982b) estimated total prey
consumed by trout and by predaceous stoneflies at three study
sites. These calculations were derived from gut analysis to
estimate meal size, and estimates of gut clearance rate (Elliott
and Persson, 1978, provide a clear discussion of methods and
assumptions; Hyslop, 1980, evaluates various techniques of
stomach content analysis). Gut clearance rate is only crudely
known for stoneflies, and little is known about feeding rates at
near-0° temperatures, so these are particular weaknesses in the
analysis. At the highest site (3350 m), where trout were absent,
stoneflies consumed only about twice the standing crop of prey.
This suggests that prey were abundant relative to consumption.
The dominant stonefly _Megarcys signata_ was observed to grow at a
nearly constant 5% per day (dry wt.) throughout summer and fall,
which also suggests an abundance of food. At two other sites,
where brook trout (3050 m) or brown trout (2740 m) were present,
consumption by trout and stoneflies combined was 8 to 9 times
mean annual standing crop biomass of prey. However, when the
comparison was made to standing crop of Ephemeroptera and
Diptera, which were the predominant prey at least at the brook
trout site based on stomach analysis (Allan, 1981), the ratios
were higher, 15 to 25 times the standing crop.

Allan (1982b) suggested that this discrepancy, and that from
his recalculation of the Horokiwi data, may be reconciled through
improvements in both estimates. That is, prey consumption may be
over-estimated. In the recalculated Horokiwi estimate, the value
generated was the maximum feeding rate under laboratory
conditions. Additionally, prey standing crop may be
systematically under-estimated. This could be due in part to
failure to sample adequately the hyporheic zone (e.g., Radford
and Hartland-Rowe, 1971; Hynes et al., 1976), and in part to a
tendency for invertebrates to wash around or escape sampling
devices. The upshot is that we can plausibly postulate enough
"fudge factors" at present to bring the ratio of food consumed to
food available low enough to imply minimal grazing. As a result,
final conclusions cannot be drawn.

Benke (1976, 1978) conducted the most extensive comparable
study of predaceous invertebrates, utilizing dragonfly larvae in
a South Carolina farm pond. The standing crop biomass of
odonates was roughly 3 times that of their principal prey
(chironomids and _Caenis_ mayflies), suggesting considerable
potential for control of prey abundance and composition. Benke
(1976) calculated annual production of odonates directly, then
back-calculated the total annual production of prey necessary to

sustain the observed production of predators. Since standing
crop of prey was known, turnover rate of prey could be inferred.
Values in excess of 30 were necessary to account for observed
odonate production. Various sources of error were considered,
and in this instance, under-estimation of prey standing crop did
not seem likely. As the study site had a relatively warm
temperature regime and the prey were mostly short-lived
dipterans, very high turnover rates seem a plausible resolution
of "Allen's paradox" in this instance. Benke suggested that this
explanation may apply to other instances where prey consumption
appears to exceed prey production, while Allan (1982b) argued
that under-estimation of prey standing crop was more likely in
stony-bottom, cool trout streams.

Benke (1978) undertook a field experiment to investigate
whether a reduction in larval odonate density would result in
increased abundance of prey. The majority of early or late
(adult flight period) dragonflies could be excluded from
ovipositing in a series of 4m x 4m pens by placing 1" chickenwire
over the tops of pens during the emergence of particular species.
Significant treatment effects were observed; however results were
not consistent with release from predation. Pens that excluded
the majority of odonates typically had intermediate or low prey
density biomass. This is partially explained by the fact that
screens intended to exclude only dragonflies evidently excluded
or reduced the number of ovipositing prey as well. Compared to
control pens, experimental pens had fewer numbers of both
predators and prey. Certain comparisons were suggestive of a
weak effect of odonates on prey standing stock, however.

Benke (1978) argued that while odonate larvae were
potentially capable of annihilating their prey, refuges prevent
this from occurring. He further noted the general absence of
evidence that prey abundance responds to changing predator
levels, and suggested that the main effect may be an alteration
or adjustment of production. However, this second argument
presents difficulties. If one assumes that natural predation
levels are high, reduction in predation ought to allow greater
survival of prey. It may be that increased prey abundance will
lead to crowding and reduced growth, so that total prey
production changes little or actually declines. Nonetheless, any
"production compensation" should occur only after an increase in
prey density due to reduced predation. The fact that we seldom
observe an increase in prey numbers following release from
predation argues that predators are cropping a surplus and not
limiting prey numbers. Turning the focus from numbers to biomass
production does not change this interpretation.

Warren et al. (1964) manipulated food supply to cutthroat
trout (Salmo clarki) in a small Oregon stream via a sucrose

enrichment that stimulated growth of a bacterial (Sphaerotilus) slime. Benthic sampling showed that annual mean biomass increased in all invertebrate groups, but especially in the Chironomidae. Relative to control sections, trout in the sucrose enrichment section consumed more aquatic invertebrates, especially Chironomidae, Nemouridae, and Perlidae, and trout production was enhanced. In two control sections, Warren et al. calculated total consumption/mean standing crop to be 11.5 and 4.6, while two enriched sections provided estimates of 2.3 and 2.8. As terrestrial items were determined to constitute from 29-58% of trout diet, these figures are well within conservative estimates of P/B. This experiment documented that an increase in food supply can stimulate greater trout production. Small (7g) trout were used, which may explain why an increase in chironomid prey was influential. It also should enhance the production response, as young trout typically suffer the greatest mortality and are capable of the greatest relative growth.

Warren and his colleagues also utilized laboratory streams to investigate the effects on invertebrates of trout, sculpins (Cottus perplexus) and a perlid stonefly (Acroneuria pacifica) (Davis and Warren, 1965; Brocksen et al., 1968). Perhaps the most striking result, relative to the question of trout predation in streams, was that stoneflies appeared more capable than sculpins of cropping insect prey, and stoneflies plus sculpins combined were more effective in reducing prey than were trout. A strong correlation was found between increasing biomass of sculpins and stoneflies, and decreasing biomass of benthos. Thus interstitial and bottom feeders (stonefly and sculpin) were more effective in controlling invertebrate density than was the water column feeder (trout). One qualification is that food consumption was estimated indirectly, using growth rates of the fish, and food conversion efficiencies of the stoneflies. The latter in particular may be questioned, as they were based on variable laboratory estimates which showed stoneflies to have low food conversion efficiencies. Moreover, the laboratory environment might be considered "biased" against trout, as chironomids were much more abundant than larger invertebrates. Certainly stoneflies, and probably sculpins, are more effective than trout at utilizing chironomid larvae in the benthos, while trout rely more on drifting larvae, pupae and adults (whose rate of supply trout cannot control). Nonetheless, for the system under study it seems clear that stoneflies and sculpins can depress prey abundance, while trout cannot. Since young trout are very dependent upon chironomids as prey (Allen, 1941; Allan, 1981), these studies certainly suggest an important competitive bottleneck through which trout must pass.

In summary, comparisons of food consumed to food available generally indicate very heavy utilization of the invertebrate fauna by major predators. While this evidence is circumstantial, it certainly leads one to expect that removal of predators would greatly benefit the prey community. Thus, these results are at odds with results from manipulation experiments that do not indicate any change in stream communities following predator removal. However, various errors associated with estimating food consumed and prey available limit the confidence we may place in conclusions obtained by such calculations.

ADAPTATIONS OF PREDATOR AND PREY

Predators: Mechanisms of Choice

All predators exhibit some degree of what the investigator may term "choice", if only in eating prey within a certain size range as a consequence of the predator's own size. Whether some more complex pattern of choice is exhibited can be quite difficult to establish. This is particularly true when conclusions are drawn from a comparison of what is in the predator's guts vs. what is captured in the investigator's nets. Any difference may be as easily attributed to biased collecting by the biologist, as to selection by the predator. Nonetheless, the topic is one of interest for several reasons. It relates to the foraging tactics of the predator, which may be evaluated against various theoretical expectations (e.g., Pyke et al., 1977). Differences in prey consumption between individual species or major groups of predators bear on the potential for competitive interactions. Finally, should particular categories of prey (species, functional groups, size classes) be more heavily preyed upon than others, it is of interest to investigate what characteristics of prey render them more or less vulnerable to predation. I shall first discuss foraging tactics and prey choice by different types of predators, and then turn to the question of differential vulnerability among prey.

Vertebrate predators. Fish, and other vertebrate predators such as amphibian larvae, tend to show increasing preference with increasing prey size. Zaret (1980) discusses this extensively for lake planktivores, and suggests that electivity (frequency of prey category in gut - frequency in the environment) increases monotonically (Fig. 3) with increasing visibility of prey. While body size is perhaps the most obvious correlate of visibility, any aspect of pigmentation, motion, etc., that increases the conspicuousness of a potential prey must also be considered. Several studies have investigated factors affecting prey choice in fishes that occur in running waters (e.g., Salmo gairdneri, Ware, 1971, 1972; S. trutta, Ringler, 1979). Size, contrast, and motion have all been implicated. Relatively little analysis of prey ingested vs. prey available has been conducted in running

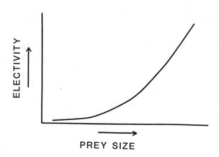

Fig. 3. Electivity curve for fish and other vertebrates such as
         amphibian larvae and salamanders feeding on aquatic
         invertebrates; predators which are large relative to
         their prey. Redrawn from Zaret (1980).

water. One of the earliest such studies is Allen's (1941)
investigation of the food of young salmon (S. salar) prey in
England and Scotland. He computed an availability factor for
each prey taxon (AF = E' of Ivlev (1961), as defined above).
Generally, large taxa showed higher values of AF than did small
taxa, indicating a preference for large prey by young salmon.
Elliott (1967) conducted an analysis of brown trout diet, with
similar results.

To my knowledge, only Metz (1974) and Allan (1978) provide
direct comparisons, based on size categories, between prey
available and prey eaten. Metz investigated rainbow trout (Salmo
gairdneri) feeding on surface drift of emerging aquatic insects
(Chironomidae and Ephemeroptera). Trout appeared to select prey
exclusively on the basis of size. Log E' was highly correlated
with mean body length of insect prey, and relative abundance did
not affect selection. Allan observed that small size classes of
a mayfly nymph occurred in the stomachs of brook trout
(Salvelinus fontinalus) far less frequently than would be
expected on the basis of their frequency in the drift, while
large size classes were over-represented in the diet. This was
particularly apparent for comparisons based on trout collected in
late afternoon and containing prey which showed little evidence
of digestion, i.e., for fish feeding by day. The pattern based
on fish collected just before dawn, hence feeding during the
dark, was less pronounced. This is consistent with size
selection based on vision.

These studies demonstrate that, at least under some
circumstances such as drift-feeding, large prey items are more
vulnerable to predators than are small prey. However, the

numerous comparisons between fish stomach contents and
collections of prey from the environment (e.g., Allen, 1941;
Mundie, 1969; Elliott, 1970; Jenkins et al., 1970; Bryan and
Larkin, 1972; Hunt, 1975; Bisson, 1978; Tippets and Moyle, 1978)
can not be explained simply by this one factor. Fish usually are
termed opportunistic or generalists in these studies, which at
least is consistent with the bewildering diversity of prey
ingested.

Allan (1981) examined the composition (by numbers) of brook
trout diet at different times of the year, to determine whether
relative abundance or prey size were better predictors of what a
trout eats. On each of the five study dates there was a
significant tendency for taxa common in the drift to be common in
trout diet, and vice-a-versa. In addition, however,
large taxa tended to be somewhat or substantially more common in
the diet than the drift, and small taxa less common in the diet
than the drift. Species that grew rapidly between sampling dates
tended to change from under- to over-representation in trout diet
(Fig. 4). Thus from a predictive point of view, relative
abundance, then body size, seemed to explain diet choice of brook
trout. It should be kept in mind, however, that large species
are much less common than small species. From the trout's
perspective, prey size may be more critical than prey abundance
in ranking prey, but the rarity of large prey makes it difficult
or impossible to subsist on large prey alone.

The importance of prey size (or in a broader sense,
visibility) is further born out by laboratory experiments. Ware
(1971, 1972, 1973) studied prey choice by wild-caught rainbow
trout brought into the laboratory and offered light or dark liver
bits. Trout reactive distance increased with size, contrast and
movement of the "prey" and with experience of the trout.
Experience with one type of prey (blanched liver bits) did not
appear to transfer to other prey types (dark liver bits). This
last point is consistent with the substantial individual
specialization often reported from analyses of fish stomachs
(e.g., Allen, 1941). Bryan and Larkin (1972) examined food
specialization of individual marked trout on a series of dates,
using a stomach flushing technique. Particular individuals
showed specialization in diet (e.g., a propensity to eat snails
vs. salamanders) that persisted over time, even if the individual
trout was caught elsewhere (i.e., this result did not appear to
be due to habitat specialization).

Ringler (1979) also examined prey selection under laboratory
conditions, using brown trout feeding on various size classes of
drift. Trout showed strong size selection, as expected. Perhaps

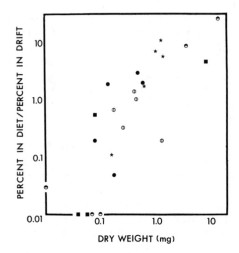

Fig. 4.   The ratio percent in diet/percent in drift (AF of Allen,
          1941, E' of Ivlev, 1961) increases with increasing body
          size for five prey taxa which were small in size on some
          dates, large in size on other dates.   Individuals $\leq$ 1 mg
          dry wt. generally are consumed in proportions equal to
          or less than their occurrence.   Star, Cinygmula sp.;
          square, Rhithrogena hageni and R. robusta; solid circle,
          Zapada haysi; half-circle, Ephemerella doddsi; divided
          circle, Alloperla spp.   From Allan, 1981.   Permission
          granted from Canadian Journal of Fisheries and Aquatic
          Sciences.

more surprising are the observations that trout required 4-6 days
of experience to show substantial improvement in the energy
return of their diet (compared to random feeding), and trout
never stopped "sampling" small prey.

     In summary, studies of the feeding of salmonids in streams
allow the following generalizations.   Diet is diverse and changes
with age and experience.   Fry eat very small prey such as
chironomids (Mundie, 1969; Miller, 1974).   Older fish feed mainly
on the drift, and individuals > 2 years may make more use of
surface drift and especially of the benthos (and large, cased
caddis larvae) than do 0^{+} and 1^{+} individuals (Allen, 1941;
Elliott, 1967, 1970; Tippets and Moyle, 1978).   As aquatic prey
become rarer in late summer, reliance on terrestrial prey
increases (Hunt, 1975).   Different species of salmonids have been
observed to differ in their use of habitats and resulting diets
(Schultz and Northcote, 1972; Allan, 1975b).   In addition to
these differences due to age, species of fish, and availability

of prey, individual fish specialize to varying degrees. Clearly there is much to learn yet about the mechanisms of prey choice in even such well-studied fishes as the Salmonidae, and further mechanistic studies such as those of Ware and Ringler are especially needed. The implications of selection for large size, or other aspects of prey which enhance vulnerability to predation, will be discussed later from the prey's perspective.

While the feeding of other stream fishes has received some investigation, much less is known of their feeding compared to the Salmonidae. For example, sculpins (Cottidae) have been reported to feed primarily on immature aquatic insects (Dineen, 1951; Daiber, 1956; Novak and Estes, 1974; Petrosky and Waters, 1975). Novak and Estes (1974) calculated forage ratios for various prey, obtained a wide range of values and suggested that prey habitat and availability were important factors determining forage ratio. Other stream fishes including Fundulus (McCaskill et al., 1977), minnows (Cyprinidae, Mendelson, 1975) and darters (Percina and Etheostoma) also are known to feed on aquatic invertebrates, but little investigation of prey selection has been undertaken.

Invertebrate predators. Virtually every group of aquatic insects includes at least some species that are predaceous on other invertebrates. Larval odonates (Koslucher and Minshall, 1973) and megalopterans (Steward et al., 1973; Hildrew and Townsend, 1976), setipalpian stonefly nymphs (Siegfried and Knight, 1976; Fuller and Stewart, 1977), rhyacophilid (Thut, 1969) and polycentropodid caddis larvae (Hildrew and Townsend, 1976), some hydropsychid caddis larvae (Wallace, 1975), and various chironomid groups (Tanypodinae, some Chironomini, Merritt and Cummins, 1978) are carnivorous during much of their life cycle. In some groups (e.g., Ephemeroptera) the predaceous habit has been adopted in only a few species (e.g., the mayfly Dolania, Tsui and Hubbard, 1979). However, it may be that many species not characterized as predators do capture animal prey at some point in their life cycle (Anderson, 1976). Thus the potential range of invertebrate predators is very considerable.

Most of what is known about the feeding habits of invertebrate predators comes from inspection of gut contents. Generally, feeding is considered to be opportunistic and in proportion to foods present (e.g., Koslucher and Minshall, 1973). Some recent studies that describe prey ingestion are summarized in Table 2. Dipterans (especially chironomids) and mayflies typically are the most common prey, although larger invertebrates often consume additional taxa. The most obvious exception is the dragonflies in Deep Creek, Idaho-Utah, which ate no chironomids and few mayflies (Koslucher and Minshall, 1973).

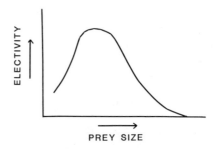

Fig. 5.   Electivity curve for invertebrate predators, which tend
          to be limited by their own size in the largest prey that
          they ingest. Redrawn from Zaret (1980).

     Zaret (1980) characterized invertebrate predators by a
humped electivity curve (Fig. 5). Larger prey are preferred, but
as size of prey increases it becomes more difficult for the
predator to subdue the prey, and handling time increases as well.
Inevitably the curve descends to the right as prey become large
relative to the predator. Zaret developed this relationship for
invertebrate predation within the freshwater zooplankton.
Thompson (1975) provides an excellent demonstration of its
validity for dragonflies that ambush zooplankton prey. Its
validity for predaceous invertebrates in streams is unknown at
present; one possibility is that the curve is fairly flat over
the range of prey sizes normally encountered, so that selection
is rarely in evidence.

     The most extensive efforts to study feeding specificity of
invertebrate predators in streams all have centered on perlid and
perlodid stoneflies. Fuller and Stewart (1977, 1979) and
Siegfried and Knight (1976) computed electivities for various
size classes of several prey species. A preference for larger
prey usually, but not invariably, was observed. It is difficult
to elucidate mechanisms from field studies in which prey
available are compared to prey eaten, since the size of both the
predator and the prey are not controlled variables. Furthermore,
sizes usually are expressed as head width rather than weight, and
so are not comparable across taxa or between studies. One
investigation might be based on a small species of predator
co-occurring with relatively large prey, hence be relevant only
to the descending side of Fig. 5, and suggest preference for prey
smaller than the average. A converse set of comparisons might
generate the opposite conclusion. Without a laboratory (or very
fortuitous field) experimental design, it is extremely difficult
to generalize on prey choice by invertebrate predators. An
excellent model for investigation of predator-prey relations,

Table 2. Food habits of invertebrate predators in streams, based
          on gut analysis. Only the animal portion of the diet
          is summarized here. Studies have been selected to
          illustrate main trends.

Predator	Most Frequent Prey (in order)	Reference
Odonata	Hyalella, Hydropsyche, Simulium	Koslucher and Minshall, 1973
Plecoptera (Setipalpia)	Chironomidae, Baetidae, Simuliidae, Heptageniidae, Trichoptera	Sheldon, 1969, 1972; Siegfried and Knight, 1976; Fuller and Stewart, 1977
Trichoptera (Rhyacophilidae)	Chironomidae	Thut, 1969; Devonport and Winterbourn, 1976
(Polycentro-podidae)	Chironomidae, Plecoptera, Microcrustacea	Townsend and Hildrew, 1977
(Hydropsychidae)	Chironomidae, Simuliidae	Mecom, 1972; Wallace, 1975; Wallace et al., 1977
Ephemeroptera (Dolania)	Chironomidae	Tsui and Hubbard, 1979
Megaloptera	Chironomidae, Ephemeroptera	Devonport and Winterbourn, 1976; Steward et al., 1973; Hildrew and Townsend, 1976
Chironomidae	Chironomidae, micro-crustacea, Oligochaeta	Roback, 1968; Armitage, 1968

based mainly on terrestrial arthropods, is provided by Hassell
(1978).

A number of studies report that diet of predaceous
invertebrates tends to change during development. Setipalpian
stoneflies tend to ingest diatoms and chironomids in early
instars, eat virtually a pure chironomid diet over much of their
life cycle, and finally consume a broader diet in which mayflies,
simuliids, and trichopterans supplement chironomids as prey
(Mackereth, 1957; Sheldon, 1969; Siegfried and Knight, 1976;
Fuller and Stewart, 1977, 1979). This is illustrated in Fig. 6
from a study by Allan (1982d), who inspected the foreguts of
three species of stoneflies at each of three sites. Any two
predator individuals of the same size but different species were
more similar in their diet than two individuals of the same
species but different size. Some differences occurred from site
to site, due to differing availability of prey. Small predator
species tend to have less diverse diets than larger species
because they don't reach a sufficient size to capture those prey
that are larger and more agile than midge larvae.

A positive correlation between size (head capsule width) of
predator and prey has been reported several times (e.g., Sheldon,
1972). This is consistent with the broadening of diet to include
larger taxa as stoneflies grow, illustrated in Fig. 6. Obviously
there is a tendency for larger invertebrate predators to ingest
larger prey, but the reasons for this are not clear. Allan
(1982d) computed the average size (mg dry wt.) of each of two
prey groups, Baetis and Chironomidae, ingested by each of three
stonefly predators (Fig. 7). The similarity of the relationships
suggests a size-based phenomenon. Acutal choice is not
demonstrated, however. In cold-water streams, many species are
univoltine and so at any one time, only a narrow range of sizes
of predator and prey can be observed. Results such as Fig. 7 are
obtained by pooling data over several seasons. Allan (1982d)
compared size of Baetis ingested to size available for a number
of dates where this bivoltine mayfly was represented by a wide
range of sizes. Out of 10 possible comparisons, eight showed no
selection and two showed preference for prey smaller than
average. This result must be viewed in light of several
cautionary points already mentioned: the investigator, not the
predator, may be the biased collector; and only a particular
range of predator and prey co-occurred for comparison. Nonetheless, it is sufficient to cause serious questioning of the role
of active prey preference in generating results such as Fig. 7.
More likely, the cause is juxtaposition of predator and prey life
cycles, so that larger size classes of predators tend to coincide
with larger classes of prey (Sheldon, 1972; Allan, 1982d).

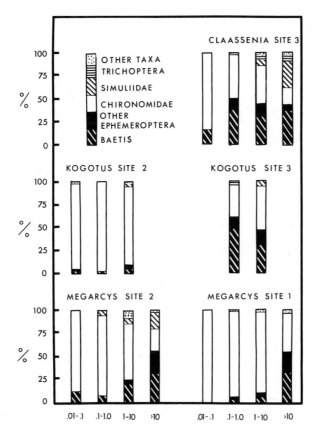

Fig. 6. Taxonomic composition of diet for various size classes
        of predaceous stoneflies. Three species of stoneflies
        were examined at 3 sites. From Allan (1982d).
        Copyright© (1982), Duke University Press (Durham, N.C.).

Prey: Mechanisms for Minimizing Risk of Predation

Prey species may be expected to evolve mechanisms that
reduce their risk of predation. Ability to detect a predator via
visual, chemical, tactile or other cues, escape and avoidance
responses, and a body size that does not correspond to an
electivity peak (Fig. 5) are some possible adaptations to
predation. As prey evolve the ability to foil predators,
predators may counter with their own advances, resulting in
co-evolutionary "arms races" (Vermeij, 1976; Dawkins and Krebs,
1979).

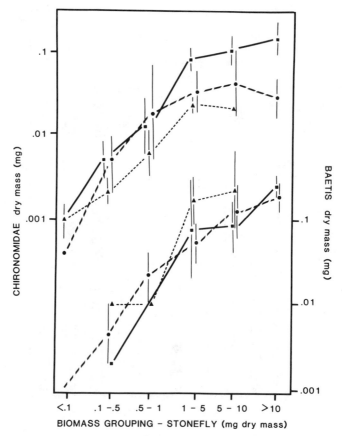

Fig. 7. Average size of prey ingested by various size classes of
each of three species of predaceous stoneflies; _Megarcys_
signata (sqares), _Kogotus_ _modestus_ (triangles) and
Claassenia _sabulosa_ (circles). Mean and 95% CL are
shown for two major prey, _Baetis_ and Chironomidae.

An apparent visual response by invertebrates to presence of
fish has been demonstrated by Stein and Magnuson (1976) and Stein
(1977). Crayfish (_Orconectes_) reduced their walking and foraging
activity, selected substrates that afforded the most protection
and increased their defensive activity in the presence of
smallmouth bass (_Micropterus_ _dolomieui_). Field distribution
appeared to be modified in nature where fish predation was
highest, so as to reduce risk of predation (Stein, 1977).
Charnov et al. (1976) reported that mayflies (Baetidae) in

laboratory tanks tended to aggregate in dark, corner regions
where they were less susceptible to fish when kokanee
(Oncorhynchus nerka) were present. Near dusk, when mayflies move
to the surface of stones, a human observer can elicit rapid
escape movements by simply passing one's hand over a motionless
individual (Allan, pers. observ.). Peckarsky (1980) tested the
response of mayflies to a predaceous stonefly in a glass test
tube, and found no evidence of visual detection of an
invertebrate predator. In contrast, juvenile Notonecta
hoffmanni, given the opportunity to forage where prey are
abundant but risk of predation is great (from adult N.
hoffmanni), prefer to forage in low prey, but also low risk,
habitats (Sih, 1980). This may indicate visual avoidance of
predators, as prey are located visually by Notonecta; however,
prey detection is enhanced by movement (Sih, 1979).

    Chemical detection of predator and prey by one another have
been reported for various marine invertebrates (Feder, 1972).
Water collected from aquaria containing a single starfish
(Pycnopodia helianthoides) was sufficient to trigger defensive or
escape responses in the urchin Strongylocentroides purpuratus and
the snail Tegula funebralis (Phillips, 1978). Well-developed
behavioral escape responses have been documented for several
marine molluscan species (Fishlyn and Phillips, 1980; Hadlock,
1980; Ordzie and Garofalo, 1980). The limpet Notoacmea paleacea
appears to camouflage its scent by incorporating into its shell a
pigment of the plant on which it commonly rests. Predatory sea
stars fail to detect its presence in most encounters. The limpet
shows no behavioral escape response, unlike other co-occurring
species that are not chemically camouflaged. Vermeij (1978)
suggests that the relative speed of prey and predator may
influence the usefulness of escape responses; slow-moving prey
may use escape effectively only against slow-moving predators.
Many species of freshwater gastropods exhibit an alarm response
to crushed extract of conspecifics, usually self-burial or
crawling away (Snyder, 1967). Species which responded by
self-burial did so in daylight but only weakly or not at all in
the dark, suggesting an evolved response to visually searching
predators. In a limited number of trials, only a few species
responded to water from predator tanks in Snyder's experiments.

    Peckarsky (1980) investigated the ability of several mayfly
species to detect and avoid stonefly predators by non-contact
chemical stimuli, and a number of tests showed a significant
avoidance response. While lack of consistency in response among
species indicates the need for further studies, it seems clear
that some mayflies can distinguish invertebrate predators by
chemical cues. There may be considerable specificity to these
responses. Peckarsky determined that a large detritivorous
stonefly failed to elicit avoidance via chemical cues.

Perhaps surprisingly, since prey evidently sense their predators, Peckarsky and Dodson (1980b) reported that stoneflies did not discriminate between cages with high initial prey density and cages with zero initial prey density. Either they lack the ability to detect prey chemically, or perhaps aggregations are inhibited by intraspecific interactions (Walton et. al., 1977).

Relatively little is known of chemical detection of fish by stream invertebrates. Williams and More (1982) observed drift behavior of Gammarus (Amphipoda) in a laboratory stream with and without rainbow trout. Addition of trout suppressed drift almost completely.

Drift of late instar nymphs of the mayflies Baetis tricaudatis and Leptophlebia cupidae in an artifical stream was enhanced by addition of the stonefly Isogenoides elongatus (Corkum and Clifford, 1980). While this may be a chemically mediated response, it could be due simply to contact (tactile) responses. Predaceous stoneflies presumably use their long antennae in prey detection, and Peckarsky (1980) has reported on the response of several mayfly species to such contact. Some showed very pronounced locomotory responses, including species such as Cinygmula sp. which did not appear to respond to chemical cues. Others, such as Ephemerella, which did respond to non-contact chemical stimuli, remained motionless.

In summary, evidence currently exists to suggest that invertebrate prey in streams have evolved defensive or escape responses to visual cues from fish but perhaps not commonly to invertebrate predators, chemical cues from both, and contact cues from invertebrate predators. Few studies have been conducted, however, and this remains a fertile area for further research (see Peckarsky, this volume).

Not all prey are equally vulnerable to predators. The most obvious determinant of risk of predation by fish, for invertebrates, is their visibility and in particular their size. Several studies document that invertebrates change their behavior coincident with changing size and changing risk of predation. Stein and Magnuson (1976) reported that small crayfish, which were more vulnerable to bass than larger crayfish, showed the greatest restriction in activity. Those life stages most vulnerable to fish appeared to modify their distribution in the field so as to reduce predation risk (Stein, 1977). Peckarsky (1980) suggested that the curved-tail ("scorpion") posture in Ephemerella serves in defense against predaceous stoneflies, and is less pronounced in young Ephemerella because it would be less effective. Terrestrial insects including Lepidoptera (Iwao and Wellington, 1970) and grasshoppers (Schultz, 1981) change anti-predator behavior as they grow.

It seems highly likely that the nocturnal periodicity of drift is an anti-predator adaptation shared by a majority of aquatic invertebrates. Fish strongly select large over small prey (Allan, 1978; Ringler, 1979). Hence, we may hypothesize that taxa or growth stages that are large, and hence suffer greatest predation risk, should be most strongly constrained to nocturnal behavior, while smaller taxa and stages may be aperiodic or diurnal (Allan, 1978). In a Colorado trout stream where the mayfly Baetis bicaudatus had overlapping generations, hence an array of sizes present at any one time, Allan (1978) found that the ratio drift by night : drift by day increased with body size (Fig. 8). The only exception was the largest size class on one day, where it appeared that increased afternoon activity associated with adult emergence reversed the trend for the largest sizes. Relatively few studies of drift activity have explicitly considered body size. However, some verification of this pattern with other species is possible (Table 3).

Edmunds and Edmunds (1980) provide a further example of the apparent modification of behavior in mayflies subject to high predation. Lowland tropical species from lineages which have relatively long adult lives (i.e., > 2 hr) emerge as sub-imagos in the first 2 hours of darkness, transform to imagos before dawn, and mate and oviposit by mid-morning. Edmunds and Edmunds suggest that this is an adaptation to minimize heavy predation by birds and odonates in the lowland tropics.

Otto and Svensson (1980) argue that the cases of caddis larvae are primarily anti-predator devices. Cases may provide camoflage, although Otto and Svensson place less emphasis on this point; make the larvae too bulky to attack, especially by lateral enlargements and projecting sticks; and increase resistance to crushing. In a series of laboratory experiments with Potamophylax cingulatus, they demonstrate that young larvae, which are autumnal and utilize leaf discs in case building, have cases that are less resistant to crushing but less costly to make. As the larvae grow, first bark, then mineral material is used. More secretion is needed to cement the many fine particles together, but resistance to crushing increases many fold. Greater protection from trout predation is the inferred benefit.

Why Aren't Streams "Structured" by Predation?

It seems that continuing efforts to understand the basis of prey choice by predators, and adaptive responses on the part of the prey to reduce risk of predation, will repay the investigator with further evidence of co-adaptation. Why, then, given this line of evidence of the importance of predation, do stream communities commonly not respond to manipulations of predator level? Since freshwater plankton (Zaret, 1980) and the marine

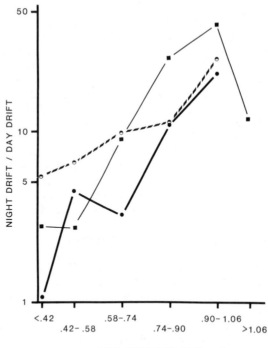

SIZE GROUPING (HW, mm)

Fig. 8.   Ratio of total nighttime drift to total daytime drift in
Baetis as a function of body size on each of 3 dates.
Even the smallest nymphs show a bias towards nocturnal
activity, but this bias increases with size (and risk of
predation).   Data from Allan (1978).   Squares, 18–19
July 1976; half-circle, 2–3 August 1976; solid circle,
11–12 September 1976.

intertidal community (Paine, 1980) respond so strikingly to such
manipulations, there is cause to be skeptical of a limited number
of experiments that conclude running waters to be the exception.
However, this review of field manipulations poses the possibility
that the freshwater benthos is different.

First, it is hardly plausible that removal of any species
chosen at random, even any predator, will have detectable effects
on much of the community.   Many species may be "weak interactors"
in the terminology of Paine (1980).   In contrast, a "strong
interactor" is one whose removal results in pronounced changes in
other species.   Such an effect is especially likely when (1) a
predator plays a key role in controlling a particular prey that
(2) is competitively superior to other species in the food web.

Table 3.   Summary of information supporting the hypothesis that
           smaller taxa or stages should be less strongly
           constrained to drift by night, relative to larger taxa
           or stages.

1)	Average weight of individuals is greater in samples collected during the night, compared to samples collected during the day.	<u>Baetis</u> <u>Capnia</u>, Chironomidae, <u>Paraleptophlebia</u>	Anderson and Lehmkuhl, 1968
		Ephemeroptera, possibly Simuliidae, but not Chironomidae	Steine, 1972
2)	Taxa that are small tend to drift by day or be aperiodic	mites	Elliott and Minshall, 1968
		Chironomidae (drift pattern) variable)	Waters, 1972
3)	Moonlight suppresses nocturnal drift of large individuals more than it suppresses small individuals	Ephemeroptera, Plecoptera and Simuliidae	Anderson, 1966
4)	Within a taxon, early instars show the greatest tendency to drift by day, compared to later stages.	<u>Baetis</u> <u>bicaudatus</u>	Allan, 1978
		<u>Deleatidium</u> sp.	Devonport and Winterbourn, 1976
		<u>Polypedilum</u> <u>haterale</u>	Colwell and Carew, 1976
		<u>Amiocentrus</u> <u>apilus</u>	Anderson, 1967
		<u>Helicopsyche</u> <u>borialis</u>	Bishop and Hynes, 1969
		<u>Rhyacophila</u> <u>nubila</u>	Ejellheim, 1980

Thus two premises are involved, neither of which has received strong support at present.

Second, a number of experimental manipulations of predators of stream benthos (Zelinka, 1974; Allan, 1982a; Reice, 1982) and pond benthos (Hall et. al., 1970; Benke, 1978; Thorp and Bergey, 1981) have shown a weak or no community response. At most, an individual prey group has shown a small benefit from release from predation. Thorp and Bergey (1981) present several reasons why the freshwater benthos may differ in response relative to the marine intertidal. These include a lesser tendency for one species to dominate a major resource such as space and a greater mobility of organisms, which results in rapid re-distribution of organisms into areas made available by past predation. Macan (1977), Benke (1978), and Thorp and Bergey (1981), among many, discuss the importance of refuges for prey in preventing their annihilation. It may be that the importance of predation will be found to depend on substrate, which is the major refuge of benthic prey. Brusven and Rose (1981) found that sculpins (Cottus rhotheus) in laboratory streams consumed more insect prey on a sand substrate, compared to pebbles and cobbles. Many of the top predators in streams, such as salmonids, have their prey "delivered to them" as drifting invertebrates. Because these predators do not actively seek out which prey species they eat, but feed from what is presented, the likelihood is reduced of their having a dominating influence on community structure. Those predators most capable of entering the substrate, and those substrates offering the least refuge to prey, seem likely candidates to offer evidence contrary to that acquired to date.

Third, in the plankton of lakes and ponds, and to some extent in the rocky intertidal, situations exist where predators are predictably absent or by chance will not be encountered during the vulnerable stages of an organism's development. The freshwater zooplankton includes species which are adapted to either low or high fish predation, hence communities respond dramatically to fish manipulation. Indeed, a zooplankton specialist can predict whether fish are present in a pond without ever looking for fish. In streams, by way of contrast, there does not appear to be one discernible faunal assemblage adapted to the absence of fish, and another to their presence. This may be because streams, by their anastomosing nature, never provide refuges where major predators are predictably absent. Very likely the fauna of streams has evolved in the presence of fish predation, resulting in appropriate adaptations. Tillyard (1926) noted that the mayfly fauna of Australia and New Zealand was not specialized to withstand trout predation, since most were active swimmers rather than burrowers. Apparently, introduction of trout severely diminished this once abundant fauna. Other observations consistent with this argument include well-developed predator-avoidance behaviors such as nocturnal drift and chemical detection of predators. The preponderance of small size and cryptic coloration may be an evolved response to predation, as may be life cycles in which adult emergence occurs before summer temperatures encourage high fish feeding rates. Linking the individual and community levels of investigation, then, it seems reasonable to suggest that a fauna well-adapted to co-occurring with predators should not respond significantly to predator manipulations.

Final answers still are elusive, of course, partly for lack of sufficient studies and partly due to the difficulty of obtaining unambiguous results. The apparent lack of response of the stream benthos to varying levels of predation is difficult to reconcile with evidence that predators heavily graze their prey. Problems with measurement and with experimental field techniques present major challenges (cf. Allan, 1982c). As the role of predators in streams is of considerable practical as well as theoretical interest, a critical need exists for the next generation of studies to develop improved approaches to answer such questions.

ACKNOWLEDGMENTS

Numerous individuals have aided me in my field research and through discussion of ideas. I am especially grateful to the Colorado Division of Wildlife, and its staff members, Walt Burkhardt and Rick Sherman, who were generous in their assistance to my field work, and Jim Kuhar and Doug Robison for their hard work and good companionship. The manuscript was improved by anonymous reviewers and the suggestions of Alex Flecker, Randy Fuller, Mark Gardner, and Bobbi Peckarsky. Research was supported by NSF grants DEB-75-03396, DEB-77-11131, and computer funds from the University of Maryland.

REFERENCES

Allan, J. D. 1975a. The distributional ecology and diversity of benthic insects in Cement Creek, Colorado. Ecology 56:1040-53.

Allan, J. D. 1975b. Diet of brook trout (Salvelinus fontinalis Mitchell) and brown trout (Salmo trutta L.) in an alpine stream. Verh. Internat. Verein. Limnol. 20:2045-50.

Allan, J. D. 1978. Trout predation and the size composition of stream drift. Limnol. Oceanog. 23:1231-1237.

Allan, J. D. 1981. Determinants of diet of brook trout (Salvelinus fontinalis) in a mountain stream. Can. J. Fish. Aquat. Sci. 38:184-192.

Allan, J. D. 1982a. The effects of reduction in trout density on the invertebrate community of a mountain stream. Ecology 63:1444-1455.

Allan, J. D. 1982b. Food consumption by trout and stoneflies in a Rocky Mountain stream, with comparison to prey standing crop. In: T. Fontaine and S. Bartell (eds.), Dynamics of lotic ecosystems. Ann Arbor Science, Ann Arbor, Mich.

Allan, J. D. 1982c. Hypothesis testing in ecological studies of the freshwater benthos. In: V. H. Resh and D. M. Rosenberg (eds.), Ecology of aquatic insects. Praeger Scientific.

Allan, J. D. 1982d. Life cycle, feeding habits and prey consumption of three setipalpian stoneflies (Plecoptera) in a mountain stream. Ecology 63:26-34.

Allen, K. R. 1941. Studies on the biology of the early stages of the salmon (Salmo salar): 2. Feeding habits. J. Anim. Ecol. 10:47-76.

Allen, K. R. 1951. the Horokiwi Stream: a study of a trout population. New Zealand Marine Dept. Fisheries Bull. 10, 238 pp.

Anderson, N. H. 1966. Depressant effect of moonlight on activity of aquatic insects. Nature 209:319-320.

Anderson, N. H. 1967. Biology and downstream drift of some Oregon Trichoptera. Can. Entomol. 99:507-521.

Anderson, N. H. 1976. Carnivory by an aquatic detritivore, Clistoronia magnifica (Trichoptera: Limniphilidae). Ecology 57:1081-85.

Anderson, N. H. and D. H. Lehmkuhl. 1968. Catastrophic drift of insects in a woodland stream. Ecology 49:198-205.

Armitage, P. D. 1968. Some notes on the food of the chironomid larvae of a shallow woodland lake in South Finland. Ann. Zool. Fenn. 5:6-13.

Benke, A. C. 1976. Dragonfly production and prey turnover. Ecology 57:915-27.

Benke, A. C. 1978. Interactions among coexisting predators--a field experiment with dragonfly larvae. J. Anim. Ecol. 47:335-50.

Bishop, J. E. and H. B. N. Hynes. 1969. Downstream drift of invertebrate fauna in a stream ecosystem. Arch. Hydrobiol. 66:56-90.

Bisson, P. A. 1978. Diel food selection by two sizes of rainbow trout (Salmo gairdneri) in an experimental stream. J. Fish. Res. Board Can. 36:392-403.

Brocksen, R. W., G. E. Davis and C. E. Warren. 1968. Competition, food consumption, and production of sculpins and trout in laboratory stream communities. J. Wildl. Manag. 32:51-75.

Brooks, J. L. and S. I. Dodson. 1965. Predation, body size, and composition of the plankton. Science 150:28-35.

Brusven, M. A. and S. T. Rose. 1981. Influence of substrate composition and suspended sediment on insect predation by the torrent sculpin, Cottus rhotheus. Can. J. Fish. Aquat. Sci. 38:1444-48.

Bryan, J. E. and P. A. Larkin. 1972. Food specialization by individual trout. J. Fish. Res. Board Can. 29:1615-1624.

Chapman, D. W. 1978. Production in fish populations. pp. 5-25. In: S. D. Gerking (eds.), Ecology of freshwater fish production. Blackwell Scientific Publications, Oxford, England.

Charnov, E. L., G. H. Orians and K. Hyatt. 1976. Ecological implications of resource depression. Amer. Natur. 110:247-59.

Colwell, B. C. and W. C. Carew. 1976. Seasonal and diel periodicity in the drift of aquatic insects in a subtropical Florida stream. Freshw. Biol. 6:587-594.

Connell, J. H. 1975. Some mechanisms producing structure in natural communities. pp. 460-491. In: M. L. Cody and J. M. Diamond (eds.), Ecology and evolution of communities. Harvard Univ. Press, Cambridge.

Corkum, L. D. and H. F. Clifford. 1980. The importance of species associations and substrate types to behavioral drift. pp. 331-341. In: J. F. Flannagan and K. E. Marshall (eds.), Advances in Ephemeroptera biology. Plenum Press, New York.

Daiber, F. C. 1956. A comparative analysis of the winter feeding habits of two benthic fishes. Copeia 1956:414-51.

Davis, G. E., and C. E. Warren. 1965. Trophic relations of a sculpin in laboratory stream communities. J. Wildl. Manag. 29:846-71.

Dawkins, R. and J. R. Krebs. 1979. Arms races between and within species. Proc. R. Soc. Lond. B. 205:489-511.

Devonport, B. F. and M. J. Winterbourn. 1976. The feeding relationships of two invertebrate predators in a New Zealand river. Freshw. Biol. 6:167-76.

Dineen, C. F. 1951. A comparative study of the food habits of Cottus bairdi and associated species of Salmonidae. Amer. Midl. Nat. 46:640-645.

Elliott, J. M. 1967. Invertebrate drift in a Dartmoor stream. Archiv. Hydrobiol. 63:202-237.

Elliott, J. M. 1970. The daily activity patterns of caddis larvae (Trichoptera). J. Zool. (Lond.) 160:279-290.

Elliott, J. M. 1975. Number of meals in a day, maximum weight of food consumed in a day, and maximum rate of feeding for brown trout, Salmo trutta L. Freshwat. Biol. 5:287-303.

Elliott, J. M. 1977. Feeding, metabolism and growth of brown trout. Freshwater Biological Association, U.K. Annual Report 45:70-77.

Elliott, J. M. and G. W. Minshall. 1968. The invertebrate drift in the River Duddon, English Lake District. Oikos 19:39-52.

Elliott, J. M. and L. Persson. 1978. The estimation of daily rates of food consumption for fish. J. Anim. Ecol. 47:977-991.

Edmunds, G. F., Jr. and C. H. Edmunds. 1980. Predation, climate and mating of mayflies. pp. 277-86. In: J. F. Flannagan and K. E. Marshall (eds.), Advances in Ephemeroptera biology. Plenum Press, New York.

Feder, H. M. 1972. Escape responses in marine invertebrates. Sci. Amer. 227:92-100.

Fishlyn, D. A. and D. W. Phillips. 1980. Chemical camoflaging and behavioral defenses against a predatory seastar by three species of gastropods from the surfgrass Phyllospadix community. Biol. Bull. 158:34-48.

Fjellheim, A. 1980. Differences in drifting of larval stages of Rhyacophila nubila (Trichoptera). Holarc. Ecol. 3:99-103.

Fuller, R. L. and K. W. Stewart. 1977. The food habits of stoneflies in the Upper Gunnison River, Colorado. Env. Ent. 6:293-302.

Fuller, R. L. and K. W. Stewart. 1979. Stonefly (Plecoptera) food habits and prey preference in the Delores River, Colorado. Amer. Midl. Nat. 101:170-181.

Gerking, S. D. 1962. Production and food utilization in a population of bluegill sunfish. Ecol. Monogr. 32:31-78.

Giguere, L. A. and L. M. Dill. 1979. The predatory response of Chaoborus larvae to acoustic stimuli, and the acoustic characteristics of their prey. Z. Tierpsychol. 50:113-23.

Griffith, R. W. 1981. The effect of trout predation on the abundance and production of stream insects. M. S. Thesis, University of British Columbia, Vancouver.

Hadlock, R. P. 1980. Alarm response of the intertidal snail Littorina littorea (L.) to predation by the crab Carcinus maenas (L.). Biol. Bull. 159:269-279.

Hall, D. J., W. E. Cooper, and E. E. Werner. 1970. An experimental approach to the production dynamics and structure of freshwater animal communities. Limnol. Oceanogr. 24:131-37.

Hassell, M. P. 1978. The dynamics of arthropod predator-prey system. Monographs in Population Biology 13. Princeton University Press, 237 pp.

Hildrew, A. G. and C. R. Townsend. 1976. The distribution of two predators and their prey in an iron rich stream. J. Anim. Ecol. 45:41-57.

Holey, M., B. Hollender, M. Inhof, R. Jesian, R. Konopacky, M. Toneys and D. Coble. 1979. Never give a sucker an even break. Fisheries 4:2-6.

Horton, P. A. 1961. The bionomics of brown trout in a Dartmoor stream. J. Anim. Ecol. 30:311-338.

Horwitz, R. J. 1978. Temporal variability patterns and the distributional patterns of stream fishes. Ecol. Monogr. 48:307-321.

Hunt, R. L. 1969. Effects of habitat alteration on production, standing crops, and yield of brook trout in Lawrence Creek, Wisconsin. pp. 281-312. In: T. G. Northcote (ed.), Symposium on salmon and trout in streams. University of British Columbia, Vancouver, Canada.

Hunt, R. L. 1971. Responses of a brook trout population to habitat development in Lawrence Creek. Tech. Bull. 48, Dept. of Nat. Res., Madison, Wisc.

Hunt, R. L. 1975. Food relations and behavior of salmonid fishes. pp. 137-151. In: A. D. Hasler (ed.), Coupling of land and water systems. Springer-Verlag, New York.

Hynes, H. B. N. 1970. The ecology of running waters. Liverpool Univ. Press. 555 pp.

Hynes, H. B. N., D. D. Williams and N. E. Williams. 1976. Distribution of the benthos within the substratum of a Welsh mountain stream. Oikos 27:307-310.

Hyslop, E. J. 1980. Stomach content analysis--a review of methods and their application. J. Fish. Biol. 17:411-429.

Ivlev, V. S. 1961. Experimental ecology of the feeding of fishes. Yale University press, New Haven CT. 302 p.

Iwao, S. and W. G. Wellington. 1970. The influence of behavioral differences among tent-caterpiller larvae on predation by a pentatomid bug. Can. J. Zool. 48:896-898.

Jacobi, G. Z. 1979. Ecological comparisons of stream sections with and without fish populations, Yellowstone National Park. National Park Service Research Center, Univ. of Wyoming, Laramie. 20 pp.

Jenkins, T. M., Jr., C. R. Feldmeth, and G. V. Elliott. 1970. Feeding of rainbow trout (Salmo gairdneri) in relation to abundance of drifting invertebrates in a mountain stream. J. Fish. Res. Board Can. 27:2356-2361.

Koslucher, D. G. and G. W. Minshall. 1973. Food habits of some benthic invertebrates in a northern cool-desert stream (Deep Creek, Curlew Valley, Idaho-Utah). Trans. Amer. Micros. Soc. 92:441-452.

LeCren, E. D. 1969. Estimates of fish populations and production in small streams in England. pp. 269-280. In: T. G. Northcote (ed.), Symposium on salmon and trout in streams. University of British Columbia, Vancouver, Canada.

Macan, T. T. 1958. Causes and effects of short emergence periods in insects. Verh. Internat. Verein. Limnol. 13:845-849.

Macan, T. T. 1966. The influence of predation on the fauna of a moorland fishpond. Arch. Hydorbiol. 61:432-452.

Macan, T. T. 1977. The influence of predation on the composition of freshwater animal communities. Biol. Rev. 52:45-70.

Mackereth, J. C. 1957. Notes on the Plecoptera from a stony stream. J. Anim. Ecol. 19:159-174.

Mann, K. H. 1965. Energy transformation by a population of fish in the River Thames. J. Anim. Ecol. 34:253-75.

Mann, K. H. 1978. Estimating the food consumption of fish in nature. pp. 250-278. In: S. D. Gerking (ed.), Ecology of freshwater fish production. Blackwell Scientific Publications, Oxford, England.

McCaskill, M. L., J. E. Thomeison, and P. R. Mills. 1972. Food of the northern studfish, Fundulus catenatus, in the Missouri Ozarks. Trans. Amer. Fish Soc. 101:375-377.

Mecom, J. R. 1972. Feeding habits of Trichoptera in a mountain stream. Oikos 23:401-407.

Mendelson, J. 1975. Feeding relationships among species of Notropis (Pisces: Cyprinidae) in a Wisconsin stream. Ecol. Mongr. 45:199-230.

Merritt, R. W. and K. W. Cummins, eds. 1978. An introduction to the aquatic insects of North America. Kendall/Hunt, Dubuque, Iowa. 441 pp.

Metz, J. P. 1974. Die Invertebratendrift an der Oberflache eines Voralpenflusses and ihre selektive Ausnutzung durch die Regenborgen forellen (Salmo gairdneri). Oecologia 14:247-267.

Miller, M. M. 1974. The food of brook trout Salvelinus fontinalis (Mitchell) from different subsections of Lawrence Creek, Wisconsin. Trans Amer. Fish. Soc. 103:130-134.

Moyle, P. B. 1977. In defense of sculpins. Fisheries 2:20-23.

Mundie, J. H. 1969. Ecological implications of the diet of juvenile coho in streams. pp. 135-152. In: T. G. Northcote (ed.), Symposium on salmon and trout in streams. University of British Columbia, Vancouver, Canada.

Mundie, J. H. 1974. Optimmization of the salmonid nursery stream. J. Fish. Res. Board Can. 31:1827-1837.

Novak, J. K. and R. D. Estes. 1974. Summer food habits of the black sculpin, Cottus baileyi, in the Upper South Fork Holston River Drainage. Trans. Amer. Fish. Soc. 103:270-276.

Ordzie, C. J. and D. C. Garofolo. 1980. Behavioral recognition of molluscan and echinoderm predators by the bay scallop Argopecten irradians (Lamarck) at two temperatures. J. Exp. Mar. Biol. Ecol. 43:29-37.

Otto, C. and B. S. Svensson. 1980. The significance of case material selection for the survival of caddis larvae. J. Anim. Ecol. 49:855-866.

Paine, R. T. 1966. Food web complexity and species diversity. Amer. Natur. 100:65-75.

Paine, R. T. 1971. A short-term experimental investigation of resource partitioning in a New Zealand rocky intertidal habitat. Ecology 52:1096-1106.

Paine, R. T. 1980. Food webs: linkage, interaction strength, and community structure. J. Anim. Ecol. 49:667-685.

Peckarsky, B. L. 1980. Predator-prey interactions among
    stoneflies and mayflies: behavioral observations within
    streams. Ecology 61:932-943.

Peckarsky, B. L. and S. I. Dodson. 1980a. Do stonefly predators
    influence benthic distributions in streams? Ecology
    61:1275-1282.

Peckarsky, B. L. and S. I. Dodson. 1980b. An experimental
    analysis of biological factors contributing to stream
    community structure. Ecology 61:1283-1290.

Pentland, E. S. 1930. Controlling factors in the distribution
    of Gammarus. Trans. Amer. Fish. Soc. 60:89-94.

Petrosky, C. E. and T. F. Waters. 1975. Annual production by
    the slimy sculpin population of a small Minnesota trout
    stream. Trans. Amer. Fish. Soc. 104:237-244.

Phillips, D. W. 1978. Chemical mediation of invertebrate
    defensive behaviors and the ability to distinguish between
    foraging and inactive predators. Mar. Biol. 49:237-243.

Pyke, G. H., H. R. Pulliam and E. L. Charnov. 1977. Optimal
    foraging: a selective review of theory and tests. Quart.
    Rev. Biol. 201:1-18.

Radford, D. S. and R. Hartland-Rowe. 1971. Subsurface and
    surface sampling of benthic invertebrates in two streams.
    Limnol. Oceanogr. 16:114-120.

Reice, S. R. 1983. Predation and substratum: factors in lotic
    community structure. In: T. Fontaine and S. Bartell
    (eds.), Dynamics of lotic ecosystems. Ann Arbor Science,
    Ann Arbor, Michigan.

Ricklefs, R. E. 1973. Ecology. Chiron Press, Newton, Mass.
    861 pp.

Ringler, N. H. 1979. Prey selection by drift feeding brown
    trout (Salmo trutta). J. Fish. Res. Board Can. 36:392-403.

Roback, S. S. 1968. Notes on the food of Tanypodinae larvae.
    Ent. News 80:13-18

Schultz, J. C. 1981. Adaptive changes in antipredator behavior
    of grasshopper during development. Evolution 35:175-179.

Schutz, D. C., and T. G. Northcote. 1972. An experimental study
    of feeding behavior and interaction of coastal cutthroat
    trout (Salmo clarki clarki) and Dolly Varden (Salvelinus
    malma). J. Fish. Res. Board Can. 29:555-565.

Sheldon, A. L. 1969. Size relationships of Acroneuria
    california (Perlidae: Plecoptera) and its prey.
    Hydrobiologia 34:85-94.

Sheldon, A. L. 1972. Comparative ecology of Arcynopteryx and
    Diura (Plecoptera) in a California stream. Arch. Hydrobiol.
    69:521-546.

Siegfried, C. A. and A. W. Knight. 1976. Prey selection by a
    setipalpian stonefly nymph, Acroneuria (Calineuria)
    californica Banks (Plecoptera: Perlidae). Ecology
    57:603-608.

Sih, A. 1979. Stability and prey behavioral responses to predator density. J. Anim. Ecol. 48:79-89.

Sih, A. 1980. Optimal behavior: can foragers balance two conflicting demands? Science 210:1041-1043.

Snyder, N. F. R. 1967. An alarm reaction of aquatic gastropods to intraspecific extract. Cornell Univ. Agr. Exp. Stat. Mem. 403.

Snyder, N. F. R. and H. A. Snyder. 1971. Pheromone-mediated behaviour of Fasciolaria tulipa. Anim. Behav. 19:257-268.

Stein, R. A. 1977. Selective predation, optimal foraging, and the predator-prey interaction between fish and crayfish. Ecology 58:1237-1253.

Stein, R. A. and J. J. Magnuson. 1976. Behavioral response of crayfish to a fish predator. Ecology 57:571-761.

Steine, I. 1972. The number and size of drifting nymphs of Ephemeroptera, Chironomidae, and Simuliidae by day and night in the River Stranda, Western Norway. Norw. J. Ent. 19:127-131.

Stewart, K. W., G. P. Friday and R. E. Rhane. 1973. Food habits of hellgrammite larvae, Corydalus cornutus L. (Megaloptera: Corydalidae) in the Brazos River, Texas. Ann. Entomol. Soc. Am. 66:959-963.

Straskraba, M. 1965. The effect of fish on the number of invertebrates in ponds and streams. Mitt. Internat. Verein. Limnol. 13:106-127.

Thompson, D. J. 1975. Towards a predator-prey model incorporating age structure: the effects of predator and prey size on the predation of Daphnia magna by Ischnura elegans. J. Anim. Ecol. 44:907-916.

Thorp, J. H., and E. A. Bergey. 1981. Field experiments on responses of a freshwater benthic macroinvertebrate community to vertebrate predators. Ecology 62:365-375.

Thut, R. N. 1969. Feeding habits of larvae of seven Rhycophila species with notes on other life history features. Ann. Entomol. Soc. Am. 62:894-898.

Tillyard, P. 1926. The insects of Australia and New Zealand. Angus and Robertson Ltd., Sydney. 560 p.

Tippets, W. E. and P. B. Moyle. 1978. Epibenthic feeding by rainbow trout (Salmo gairdneri) in the McCloud River, California. J. Anim. Ecol. 47:549-559.

Townsend, C. F. and A. G. Hildrew. 1977. Predation strategy and resource utilization by Plectrocnemia conspersa (Curtis) (Trichoptera: Polycentropodidae). Proc. 2nd Int. Symp. Trichoptera: 283-291.

Tsui, Ph. T. P. and M. D. Hubbard. 1979. Feeding habits of the predaceous nymphs of Dolania americana in northwestern Florida (Ephemeroptera: Behningiidae). Hydrobiologia 67:119-124.

Vermeij, G. J. 1976. Interoceanic differences in vulnerability of shelled prey to crab predation. Nature (Lond.) 260:136-36.

Vermeij, G. J. 1978. Biogeography and adaptation. Harvard Univ. Press, Cambridge, Mass. 332 pp.

Wallace, J. B. 1975. The larval retreat and food of Arctopsyche; with phylogenetic notes on feeding adaptations in Hydropsychidae larvae (Trichoptera) Ann. Entomol. Soc. Am. 68:167-173.

Wallace, J. B., J. R. Webster and W. R. Woodall. 1972. The role of filter feeders in flowing waters. Arch. Hydrobiol. 79:506-532.

Walton, O. E., Jr., S. R. Reice and R. W. Andrews. 1977. The effects of density, sediment particle size, and velocity on drift of Acroneuria abnormis (Plecoptera). Oikos 28:291-298.

Ware, D. M. 1971. Predation by rainbow trout (Salmo gairdneri): the effect of experience. J. Fish. Res. Board Can. 28:1847-1852.

Ware, D. M. 1972. Predation by rainbow trout (Salmo gairdneri): the influence of hunger, prey density, and prey size. J. Fish. Res. Board Can. 29:1193-1201.

Ware, D. M. 1973. Risk of epibenthic prey to predation by rainbow trout (Salmo gairdneri). J. Fish. Res. Board Can. 37:787-797.

Warren, C. E., J. H. Wales, G. E. Davis, and P. Doudoroff. 1964. Trout production in an experimental stream enriched with sucrose. J. Wildl. Manag. 28:617-660.

Waters, T. F. 1966. Production rate, population density, and drift of a stream invertebrate. Ecology 47:595-604.

Waters, T. F. 1972. The drift of stream insects. Ann. Rev. Entomol. 17:253-272.

Waters, T. F. 1977. Secondary production in inland waters. pp. 91-164. In: A. Macfadyen (ed.), Adv. Ecol. Res., Volume 10. Academic Press, N.Y.

Waters, T. F. 1979. Benthic life histories: summary and future needs. Can. J. Fish. Res. Board Can. 36:342-345.

White, R. J. 1975. Trout population responses to streamflow fluctuation and habitat management in Big Roche-a-Cri Creek, Wisconsin. Verh. Internat. Verein. Limnol. 19:2469-2477.

Williams, D. D. and K. A. Moore. 1982. The effect of environmental factors on the activity of Gammarus pseudolimnaeus (Amphipoda). Hydrobiologia (in press).

Zaret, T. M. 1980. Predation and freshwater communities. Yale University Press, New Haven. 187 pp.

Zelinka, M. 1974. Die Eintagsfliegen (Ephemeroptera) in Forellenbachen der Beskiden. III. Der Einfluss des verschiedenen Fischbestandes. Vestnik Ceskoslovenske spolecnesti Zoologicje 38:76-80.

MACROSCOPIC MODELS OF COMMUNITY ORGANIZATION:

ANALYSES OF DIVERSITY, DOMINANCE, AND STABILITY

IN GUILDS OF PREDACEOUS STREAM INSECTS

D. A. Bruns[1] and G. W. Minshall

Biology Department
Idaho State University
Pocatello, Idaho 83209

INTRODUCTION

   Two approaches have been evident in empirical tests of
theoretical questions in community organization. One approach
focuses on within-community patterns of resource partitioning
between coexisting species. This "microscopic" approach
emphasizes detailed information about the ecology of individual
species relative to specified niche dimensions such as space,
food, or time. The objective here is to concentrate on the ways
species divide resources so as to avoid competition. Thus,
intra-community models of resource partitioning may predict
regular (non-random) spacing of species along a single resource
axis or patterns of niche complementarity whereby high overlap in
one dimension is accompanied by low overlap in another (see e.g.,
Schoener, 1974).

   Alternatively, a "macroscopic" or holistic approach to
community organization emphasizes qualitative relationships
between macroscopic variables that describe whole-system
properties (Levins, 1975, 1977). Variables of concern in this
approach may include species diversity, average niche breadth and
overlap, and the eigenvalues of the community matrix. Here, the
objective is to find predictable patterns among these variables
that reveal underlying ecological relationships. This
macroscopic approach allows for more general treatment of niche
measurement and facilitates comparisons among several different
communities since the focus is no longer on the detailed ecology

[1]Present address: Division of Life Sciences, Geosciences, and
Geography, Sam Houston State University, Huntsville, TX 77341.

of individual species. We have chosen this macroscopic approach
in our analysis of community organization in guilds of predaceous
lotic insects in the Salmon River, Idaho.

The objective of this paper is to outline three related
macroscopic models of community organization from the literature,
and to test these with spatial niche data on predaceous lotic
insects. Competition models of community structure require
assumptions of resource limitation and equilibrium populations
(see Wiens, 1977). As discussed by Bruns (1981), seasonal
changes in species diversity, spatial resource partitioning, and
general resource abundance indicate that these assumptions are
most likely to be met within the upper Salmon River basin in
winter. Therefore, we consider only data from this season to
examine aspects of equilibrium models of community organization
here.

The models that we examine are: 1) MacArthur's (1972)
theory of species diversity and the pattern of resource
partitioning, 2) McNaughton and Wolf's (1970) model of dominance
and the niche in ecological systems, and 3) a cybernetic model of
stability and complexity in ecological communities (see May,
1972; McNaughton, 1978; Lawlor, 1980). Major hypotheses of all
models are outlined first. Then we describe methods employed to
estimate macroscopic spatial niche parameters. And finally, we
present and discuss applications of the lotic niche data to the
various models.

THE MODELS

Species Diversity and Resource Partitioning

MacArthur (1972) developed a theory of species diversity
that is related to the general pattern of resource partitioning
among coexisting species. The essential features of this model
suggest that the number of species in a community or guild can
increase by any of the following: 1) an increase in the range of
resources available to and used by a community, 2) a decrease in
the range of resources used by an average species, and 3) an
increase in the average overlap in resource use between species.

To compensate for non-uniform distribution of resources,
differential abundance among species, and non-uniform utilization
of resources by species, MacArthur (1972) developed the following
equation which summarizes the essential features of the model:

$$D_s = \frac{D_r}{D_u} (1 + C \bar{\alpha}) \tag{1}.$$

D_s is species diversity, D_r is the diversity of resources used by
the whole community, D_u is the diversity of resources used by an
average species, and $\bar{\alpha}$ represents niche overlap between an
average pair of species; C denotes the number of neighbors in
niche space. MacArthur recommended the inverse of Simpson's
index as an appropriate measure of diversity in the above
equation. Thus, an increase in species diversity may be
proportional to the diversity of resources used by the community
or the average degree of niche overlap, and inversely
proportional to the diversity of resources used by an average
species.

Dominance and Niche

An empirical model of dominance and niche in ecological
systems was proposed by McNaughton and Wolf (1970). They were
concerned primarily with development of a community framework
that integrated concepts of dominance, relative abundance,
species niche breadth, and species diversity. These
investigators argued that dominant species, those with high
relative abundance, may so pervade certain ecosystems that they
may effectively modify the ecology of other coexisting species.
Thus, the role of dominance may be an important aspect of
community organization.

McNaughton and Wolf (1970) analyzed the importance of
dominance in community organization through a series of six
related hypotheses. First, from a general perspective, if
dominance does play a primary role in community organization, one
may expect that the species relative abundance distribution would
not fit a lognormal distribution. In a lognormal distribution,
most species demonstrate relative abundances within an
intermediate range with fewer species being relatively rare or
abundant. A fit to a lognormal distribution implies that a
variety of factors are important in determining the abundance of
species (May, 1975). Thus, although rejection of the lognormal
distribution does not of itself indicate that dominance is of
primary importance, a fit to the lognormal would imply that, at
best, dominance is only one of several significant factors.

Given rejection of the lognormal distribution, the second
hypothesis specifically addresses the nature of the deviation
from this distribution. The concept of dominance implies that
only a few species will be extremely abundant at the expense of
the less abundant species. Therefore, relative to a lognormal
distribution, the frequency of species in octaves above the mode
(abundant species) will be impoverished, while the frequency
below the mode (less abundant species) would be enriched (see
Figure 1a). McNaughton and Wolf (1970) examined this pattern of

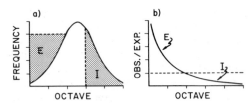

Fig. 1.   a) Generalized lognormal distribution of species
          relative abundances. Observed or empirical data that
          fall within shaded E portion denote enriched octaves
          (below the mode) relative to the lognormal distribution.
          Empirical data that fall within shaded I portion denote
          impoverished octaves (above the mode) relative to the
          lognormal distribution. b) Ratio of observed (N) to
          expected (N_r) frequencies vs. the upper limit of

          relative abundance octave. Solid curve delineates
          regular and predictable deviations from the lognormal
          distribution due to dominance relationships. Dotted
          horizontal line denotes where $N/N_r = 1$ (the mode).

          Points along solid curve below dotted line indicate
          impoverished octaves; those above, enriched octaves.
          Octaves represent relative abundance categories, where
          the highest octave is an abundance class from 50 to
          100%, the next highest, 25 to 50%, then 12.5 to 25%,
          etc. (figure modified after McNaughton and Wolf, 1970).

dominance by correlating the ratio of observed (N) to expected
(N_r) frequencies in a relative abundance octave against the upper

limit of that abundance octave. Their predicted curve
represented a negative logarithmic relationship (see Figure 1b).
The region of the curve below the horizontal line of $N/N_r = 1$

delineates octaves which have a progressively lower than expected
frequency of dominant species while that part of the curve above
the line represents octaves that have an increasingly greater
than expected frequency of rare species.

    The dominance relationships in Figure 1 suggest that
abundant species may be appropriating potential niche space of
certain subordinate species. This would imply that the more
dominant species should exhibit broader niche breadths. To
examine this idea, McNaughton and Wolf (1970) measured dominance
for a species as its relative abundance in the community where it
makes its greatest contribution to community composition. They
suggest that this index of dominance (as opposed to absolute
abundance) reflects ecological efficiency since it measures a
species response under optimal conditions. Thus, the third

hypothesis predicts a positive correlation between this measure of dominance and niche breadth.

The fourth hypothesis deals with the relationship between dominance in a community and position of the community on a habitat gradient. McNaughton and Wolf measured community dominance as the combined relative abundance of the two most prevalent species. Environmental positions of sites were ordered according to site moisture from mesic to dry along a relative scale of 1 to 10. This was intended as a measure of environmental harshness. This component of the model predicts that community dominance should be correlated with the degree of environmental harshness largely because of low diversity in harsh environments.

The last two hypotheses of the dominance model examine the ways in which additional species can be accommodated within a community. McNaughton and Wolf present two alternative strategies: either species specialize more in their use of resources, or the community as a whole expands its niche space. These two predictions of community structure are essentially the same as hypothesis 1 and 2 in the previous model of species diversity and resource partitioning.

Stability and Complexity

The relationship between stability and complexity in ecological communities has been debated for at least the last ten years. The traditional view in ecology, based in part on empirical observation, is that more diverse communities are more stable (e.g., MacArthur, 1955; Elton, 1958; Odum, 1971). In contrast, theoretical considerations of cybernetic models (e.g., Gardner and Ashby, 1970; May, 1972) have shown that greater diversity and complexity reduce stability. However, these models often are analyzed on the basis of randomly generated matrices of communities which may not realistically represent the structure of natural ecosystems (Lawlor, 1980). Thus, as McNaughton (1978) has indicated, more diverse natural communities feasibly may be more stable provided other aspects of community organization are adjusted accordingly.

In this paper, community stability is defined as resilience, or the rate of return of a system to equilibrium after a small perturbation. This aspect of stability typically has been measured in terms of the eigenvalues of the community matrix of niche overlaps in both theoretical (May and MacArthur, 1972; May, 1974) and empirical (Lane et al., 1976; Lawlor, 1980) analyses. In particular, the minimum eigenvalue of the community matrix sets the stability criteria since it represents the theoretical rate at which a community's population would return to

equilibrium levels after a perturbation (see May, 1974). We
followed Lane et al. (1976) and Lawlor (1980) and factored a (-1)
from each element of the community matrices. Thus, stability is
measured in terms of the minimum positive eigenvalue, with larger
values being indicative of more stable systems.

A general theoretical model outlined by May (1972, 1974,
1979) indicates that an increase in any one of three different
components of community complexity may diminish system stability.
These components are the number of species (n), the average
strength of interaction (i, average of non-zero niche overlaps),
and connectance (c, percentage of significant or non-zero niche
overlaps). May summarized these relationships by

$$i \sqrt{(nc)} < 1$$

where a community will be stable if this inequality is satisfied.
Thus, an increase in any one of these three parameters may act to
depress stability.

We have demonstrated qualitative relationships between each
of the above parameters and community stability in Figures 2-4.
These figures show how either species richness, average strength
of interaction, or connectance is predicted to be inversely
related to community stability. Although these results show that
more diverse communities may be less stable (Figure 2), they also

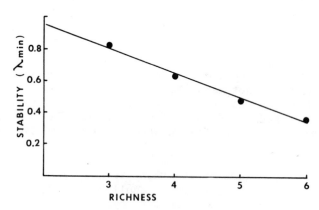

Fig. 2.   Hypothetical relationship between community stability
          (minimum eigenvalue, λmin) and richness based on the
          community matrix model. Average strength of interaction
          (0.50) and connectance (∿0.77) held constant. r = 0.99.

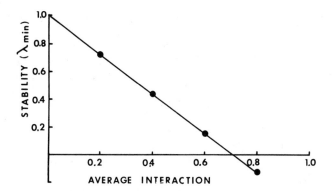

Fig. 3. Hypothetical relationship between community stability (minimum eigenvalue, λmin) and average strength of interaction. Richness (3) and connectance (0.78) held constant. r = 1.00.

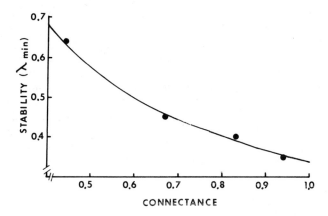

Fig. 4. Hypothetical relationship between community stability (minimum eigenvalue, λmin) and connectance. Richness (6) and average strength of interaction (~0.50) held constant. r = 0.99.

suggest how enriched communities may be structured so that stability is enhanced (Figures 3 and 4). Thus, if the average strength of interaction, or the degree of connectance, or both, progressively decrease in communities with higher numbers of species, they may potentially override the negative effect of high diversity and promote stability. Therefore, biological interaction could act to organize communities in such a way that more diverse communities are actually more stable.

It should be pointed out that the relationships depicted in
Figures 2-4 were not derived from standard numerical techniques
on random matrices (e.g., as in Gardner and Ashby, 1970).
Instead, matrices were systematically constructed so as to keep
two community parameters constant while varying the third. The
minimum eigenvalues were extracted from these matrices (IBM
Scientific Subroutine Package) and plotted as a function of each
aspect of complexity. Thus, the results in Figures 2-4 are
intended only to demonstrate the qualitative behavior of the
community matrix model and they should not be considered as
substitutes for numerical solutions of random communities.

METHODS

Study Area

Five sampling sites were chosen from within the upper basin
of the Salmon River in central Idaho (Figure 5). Terrestrial
vegetation in this area includes Douglas fir (Pseudostuga
menziesii), lodgepole pine (Pinus contorta), Engelmann spruce
(Picea engelmanii), and various species of sagebrush (Artemisia).
However, terrestrial vegetation has an important influence only
on the headwater sites (especially at site I). Although sampling
sites varied in regard to stream size, general habitat features
were relatively similar. Cobble-sized (64-256 mm on the
intermediate axis) substrate predominated at all sites and
current velocities were typically moderate to fast. Pools were
rare and a given reach was characterized by alternate runs and
riffles with occasional rapids. Bruns (1981) provides a fuller
description of the study area and presents information on habitat
characteristics and species composition.

Field Sampling

A variety of methods and techniques were available for
estimation of niche parameters from field data (e.g., see general
work of Whittaker and Levin, 1975). However, since the
perspective of this study is oriented toward analysis of
macroscopic community parameters (e.g., species diversity,
average niche breadth and overlap, etc.), it seemed best to
select a method appropriate to that scale. One such method views
sampling units as "resource states" over which one calculates
niche breadths and overlaps (Colwell anf Futuyma, 1971). Such
units need not be orderable and may differ from one another in
specified or unspecified ways. In this context, a generalist
would be a species that is distributed over many such sampling
units or resource states.

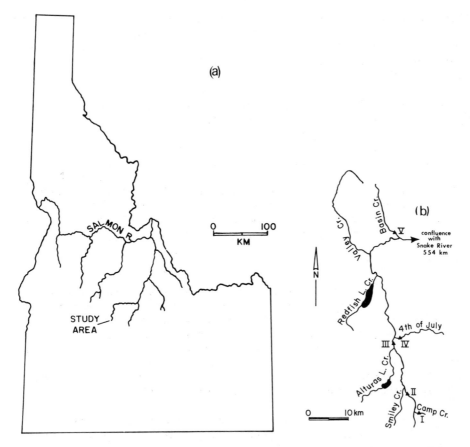

Fig. 5. a) Location of study area in Salmon River basin, Idaho.
b) Location of sampling sites within the upper basin of
the Salmon River. Sites identified by triangles.

One problem with the method, it quadrat sampling is
employed, is the arbitrary area of the sampling unit. To avoid
this, we sampled individual rocks as naturally occurring units of
the environment (Hurlbert, 1978; Yodzis, 1978). Thus, a given
rock sample represents a patch of habitat (Yodzis, 1978), or a
point in the niche hyperspace of a species' population (Hanski
and Koskela, 1977; Hanski, 1978). This technique of calculating
niche measures over specified sampling units (resource states)
has received support from both general discussions of niche
analysis (Colwell and Futuyma, 1971; Allan, 1975a; Hanski, 1978;
Hurlbert, 1978; Hutchinson, 1978; Yodzis, 1978) and particular
field investigations: birds (Cody, 1974), stream insects (Allan,
1975b), and dung-inhabiting beetles (Hanski and Koskela, 1977).
In addition, as suggested by Hutchinson (1978), the technique may
integrate as least several different niche dimensions, and thus

hopefully approximates the spatial niche as a whole. In this
study, substrate size and texture, general current velocity and
microcurrents, and the spatial distribution and abundance of prey
size and taxa would seem likely components of the niche to be
incorporated in the measurements.

At each site, 12 resource states were sampled over a variety
of rock sizes and general current regimes in lotic habitats. In
practice, a dip net (1-mm mesh) was positioned directly
downstream from a rock, the rock was placed into the net, and
sediments immediately below the rock stirred to a depth of about
5 cm (see also Bruns et al., 1982a for use of the technique). On
shore, all macroinvertebrates associated with the sample were
removed from the net and preserved in 5% formalin. The three
major axes (length, width, and depth) of each rock were measured
in order to estimate surface area.

Laboratory Analysis

Aquatic insects were sorted from detritus in the laboratory.
Species identifications were made with the aid of taxonomic keys
by Jensen (1966), Smith (1968), and Baumann et al. (1977).
Species were defined as predators on the basis of gut analyses of
selected taxa, the primary literature (e.g., Thut, 1969; Cather
and Gaufin, 1975; Sheldon, 1972) and past experience with
functional groups of macroinvertebrates from the Salmon River
(e.g., Minshall et al., 1982; Bruns et al., 1982a).

Before calculation of community parameters (see next
section), densities of all species of predators were calculated
on a per m^2 basis for each resource state (rock sample). This
was intended to minimize bias between resource states because of
differently sized rocks.

Calculation of Community Parameters

We employed the following indices to estimate niche
parameters (after Pianka, 1973):

$$\text{niche breadth} = 1/\Sigma p_i^2 \qquad (2)$$

$$\text{niche overlap} = \frac{\Sigma p_{ih} p_{jh}}{(\Sigma p_{ih}^2 \Sigma p_{jh}^2)} \qquad (3)$$

where for niche breadth, p_i is the proportion of individuals of a
given species in the ith resource state (rock sample); for niche
overlap, p_{ih} is the proportion of individuals of species i in the

hth resource state, and p_{jh} is the proportion of individuals of the species j in the hth resource state. Equation (2) (Simpson's index) also was used to estimate species diversity (see Peet, 1974; Pinanka, 1973) and community niche breadth or the diversity of resources used by the whole community (see Pianka, 1973; Lane et al., 1976). For species diversity, p_i is the proportion of all individuals in the ith species. For calculation of community niche breadth, all individuals from all species in the guild are not differentiated as to their species; p_i is simply the proportion of all individuals in the ith resource state.

RESULTS AND DISCUSSION

Species Diversity and the Pattern of Spatial Resource Partitioning

Species diversity was positively correlated with community niche breadth or the diversity of resources used by the whole guild (Figure 6). This same pattern has been observed in lotic guilds of gatherers and grazers (Bruns et al., 1982b) and is comparable to findings of Pianka (1975) in his analyses of lizard communities (see Table 1). Likewise on the basis of morphological quantification of the niche, patterns between total niche space and species diversity have been demonstrated in bat and bird communities (Findley, 1973; Ricklefs, 1979, respectively). This hypothesis also has been examined by correlating species diversity or richness with some measure of environmental productivity or heterogeneity. Thus, Brown (1975) found a significant positive correlation between rodent species diversity and an index of productivity, and Cody (1974) found a significant positive relationship between bird diversity and foliage height diversity (see also MacArthur and MacArthur, 1961; Karr and Roth, 1971; Holmes et al., 1979). Stream ecologists also have shown that aspects of habitat diversity, such as substrate heterogeneity, may enhance lotic macroinvertebrate species diversity (e.g., Allan, 1975b; de March, 1976; Hart, 1978), as well as fish species diversity (Sheldon, 1968).

A second way in which communities can accommodate additional species is through a decrease in average niche breadth. This is related to the concept of niche compression (MacArthur and Wilson, 1967) and niche breadth compensation (Cody, 1975). As new species invade a diverse community, competition may compress the range of habitat patches that an average species can exploit. Conversely, in areas of low diversity, species will tend to

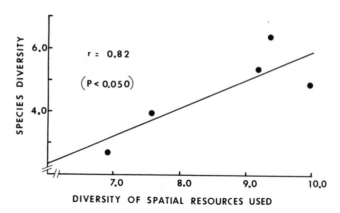

Fig. 6. Species diversity of predaceous stream insects vs.
         community niche breadth or the diversity of spatial
         resources (microhabitat patches) used by the whole
         guild. Both parameters estimated with equation (4),
         Simpson's diversity index; see text.

expand their use of habitat patches and demonstrate a broader
niche breadth. This relationship was not observed in lotic
predator guilds (Figure 7); species diversity was not correlated
with average niche breadth. Nevertheless, such patterns have
been demonstrated in other lotic guilds such as filter feeders,
grazers, and gatherers in summer (Bruns et al., 1982b), and in
various terrestrial communities (see Table 1, e.g., Wiens and
Rotenberry, 1979; Bernstein, 1979). However, in the lotic
guilds, the decrease in average niche breadth with high species
diversity or richness was associated with increased food
abundance. Thus, this pattern appeared to be related more to
resource availability rather than degree of competitive niche
compression.

     Species diversity also can be enhanced in communities where
species on the average overlap more in their exploitation of
habitat patches. In guilds of predaceous stream insects, species
diversity was significantly associated with average niche
overlap. However, these parameters are negatively correlated
(Figure 8), so that increased species packing was not a mechanism
of community species enrichment. Instead, this inverse
relationship fits predictions of "diffuse" competition theory
(Pianka, 1974). Diffuse competition represents competitive
interactions within a constellation of several species
(MacArthur, 1972) as opposed to interactions between just two
species as in classical studies of competition (e.g., Gause,
1934). Pianka (1974) has argued that in more diverse

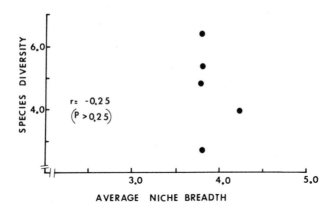

Fig. 7. Species diversity of predaceous stream insects vs.
average spatial niche breadth or the diversity of
microhabitat patches used by an average species. Both
parameters estimated by equation (4), Simpson's
diversity index; see text.

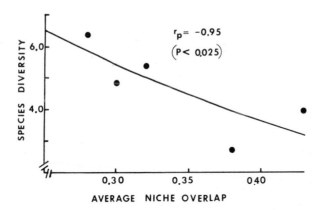

Fig. 8. Species diversity of predaceous stream insects vs.
average spatial niche overlap. Partial correlation with
average niche breadth held constant. Species diversity
estimated by equation (4), Simpson's diversity index;
average niche overlap in microhabitat patch use
estimated by equation (5); see text.

Table 1. Comparison among empirical studies of species diversity (or richness) and the pattern of resource partitioning. Legend: NA = not analyzed; + = positive association; − = negative association; 0 = no relationship.

Investigator	Community	Species Diversity vs. Diversity of Resources Used	Species Diversity vs. Average Niche Breadth	Species Diversity vs. Average Niche Overlap
This study	lotic predators	+	0	−
Bruns et al. (1982b)	lotic gatherers[1]	+	−	0
	lotic grazers[1]	+	−	+
	lotic filter feeders[1]	0	−	+
Pianka (1975)	lizards	+	0	−
Brown (1975)	rodents	+[2]	NA	+
Cody (1974)	birds	+	0	0
Findley (1973)	bats[3]	+	NA	NA
Ricklefs (1979)	birds[3]	+	NA	+
Allan (1975b)	lotic insects	+[4]	NA	NA
de March (1976)	lotic insects	+[4]	NA	NA
Hart (1978)	lotic insects	+[4]	NA	NA

Investigator	Community	Species Diversity vs. Diversity of Resources Used	Species Diversity vs. Average Niche Breadth	Species Diversity vs. Average Niche Overlap
Wiens and Rotenberry (1979)	birds	NA	-	-
Rusterholz (1981)	birds	NA	NA	-
Bernstein (1979)	ants	NA	-	NA
Muhlenberg et al. (1977a)	spiders	+[4]	0	-
M'Closkey (1978)	rodents	NA	NA	-

1: niche parameters based on body sizes
2: measured as resource productivity
3: niche parameters based on morphology
4: measured as habitat heterogeneity

communities, species potentially are exposed to a more intense competitive regime. Thus, given predictions on the upper limits to niche similarity, a higher demand for resources in diverse communities would result in decreased ecological overlap between species pairs. Related evidence of diffuse competition has been documented in communities of lizards, spiders, rodents, and birds (see Table 1).

MacArthur's (1972) theory of species diversity brings into focus the role of biotic factors in community structure. Thus, competitive interactions may affect the manner in which species partition resources. Of the several field studies explicitly aimed at analyzing this model (Cody, 1974; Pianka, 1975; Brown, 1975; Bruns et al., 1982b; see Table 1), our results and conclusions on predaceous lotic insect guilds are most similar to those concerning lizard communities (Pianka, 1975). In both cases, the diversity of resources used by the whole guild or community was positively correlated with species diversity while changes in average niche breadth were unimportant. In particular, diffuse competition played an important role in organizing both sets of communities with species diversity being negatively correlated with average niche overlap. In contrast to these patterns, other guilds of lotic insects are structured along somewhat different lines (Bruns et al., 1982b). In guilds of filter feeders, scrapers, and gatherers, species diversity or richness was negatively correlated with average niche breadth. Also, scraper and filter feeder species richness was positively correlated with average niche overlap. On the other hand, as in predaceous guilds, community niche breadth was positively correlated with diversity and richness for gatherers and scrapers, respectively. Overall, these patterns may reflect differential mechanisms in the organization of various guilds of stream insects. Competition seems more important in the guild structure of predators than in non-predators. For filter feeders, scrapers, and gatherers, changes in average niche breadth were better explained by trends in food abundance rather than competitive interactions. Although this seems a reasonable interpretation, it should be noted that these differences are due at least in part to seasonal effects since predator data are from winter and data on the other guilds are from summer. Nevertheless, as Bruns (1981) has shown, predator species diversity in summer was positively correlated with average niche overlap as in the other lotic guilds. Thus, the manner and mechanism of community organization may vary with season and these should be given careful consideration in future studies (e.g., Wiens, 1977, 1981). Bruns (1981) gives a full account of seasonal changes in community structure of lotic predator guilds along with consideration of alternate theories of species diversity.

Dominance and Niche

In the previous section we demonstrated how competitive interactions affected the way in which species of lotic predators partitioned patches of microhabitat. In this section, the role of competition in community organization is examined from a different perspective. Here, we are concerned primarily with how competitively dominant species may modify abundance relationships and aspects of niche breadth.

Patterns of species relative abundance of lotic predaceous insects agreed with predictions of the dominance and niche model. Relative abundances did not fit a lognormal distribution (Figure 9); data points tend to fall within areas that imply dominance relationships. When deviations from the lognormal are considered in detail, a significant relationship is found (Figure 10). The frequency of abundant species was less than expected and that for rare species was greater than expected: in both cases, the degree of divergence between observed and expected was progressively greater the more rare or abundant the octave class.

The pattern expressed in Figure 10 suggests that the more abundant species may occupy the potential niche space of species that are rare. Thus, the third component of the dominance and niche model predicts that a species' ability to appropriate resources should be reflected in its degree of dominance. This hypothesis was supported by data from predaceous stream insect

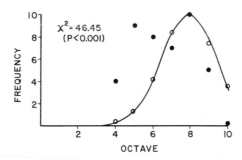

Fig. 9. Observed (solid datum points) frequencies of species relative abundances in guilds of predaceous stream insects compared with frequencies predicted (open circles) by lognormal distribution. Octaves represent relative abundance categories where the highest octave ranges from 50 to 100%, the next highest 25 to 50%, etc.

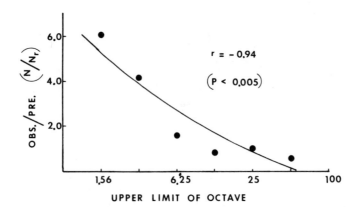

Fig. 10.   Relation between observed (N) and predicted (N_r)
            frequencies in a relative abundance octave and the
            upper limit of that octave (in percentage relative
            abundance). Octaves represent relative abundance
            categories where the highest octave ranges from 50 to
            100%, the next highest 25 to 50%, etc.

guilds: species dominance was positively correlated with niche
breadth (Figure 11). As noted earlier, dominance was measured as
the relative abundance for a species in a community where it made
its largest contribution to community composition. As such, this
estimate of dominance should reflect ecological efficiency since
it reflects optimal conditions. According to the dominance
model, unequal efficiency in the exploitation of resources among
species generates dominance and therefore distorts the lognormal
distribution in a regular and predictable way (Figure 10). Thus,
the relationship demonstrated in Figure 11 indicates that in the
Salmon River dominant species do in fact exploit a wider range of
microhabitat patches than do rare species.

     To test the fourth component of the dominance model, an
independent measure of environmental harshness is needed.
McNaughton and Wolf (1970) used a relative ranking of sites by
soil moisture content. Diel temperature flux or seasonal
variation in discharge may represent potential measures of
harshness in stream systems. However, those data are unavailable
at present for sampling sites of this study. Nevertheless, as a
first step in addressing the potential relationship between
community dominance and environmental harshness, we used a
measure that at least on theoretical grounds should reflect the
relative degree of habitat variability. This measure is the
minimum eigenvalue (min) of the community matrix of niche
overlaps. As discussed earlier, the minimum eigenvalue typically

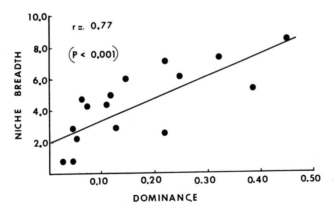

Fig. 11.   Relation between spatial niche breadth and species
           dominance in guilds of predaceous lotic insects. Niche
           breadth or diversity of microhabitat patches used
           estimated by equation (4), Simpson's diversity index;
           species dominance measured as a species maximum
           relative abundance from among all communities; see
           text.

is used in analyses of community stability (see also next
section). In addition, within the context of ecological
stability in a randomly fluctuating environment, May and
MacArthur (1972) suggested that the minimum eigenvalue should be
roughly proportional to the environmental noise level. This
association has been suggested in other guilds of lotic insects
(Bruns et al., 1982b) but has not been indicated for predator
guilds. Assuming such a relationship holds in this case also, we
found a significant positive correlation between community
dominance and the minimum eigenvalue (Figure 12). Although this
finding agrees with the predictions of the dominance model, it
should be viewed with caution. An association between the
minimum eigenvalue of the community matrix and environmental
variability has not yet been tested for lotic predators. Also,
the minimum eigenvalue was related to species richness (see next
section). Since species relative abundances are an important
aspect of diversity (e.g., see Peet, 1974), the pattern shown in
Figure 12 may be fortuitous. Nevertheless, we have presented
these results to demonstrate at least one potential approach to
the problem.

     The last two components of the dominance and niche model
indicate how additional species may be accommodated within
communities. For lotic predators, species richness was unrelated
to average niche breadth ($r = 0.24$, $P > 0.25$). As richness

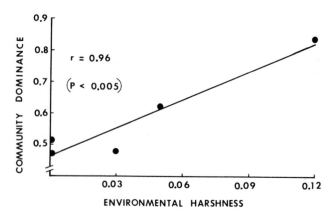

Fig. 12. Community dominance vs. environmental harshness
          (assumed to be approximated by the minimum eigenvalue
          of the community matrix) for guilds of predaceous
          stream insects. Community dominance estimated as
          combined relative abundance of the two most prevalent
          species in a community.

increased within these guilds of predaceous insects, average
niche breadth remained relatively constant. This is essentially
the same finding as when species diversity was correlated with
average niche breadth (see previous section, Figure 7). This is
to be expected given that richness was strongly correlated with
diversity (r = 0.97, P < 0.005). Likewise, species richness was
positively correlated with a measure of the range of resources
used by the entire community (Figure 13). However, in the
dominance and niche model this latter parameter was estimated in
a somewhat different manner than in the previous section. Here,
such a measure was calculated as the sum of the niche breadths of
the component species. McNaughton and Wolf (1970) called this
index "carrying capacity". Since the average niche breadths in
each guild were quite consistent, it is easy to see how this
relationship would result with so little deviation from a fitted
line. If average niche breadths had demonstrated more variation,
or if they had varied inversely with richness, the tight pattern
in Figure 13 may not have been evident or even statistically
significant.

     McNaughton and Wolf (1970) developed their model of
dominance and niche based on empirical relationships within plant
communities. They tested the generality of the model on other
plant communities and on bird communities. Although they
expressed hope that other animal systems would become available
for testing the model, few such studies have followed. Table 2

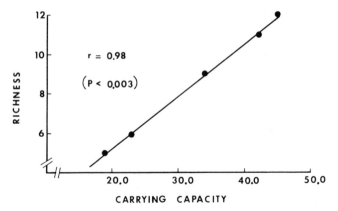

Fig. 13. Species richness of lotic predators vs. "carrying
capacity". Carrying capacity estimated as the sum of
niche breadths of individual species within a guild;
niche breadth estimated by equation (4), Simpson's
diversity index; see text.

compares our results with those of previous investigators. The
last three studies shown in the table do not make reference to
the model, nor do they test the respective hypothesis in terms of
maximum relative abundance (dominance). They were included for
strictly comparative purposes. Overall, the model is strongly
supported by data in the original paper, and by data in this
study on predaceous stream insects. Of those shown, only these
two studies analyze all aspects of the model. Between these two,
all components of the model are supported in the majority of the
cases. The most prevalent findings are the positive relationships between niche breadth and dominance, and between species
richness and carrying capacity. The former relationship, in
conjunction with results from the last three studies shown in
Table 2, indicate that correspondence between niche breadth and
some measure of species abundance is fairly common. If this
pattern continues to hold in a variety of situations, our ability
to test aspects of niche theory with already available data may
be enhanced.

     Data on predaceous stream insects supported the dominance
and niche model over five of its six components. The only
hypothesis that was not supported by data, richness vs. average
niche breadth, did not necessarily negate the model. Instead
this hypothesis tends to complement the one concerned with
richness vs. carrying capacity, and as such represents a
potential alternative in community organization. The relatively
good fit between the model and our data indicates that

Table 2. Comparison among studies of the dominance and niche model (McNaughton and Wolf, 1970). Includes related references (last 3) that do not specifically address the model. Legend: NA = not analyzed; N.S. = not significant; + = significant positive correlation; − = significant negative correlation.

Investigator	Community	Fit to Lognormal	N/N$_r$ vs. Abundance Class	Niche Breadth vs. Dominance	Community Dominance vs. Harshness	Richness vs. Ave. Niche Breadth	Richness vs. Carrying Capacity
This study	lotic insects	No	−	+	+	N.S.	+
McNaughton and Wolf (1970)	trees, grasses, and shrubs	No	−	+	N.S.	NA	NA
	−trees	NA	NA	NA	NA	−	N.S.
	−grasses and shrubs	NA	NA	NA	NA	−	+
	shrub-grass	No	−	+	+	−	+
	birds	Yes	N.S.	+	+	N.S.	+
Ricklefs (1972)	birds	NA	NA	N.S.	NA	NA	NA
Johnson (1977)	bog plants	NA	NA	+	NA	NA	+
Muhlenberg et al. (1977a)	spiders	NA	NA	+[1]	NA	NA	NA

Investigator	Community	Fit to Lognormal	N/N_r vs. Abundance Class	Niche Breadth vs. Dominance	Community Dominance vs. Harshness	Richness vs. Ave. Niche Breadth	Richness vs. Carrying Capacity
Muhlenberg et al. (1977b)	ants	NA	NA	+[2]	NA	NA	NA
Levins (1968)	fruitflies	NA	NA	+[1]	NA	NA	NA

1: tested as niche breadth vs. absolute abundance
2: tested as niche breadth vs. relative abundance

competitive interactions are an important element in shaping
species abundance relationships and the degree of niche
generalization. In conjunction with conclusions from the
previous section, competition seems to play a potentially
important role in the organization of predaceous lotic insect
guilds in winter. In the next section we examine whether these
biotic interactions may have such an influence on the manner in
which lotic guilds are organized so that stability may be
enhanced in more diverse communities.

Stability and Complexity

We partitioned community complexity into the number of
species, degree of connectance, and average strength of
interaction in order to determine whether guilds of predaceous
stream insects were structured in such a way that more diverse
guilds were more stable. To calculate connectance, it is
necessary to know the percentage of significant non-zero overlap
values for a guild. Overlap estimates used in the first section
of this paper were based on equation (3) which gives values that
may range from 0 to 1. However, this method provides no
objective criteria for choosing which measures represent
significant non-zero values. Therefore, we calculated
correlation coefficients between all possible pairs of species
based on raw distributional data within a guild. Spatial
overlaps determined in this way were strongly correlated with
values based on equation (3) (e.g. at site I, $r = 0.97$, $P <
0.001$). Thus, we defined significant non-zero overlap values as
those correlation coefficients where $P < 0.050$. This approach is
analogous to that developed by McNaughton (1978) in a similar
investigation of stability and complexity in plant communities.
Likewise, following McNaughton (1978), we defined average
strength of interaction as the mean of the significant non-zero
spatial overlaps.

Connectance tended to be higher in the more diverse guilds
of predaceous stream insects (Figure 14). However, this pattern
was not statistically significant. In contrast, species richness
was negatively correlated with the average strength of
interaction (Figure 15), a relationship that was significant.
This is to be expected given that species diversity was shown to
be inversely related to average niche overlap in a previous
section, and that average niche overlap (based on equation 3)
itself was correlated ($r = 0.91$, $P < 0.025$) with average strength
of interaction (based on significant correlation coefficients).
Thus, the two measures seem interchangeable even though the
latter is defined in a more restrictive sense.

The above results suggest that predator guilds of stream
insects may be organized in a way that more diverse guilds are

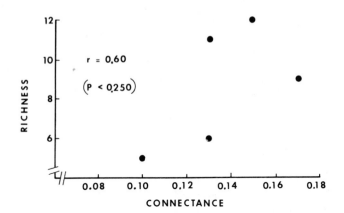

Fig. 14.   Relation between species richness and connectance in
           guilds of predaceous stream insects.   Connectance
           measured as the percentage of significant non-zero
           overlap values in the community matrix; see text.

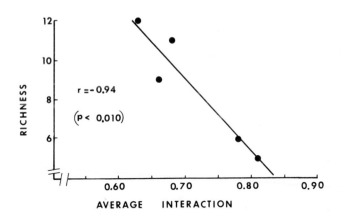

Fig. 15.   Species richness in guilds of lotic insect predators
           vs. average strength of interaction.   Average strength
           of interaction estimated as the mean of significant
           non-zero overlap values in the community matrix; see
           text.

more stable. Connectance was not significantly correlated with
species richness, and thus, it may not act to depress stability
in enriched communities. And since average strength of
interaction actually decreased with species richness, it should
enhance the stability of diverse predator guilds. To test
whether these points are plausible, we correlated species
richness against the minimum eigenvalues of the community
matrices (based on the original set of niche overlap values) and
found that theoretical stability (resilience) was inversely
related to predator richness (Figure 16). This indicates that
high connectance may have been a partial factor negatively
affecting stability conditions within the more diverse guilds.
In addition, the above result suggests that species richness
increased across these guilds at a rate higher than that for the
decrease in average strength of interaction. Closer scrutiny of
the data supports this latter point: overall, species richness
increased by a factor of almost 2.5 while average strength of
interaction decreased only by a factor of 1.3.

The results of this study agree with predictions of
theoretical models (e.g., May, 1974, 1979) that more diverse
guilds are less stable than those with fewer species. Thus, it
appears that given some kind of minor environmental perturbation,
populations within diverse guilds of predaceous stream insects
would take longer to return to their previous structural
equilibrium levels than those within species poor guilds.

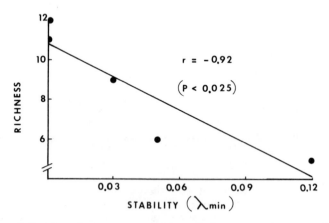

Fig. 16. Relation between species richness and community
          stability (minimum eigenvalue, λmin) in guilds of
          predaceous lotic insects.

Few attempts have been made to examine the
stability-complexity question with empirical field data although
the problem has received considerable attention in the
theoretical literature (e.g., see May, 1974, 1979). McNaughton's
(1978) study of plant communities was a notable exception, but
unfortunately his work was flawed by methodological sampling
problems (Harris, 1979; Lawton and Rallison, 1979). Recently,
Lawlor (1980) has re-examined empirical data on bird (Cody, 1974)
and lizard (Pianka, 1967) communities. In both cases, species
richness was negatively associated with the minimum eigenvalues
of community matrices (see Table 3). However, stability in these
natural communities was generally higher than that in analogous
randomized ones. Lawlor attributed this to underlying biological
organization in real communities, perhaps shaped by competitive
processes. In a more recent study, Bruns et al. (1982b) have
analyzed various aspects of the stability and complexity problem
for guilds of lotic gatherers, filter feeders, and grazers (see
Table 3). No patterns were evident for gatherers. However, for
both filter feeders and grazers, species richness was positively
associated with average strength of interaction. Additionally,
filter feeder species richness was positively correlated with
connectance. Thus, the significant inverse relationships between
richness and stability in guilds of filter feeders and grazers is
expected based on the above patterns and previously discussed
theoretical considerations. Overall, the most important and
consistent pattern established in Table 3 is that more diverse
natural communities seem to be less structurally resilient.
Thus, despite the significant ecological structure of real
communities, for example as demonstrated in the various sections
of this paper, empirical studies to date indicate that less
diverse communities should be more stable.

Although the above conclusion seems consistently supported
by empirical data, several points of precaution are in order.
First, problems of sampling technique may influence results in
unknown and unexpected ways (e.g., McNaughton, 1978; see Harris,
1979; Lawton and Rallison, 1979), even though we have tried to
minimize these by avoiding point correlation analyses of nearest
neighbor data. Second, imposition of the community matrix model
on empirical data may affect the outcome of the analyses to some
unknown degree (Lawlor, 1980). And third, as Lawlor (1980) has
indicated, elements of the community matrix derived from field
data reflect ecological similarities and are not necessarily true
competition coefficients.

Nevertheless, we agree with Lawlor's contention that it is
still valuable to examine overlap or similarity matrices as a

Table 3. Comparison among empirical studies of stability-complexity relationships. Legend: NA = not analyzed; NS = not significant; + = significant positive correlation; - = significant negative correlation.

Investigator	Community	Richness vs. Connectance	Richness vs. Average Strength of Interaction	Richness vs. Min. Eigenvalue (Stability)
This study	lotic predators	NS	-	-
Bruns et al. (1982b)	lotic gatherers	NS	NS	NS
	lotic filter feeders	+	+	-
	lotic grazers	NS	+	-
Lawlor (1980)	birds	NA	NA	-
	lizards	NA	NA	-

first, although rough, approximation to the question of community stability. Such an approach seems reasonable at this time given the proliferation of theoretical studies on the subject and the apparent absence of related empirical investigations (however, see e.g., McNaughton, 1977).

SUMMARY AND CONCLUSIONS

Three macroscopic models of community organization were analyzed with spatial niche data for guilds of predaceous stream insects. Our objectives were to determine the feasibility of using data on stream insects to test general ecological theory, and to delineate the contribution of theory to further understanding of lotic insect communities. The following models were analyzed: 1) MacArthur's (1972) theory of species diversity and resource partitioning, 2) dominance and niche (McNaughton and Wolf, 1970), and 3) stability and complexity in ecological communities (May, 1974).

In regard to the first model, species diversity was positively correlated with community niche breadth, negatively correlated with average niche overlap, and unrelated to average niche breadth. The relationship between diversity and overlap supported predictions of diffuse competition theory.

The importance of competitive interactions also was implicated in data supporting the dominance and niche model. Relative abundances of predatory species did not fit a lognormal distribution: high abundance octaves were less than predicted while low abundance octaves were enriched. In addition, spatial niche breadth of a species was positively correlated with dominance. Collectively, these relationships indicate that dominant species appropriate niche space of rare species due to differential efficiency of resource exploitation. Thus, competition may affect both patterns of relative abundance and niche specialization within predator communities.

To examine the relationship between stability (resilience) and the pattern of community organization, complexity was partitioned into three components: species richness, average strength of interaction, and connectance. Richness was correlated inversely with average strength of interaction, but was unrelated to connectance. This suggests that more diverse guilds of predators may be more stable. However, richness increased across communities more rapidly than average strength of interaction decreased. Thus, a measure of community stability (minimum eigenvalue of the community matrix) was negatively correlated with species richness, a result that sustains predictions of theoretical models.

The results of this study lead to the conclusion that the use of lotic insect data can contribute positively to evaluation of general ecological theory. This approach should provide an important dimension to an overall conceptual perspective on ecological communities since evaluation of theory depends on empirical tests over a broad range of populations, communities, and environmental conditions. However, a better conceptual foundation also should add to understanding of biotic processes in stream ecosystems. For example, the importance of competition has been analyzed extensively in terrestrial and marine communities (see reviews in Cody and Diamond, 1975; May, 1976; Anderson et al., 1979). In contrast, competition in stream systems has been addressed only recently (e.g., Hildrew and Edington, 1979; Wiley, 1981). This study provides additional support for expansion of stream community ecology to include notions of ecological interaction, especially at a macroscopic level of focus.

ACKNOWLEDGMENTS

We would like to thank Drs. J.E. Anderson, M.L. Cody, J.S. Griffith, and S.J. McNaughton for their critical reviews of the manuscript. R.T. Litke provided Figure 5a.

"The work upon which this publication is based was supported partially by Grant Number DEB-7811671 from the Ecosystems Program of the National Science Foundation. This is Contribution No. 23 from the NSF River Continuum Project."

REFERENCES

Allan, J. D. 1975a. Components of diversity. Oecologia (Berl.) 18:359-367.

Allan, J. D. 1975b. The distributional ecology and diversity of benthic insects in Cement Creek, Colorado. Ecology 56:1040-1053.

Anderson, R. M., B. D. Turner, and L. R. Taylor, eds. 1979. Population dynamics. Blackwell Scientific, Oxford.

Baumann, R. W., A. R. Gaufin, R. F. Surdick. 1977. The stoneflies (Plecoptera) of the Rocky Mountains. Mem. Amer. Ent. Soc. No. 31.

Bernstein, R. A. 1979. Evolution of niche breadth in populations of ants. Am. Nat. 114:533-544.

Brown, J. H. 1975. Geographical ecology of desert rodents, pp. 315-341. In: M. L. Cody, and J. M. Diamond (eds.), Ecology and evolution of communities. Harvard University Press, Cambridge, Mass.

Bruns, D. A. 1981. Species diversity and spatial niche relations in guilds of predaceous stream insects. Doctoral dissertation. Idaho State University, Pocatello, Idaho.

Bruns, D. A., G. W. Minshall, J. T. Brock, C. E. Cushing, K. W.
    Cummins, and R. L. Vannote. 1982a. Ordination of
    functional groups and organic matter parameters from the
    Middle Fork of the Salmon River, Idaho. Freshw. Invertebr.
    Biol. 1:2-12.
Bruns, D. A., G. W. Minshall, A. B. Hale, and T. L. La Point.
    1982b. Stability and organization of lotic insect commun-
    ities over habitat templets along a river continuum. Idaho
    State University typescript.
Cather, M. R. and A. R. Gaufin. 1975. Life history of Megarcys
    signata (Plecoptera: Perlodidae), Mill Creek, Wasatch
    Mountains, Utah. Great Basin Nat. 35:39-48.
Cody, M. L. 1974. Competition and the structure of bird
    communities. Princeton University Press, Princeton, N. J.
Cody, M. L. 1975. Towards a theory of continental species
    diversities: bird distribution over Mediterranean habitat
    gradients, pp. 214-257. In: M. L. Cody and J. M. Diamond
    (eds.), Ecology and evolution of communities. Harvard
    University Press, Cambridge, Mass.
Cody, M. L., and J. M. Diamond, eds. 1975. Ecology and
    evolution of communities. Harvard University Press,
    Cambridge, Mass.
Colwell, R. K. and D. J. Futuyma. 1971. On the measurement of
    niche breadth and overlap. Ecology 52:567-576.
de March, B. G. E. 1976. Spatial and temporal patterns in
    microbenthic stream diversity. J. Fish. Res. Board Can.
    33:1261-1270.
Elton, C. S. 1958. The ecology of invasions by animals and
    plants. Methuen, London.
Findley, J. S. 1973. Phenetic packing as a measure of faunal
    diversity. Am. Nat. 107:580-584.
Gardner, M. R., and W. R. Ashby. 1970. Connectance of large
    dynamical (cybernetic) systems: critical values for
    stability. Nature (Lond.) 228:784.
Gause, G. F. 1934. The struggle for existence. Williams and
    Wilkins, Baltimore.
Hanski, I. 1978. Some comments on the measurement of niche
    metrics. Ecology 59:168-174.
Hanski, I. and H. Koskela. 1977. Niche relations among
    dung-inhabiting beetles. Oecologia (Berl.) 28:203-231.
Harris, J. R. W. 1979. Evidence for species guilds is an
    artifact. Nature (Lond.) 279:350-351.
Hart, D. D. 1978. Diversity in stream insects: regulation by
    rock size and microspatial complexity. Verh. Internat.
    Verein. Limnol. 20:1376-1381.
Hildrew, A. G. and J. L. Edington. 1979. Factors facilitating
    the coexistence of hydropsychid caddis larvae (Trichoptera)
    in the same river system. J. Anim. Ecol. 48:557-576.

Holmes, R. T., R. E. Bonney, Jr., and S. W. Pacala. 1979. Guild structure of the Hubbard Brook bird community: a multivariate approach. Ecology 60:512-520.

Hurlbert, S. H. 1978. The measurement of niche overlap and some relatives. Ecology 59:67-77.

Hutchinson, G. E. 1978. An introduction to population ecology. Yale University Press, New Haven and London.

Jensen, S. L. 1966. The mayflies of Idaho (Ephemeroptera). Master's Thesis, University of Utah, Salt Lake City.

Johnson, E. A. 1977. A multivariate analysis of the niches of plant populations in raised bogs. II. Niche width and overlap. Can. J. Bot 55:1211-1220.

Karr, J. R. and R. R. Roth. 1971. Vegetation structure and avian diversity in several new world areas. Am. Nat. 105:423-435.

Lane, P. A., G. H. Lauff, and R. Levins. 1976. The feasibility of using a holistic approach in ecosystem analysis, pp. 111-128. In: S. A. Levin (ed.), Ecosystem analysis and prediction. 2nd edition. Proceedings of a SIAM-SIMS Conference held at Alta, Utah, July 1-5, 1974.

Lawlor, L. R. 1980. Structure and stability in natural and randomly constructed competitive communities. Am. Nat. 116:394-408.

Lawton, J. H., and S. P. Rallison. 1979. Stability and diversity in grassland communities. Nature (Lond.) 279:351.

Levins, R. 1968. Evolution in changing environments. Princeton University Press.

Levins, R. 1975. The search for the macroscopic in ecosystems, pp. 213-222. In: G. S. Innis (ed.), Simulation Councils Proceedings Series. Vol. 5. No. 2.

Levins, R. 1977. Qualitative analysis of complex systems, pp. 152-199. In: D. E. Matthews (ed.), Mathematics and the life sciences, Proceedings 1975. Lecture Notes in biomathematics. Vol. 18. Springer-Verlag, New York.

MacArthur, R. H. 1955. Fluctuations of animal populations, and a measure of community stability. Ecology 36:533-536.

MacArthur, R. H. 1972. Geographical ecology. Harper and Row, New York.

MacArthur, R. H. and J. W. MacArthur. 1961. On bird species diversity. Ecology 42:594-598.

MacArthur, R. H. and E. O. Wilson. 1967. The theory of island biogeography. Princeton University Press, Princeton, N.J.

May, R. M. 1972. Will a large complex system be stable? Nature (Lond.) 238:413-414.

May, R. M. 1974. Stability and complexity in model ecosystems. 2nd Edition. Princeton University Press, Princeton, N. J.

May, R. M. 1975. Patterns of species abundance and diversity, pp. 81-120. In: M. L. Cody and J. M. Diamond (eds.), Ecology and evolution of communities. Harvard University Press, Cambridge, Mass.

May, R. M., ed. 1976. Theoretical ecology: Principles and applications. Saunders, Philadelphia.

May, R. M. 1979. The structure and dynamics of ecological communities, pp. 385-407. In: R. M. Anderson, B. D. Turner, and L. R. Taylor (eds.), Population dynamics. Blackwell Scientific, Melbourne.

May, R. M., and R. H. MacArthur. 1972. Niche overlap as a function of environmental variability. Proc. Natl. Acad. Sci. USA 69:1109-1113.

M'Closkey, R. T. 1978. Niche separation and assembly in four species of Sonoran desert rodents. Am. Nat. 112:683-694.

McNaughton, S. J. 1977. Diversity and stability of ecological communities: a comment on the role of empiricism in ecology. Am. Nat. 111:515-525.

McNaughton, S. J. 1978. Stability and diversity in ecological communities. Nature (Lond.) 274:251-253.

McNaughton, S. J., and L. L. Wolf. 1970. Dominance and the niche in ecological systems. Science 167:131-139.

Minshall, G. W., J. T. Brock, and T. W. La Point 1982. Characterization and dynamics of benthic organic matter and invertebrate functional feeding group relationships in the Upper Salmon River, Idaho (U.S.A.). Int. Rev. ges Hydrobiol. 67:793-820.

Muhlenberg, M., D. Leipold, H. J. Mader, and B. Steinhauer. 1977a. Island ecology of arthropods. I. Diversity, niches, and resources on some Seychelles Islands. Oecologia (Berl.) 29:117-134.

Muhlenberg, M., D. Leipold, H. J. Mader, and B. Steinhauer. 1977b. Island ecology of arthropods. II. Niches and relative abundances of Seychelles ants (Formicidae) in different habitats. Oecologia (Berl.) 29:135-144.

Odum, E. P. 1971. Fundamentals of ecology. Saunders, Philadelphia.

Peet, R. K. 1974. The measurement of species diversity. Ann. Rev. Ecol. Syst. 5:285-307.

Pianka, E. R. 1967. Lizard species diversity. Ecology 48:333-351.

Pianka, E. R. 1973. The structure of lizard communities. Ann. Rev. Ecol. Syst. 4:53-74.

Pianka, E. R. 1974. Niche overlap and diffusion competition. Proc. Natl. Acad. Sci. USA 71:2141-2145.

Pianka, E. R. 1975. Niche relations of desert lizards, pp. 292-314. In: M. L. Cody, and J. M. Diamond (eds.), Ecology and evolution of communities. Harvard University Press, Cambridge, Mass.

Ricklefs, R. E. 1972. Dominance and the niche in bird
    communities. Am. Nat. 106:538-545.
Ricklefs, R. E. 1979. Ecology. 2nd Edition. Chiron Press,
    Portland, Ore.
Rusterholz, K. A. 1981. Niche overlap among foliage-gleaning
    birds: support of Pianka's niche overlap hypothesis. Am.
    Nat. 117:395-399.
Schoener, T. W. 1974. Resource partitioning in ecological
    communities. Science 185:27-39.
Sheldon, A. L. 1968. Species diversity and longitudinal
    succession in stream fishes. Ecology 49:193-198.
Sheldon, A. L. 1972. Comparative ecology of Arcynopteryx and
    Diura in a California stream. Arch. Hydrobiol. 69:521-546.
Smith, S. D. 1968. The Rhyacophila of the Salmon River drainage
    of Idaho with special reference to larvae. Ann. Ent. Soc.
    Amer. 61:655-674.
Thut, R. N. 1969. Feeding habits of larvae of seven Rhyacophila
    species with notes on other life history features. Ann.
    Ent. Soc. Amer. 62:894-898.
Whittaker, R. H., and S. A. Levin, eds. 1975. Niche, theory and
    application. Benchmark papers in Ecology, Vol. 3. Dowden,
    Hutchinson and Ross, Inc. Stroudsburg, Pennsylvania.
Weins, J. A. 1977. On competition and variable environments.
    Am. Sci. 65:590-597.
Weins, J. A. 1981. Single-sample surveys of communities: are
    the revealed patterns real? Am. Nat. 117:90-98.
Wiens, J. A., and J. T. Rotenberry. 1979. Diet niche
    relationship among North American grassland and shrubsteppe
    birds. Oecologia (Berl.) 42:253-292.
Wiley, M. J. 1981. Interacting influences of density and
    preference on the emigration rates of some lotic chironomid
    larvae (Diptera: Chironomidae). Ecology 62:426-438.
Yodzis, P. 1978. Competition for space and the structure of
    ecological communities. Lecture Notes in Biomathematics.
    Vol. 25. Springer-Verlag, New York.

INSECT SPECIES DIVERSITY AS A FUNCTION OF ENVIRONMENTAL

VARIABILITY AND DISTURBANCE IN STREAM SYSTEMS

Jack A. Stanford and James V. Ward

University of Montana Biological Station, Bigfork,
Montana 59911; Department of Zoology and Entomology,
Colorado State University, Fort Collins, Colorado 80523

INTRODUCTION

Thienemann (1954) observed that the number of species
inhabiting a given locality is highest in areas which have
persisted largely unchanged for a long time and are characterized
by diverse and predictable (i.e., not deviating substantially
from the long-term norm) environmental conditions. The rhithron
(i.e., the habitat of cold, rubble-bottom streams often found in
the middle and upper reaches of river systems) epitomizes the
ideal running water environment for co-existence of many species
(Illies, 1969), most of which are usually insects (Hynes, 1970).

Diversity patterns, which encompass both species richness
(i.e., the number of different species) and evenness (i.e., the
relative abundance of species), may be considered in the context
of two different time frames. On the one hand, diversity is a
manifestation of adaptive radiation in a slowly changing, but
predictable, environment from some original genetic stock. That
is, speciation within a particular fauna is a consequence of
biogeographical patterns and natural selection over long time
periods (Illies, 1969). On the other hand, unpredictable
disturbances superimposed upon predictable enviromental
conditions (e.g., temperature, nutrients) are short-term events,
which may also structure diversity patterns (Huston, 1979).

In this paper we discuss species diversity in stream insects
as a function of environmental heterogeneity along natural and
man-induced disturbance gradients. This is done largely from a
theoretical perspective, although ecological and zoogeographical
data from lotic stonefly (Plecoptera) assemblages will be

presented in support of our ideas. We then discuss a system-level framework for the mechanisms controlling diversity of insects in streams.

ENVIRONMENTAL VARIABILITY VERSUS DISTURBANCE

The attributes, or driving components, of any ecological system are characterized by some amount of temporal variability. Some streams (e.g., springbrooks) are characterized by very little annual variability, while in others (e.g., 5th-6th order rivers) system attributes, such as temperature and discharge, follow predictable, high-amplitude cycles. The former situation might be considered one of constancy, while the latter is a variable system. However, under natural conditions, both are very predictable over a period of 50-100 years. Aquatic insects have evolved different life history strategies for dealing with natural variations in their environment (e.g., see Ward and Stanford, 1982 for a review of temperature adaptations) and, hence, predictable diversity patterns should be expected along a gradient from conditions of constancy to some degree of variability. Occasionally disturbances occur which upset environmental predictability; any stochastic event which forces normal system environmental conditions substantially away from the mean is a convenient definition of disturbance. For example, a stream might be disturbed by an unusually heavy late-summer thunderstorm and resultant flooding, that causes departure from the normal discharge regime. Or, a fire in the drainage basin removes significant amounts of canopy, opening up a formerly shaded stream segment to solar insolation and, thus, altering the thermal regime. Severity of disturbance includes both frequency (or timing) and duration. Obviously, there exists some point at which the disturbance is severe enough to completely alter the system and attendant diversity patterns (Fig. 1). In other words, if the disturbance is severe enough to substantially reduce population size of one or more species, it becomes an unpredictable consequence; whereas, mild disturbances merely reflect normal (i.e., year-to-year) environmental variability.

Species diversity in natural systems seems to correlate with temporal and spatial environmental heterogeneity (c.f., Slobodkin and Sanders, 1969; Levin and Paine, 1974; MacArthur, 1975; Horwitz, 1978; but see McLeod et al., 1981). Attempts to explain this observed correlation have produced literature which may be synthesized as follows: the environment is a dynamic, but generally predictable, "templet for ecological strategies" (Southwood, 1977) that is partitioned by competitive interactions, given a set of "founder" species (Levins, 1968); over time the system progresses toward a dynamic equilibrium between species richness and evenness and the availability of

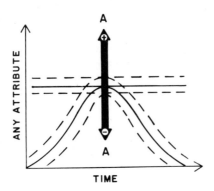

Fig. 1. Biophysical attributes that conceptually describe stream
         systems over time may vary from linear (i.e., constant)
         to periodic (i.e., cyclic), as shown by solid lines.
         Within short time intervals (ca. a few years or less)
         these attributes vary predictably; that is, deviation
         from mean amplitude is small, as shown by the broken
         lines. Disturbance is shown as a positive or negative
         vector (bold, solid line) that introduces unpredictable
         variance into the dynamic expression of system
         attributes. If severe enough, disturbance can force
         expression of any attribute toward system breakpoint
         (A), which corresponds to the boundary of ecosystem
         stability. Any feedback mechanism which prevents the
         disturbance effect from manifesting beyond A may be
         thought of as a resilience vector.

resources (including environmental variability as a resource).
However, most communities, including those in streams, do not
exist at equilibrium because population response is slow,
relative to rates of environmental change, and populations are
periodically reduced (or locally eliminated) by their
interactions (e.g., predation) and by unpredictable environmental
disturbance (Connell, 1978). Also, different populations
(species) in a community do not grow at the same rate. Huston
(1979) proposed that a stable level of diversity may be primarily
maintained by a dynamic balance between the rate of competitive
displacement and periodic population reductions. Hence,
"diversity is determined not so much by the relative competitive
abilities as by the influence of the environment on the net
outcome of their interactions" (Huston, 1979). Ward and Stanford
(1983a) argued that intermediate disturbances may determine
diversity levels in streams and that many natural streams are in
fact perturbed to the extent that competitive equilibrium cannot
be approached. However, we did not distinguish between environmental variability (predictable) and disturbance (unpredictable).

MEASURING ENVIRONMENTAL VARIABILITY AND DISTURBANCE

If we assume for the moment that insect diversity in streams is correlated with temporal environmental heterogeneity, it is necessary to select a variable, or set of variables, which measures the relative abiotic constancy and predictability (sensu Colwell, 1974) of the environment in a temporal sense. This would require data collection over a long enough period of time to accurately deduce just how predictable the variance of that particular variable(s) actually is, as well as the frequency and duration of disturbance.

Certain criteria should be considered when selecting a system attribute as a measure of environmental variability in streams:

1) it should be a major factor of biological, especially energetic, significance to lotic insect species;

2) it should be a universal variable which can be compared across geographical boundaries;

3) it should directly influence all biotic components, unlike nutrients, which directly influence only certain biotic components;

4) it should exhibit a wide range of values over time for both natural and altered stream systems; and

5) it should be easily and accurately quantified and yield data that will be comparable to existing records.

Perhaps the most useful variables which meet these conditions are temperature and discharge. Temperature is considered a driving variable in directing biotic responses in streams (Ward and Stanford, 1982), since insect energetics are closely tied to thermal patterns (see, for example, Brown and Fitzpatrick, 1978; Sweeney, 1978; Vannote and Sweeney, 1980). Discharge also directly affects lotic biota, in addition to influencing temperatures (see Ward and Stanford, 1979). In general, temperature and discharge cycles recur with predictable seasonal amplitude year after year, especially in rhithron streams. Disturbances are usually evident in time-series plots of thermal and discharge data as abnormal peaks or troughs, which skew the annual mean away from the long-term mean. Colwell's measures may be used to quantify the fluctuating pattern and relative predictability of such data (Stearns, 1981). Perturbations could be quantitatively identified in long time-series data by removing peak values in step-wise fashion until the probability statistic (P) maximizes.

Vannote et al. (1980) have conceptualized stream systems as continua that change in predictable ways from headwaters to river mouths. Well-canopied headwater springbrooks are rather constant environments; environmental conditions in middle reaches are highly variable; constancy again prevails in the higher orders. The River Continuum Concept remains to be thoroughly tested; however, temperature and discharge are attributes that seem to vary in a generally predictable manner along the continuum (Figs. 2A and B). Figure 2C, which is a simultaneous plot of annual variance in discharge and temperature, implies that the greatest amount of ecological information for these two variables occurs in the middle reaches of the stream continuum (i.e., stream orders 5 to 7). Therefore, it is not surprising that Vannote et al. (1980) predict that species diversity will maximize within the middle reaches of the stream continuum.

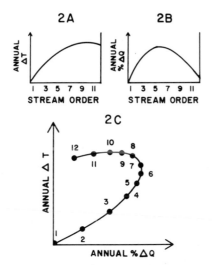

Fig. 2.   Conceptual relationships between annual variation (Δ) in temperature (T) and discharge (Q) as a function of stream order. In 2A and B, T and Q are plotted against the stream size or order. In 2C, the T and Q functions are combined over stream orders 1-12. Whereas T indicates absolute temperature change, % Q represents the percent change in discharge annually (i.e., $\frac{\text{Min Q}}{\text{Max Q}}$ x 100) within each stream order.

A pattern of generalized environmental variability emerges if different stream systems are compared in terms of the T, Q relationship (Fig. 3). A distinction is made between higher order streams in the tropics versus those in more northern latitudes, because temperature would be more constant in large, slow-moving rivers nearer the Equator. Even in the tropics, however, we expect that maximum predictable heterogeneity of environmental variables will occur in the middle reaches of river systems. Regardless of latitude, it is these same stream segments that also receive intermediate levels of disturbance.

DIVERSITY ALONG ENVIRONMENTAL GRADIENTS

River Continua

The way in which diversity pattern changes within a particular stream continuum has been measured in the Flathead River, Montana. The tributaries drain high mountains in Glacier National Park and various wilderness areas and flow through glacial valleys to Flathead Lake (see map in Hauer and Stanford, 1981). Except that alpine streams are present, the biophysical attributes of this system conform to the River Continuum Concept (Table 1). Lowland springbrooks (1_b, Table 1 and Fig. 4) are constant environments, water temperatures and flows being buffered by residence time in the ground; whereas, exposed alpine streams (1_a, Table 1 and Fig. 4) are extremely rigorous habitats,

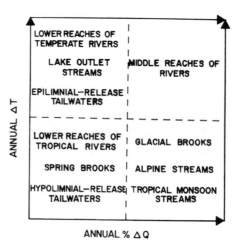

Fig. 3. Generalized relationships between annual variation (Δ) in temperature (T) and discharge (Q) for various types of streams. (No meaning is associated with placement of various stream types within a particular quadrant.)

Table 1. Biophysical attributes measured along the Flathead River continuum.

Parameter	Stream Size (order)					
	$1_{(a)}$	$1_{(b)}$	2-3	4	5	$6_{(c)}$
Mean Annual Degree Days	1070	2200	1300-1500	1900	2400	2050-2600
Annual ΔT (°C)	24	0.5	10.5	12.5	17	16-20
P/R ($X_{(n)}$)	--	$.81_{(6)}$	$.99_{(9)}$	$1.30_{(10)}$	$1.93_{(8)}$	$1.45_{(8)}$
NCPP*	--	133	178	768	607	1063
Number of Plecoptera Species	24	13	29	27	40	42

(a) tundra (alpine) stream.

(b) lowland springbrook.

(c) regulated segment (see text).

(n) number of community metabolism analyses.

* average net community primary productivity (mgC/day).

due to vagaries of climate at the high altitudes. Stream orders 2-3 are predominantly canopied, high-gradient streams. Fourth order streams are uncanopied and feed three river tributaries (5th order) on expansive valley floors. The mainstream river (6th order) is partially regulated by hypolimnial discharges from a high dam on the lower reaches of the South Fork tributary.

The plecoptera fauna will be used to examine the premises set forth in this paper. The results of nearly 30 years of collection effort and taxonomic analyses are available over all stream orders in the Flathead River system (Stanford and Gaufin, unpubl.). Due to differences in origins, the stoneflies, as a group, may not necessarily reflect overall patterns of diversity in streams. However, lotic Trichoptera and Ephemeroptera are

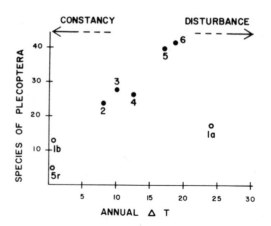

Fig. 4.   Relationship between annual temperature variation and
          the number of Plecoptera species in eight stream
          segments (orders 1-6) of the Flathead River, Montana.
          Arabic numerals designate stream order; 5_r is a
          tailwater segment below a hypolimnial release dam, 1_a is
          an exposed alpine stream, and 1_b is a springbrook.
          (See text for details.)   •

being studied in detail and follow the same general patterns
described below for the Plecoptera (Stanford et al., unpubl.).
Also, comprehensive data on Plecoptera diversity patterns are
available in other areas of the Holarctic region and in the
Neotropics.

     Stonefly diversity in the Flathead is greatest in the 6th
order segment and is a function of annual variance in
temperature, which is one measure of disturbance in a stream
continuum (Fig. 4).   Intermediate levels of predictable
variability occur at 5th-6th order, whereas the environment
characterizing headwater streams is either more rigorous (i.e.,
in alpine streams, Table 1 and Fig. 4) or more constant (i.e., in
springbrooks, Table 1 and Fig. 4).   We have argued elsewhere
(Ward and Stanford, 1983a) that constant environments (i.e.,
tailwater segments and lowland springbrooks) and extremely
variable environments (i.e., exposed, alpine streams) represent
two forms of disturbance.   This idea is implicit in the River
Continuum Concept (Vannote et al., 1980) and supported here by
the tremendous reduction of species in the tailwaters of the
hypolimnial release dam (5_r, Fig. 4).   The South Fork tailwater
reflects conditions of constancy, despite short-term flow
fluctuations induced by hydropower generation.   Enhanced thermal
constancy (5°C ± 3°) results from releasing water from deep

reservoir strata; seasonal flow constancy is a function of the controlled discharge regime. The dam has eliminated the scouring action and bedload movement normally caused by spring runoff; clearwater discharges have stabilized substrata, permitting development of a thick, uniform growth of periphytic algae. The tailwater may be thought of as severely disturbed in the sense of deviation from conditions in the unregulated tributaries. Flows from the North and Middle Forks ameliorate effects of regulation in the mainstream and, thus, the mainstream is only moderately disturbed. Even though spatial habitat heterogeneity probably is greatest in the mainstream (see Stanford and Gaufin, 1974), the intermediate disturbance of regulation over 20 years may well explain the higher richness. Population reductions are related to flow fluctuations (i.e., sudden reduced flows during cold weather have been observed to freeze thousands of teneral capniids; Stanford, 1975), which could influence inter- and intra-specific population dynamics enough to allow additional species to coexist. In fact, two Utacapnia species are found in the North Fork, while two other Utacapnia species are present in the South and Middle Forks (unregulated segments); all four Utacapnia species exist in fairly high numbers in the partially regulated mainstream segment downstream from the confluence of these tributaries.

Latitudinal Gradients in the Rocky Mountains

The Flathead system is located in an intermediate area relative to environmental variability across the latitudinal gradient. The Rocky Mountains traverse from the Yukon River (Yukon Territory-Alaska) on the north to the Rio Grande River (New Mexico) on the south. Yukon streams have smaller thermal amplitudes than do streams in the middle Rocky Mountains; and, streams in the southern Rockies have higher summer temperatures compared to Rocky Mountain streams in Montana and Wyoming. Because of the more xeric climate, streams in the southern Rockies are less well-buffered from environmental vagaries by terrestrial (riparian) vegetation than the more mesic Flathead system. Also, disturbance-level climatic events may be considered intermediate in the Flathead. Whereas the Yukon and New Mexico areas are predominately influenced by extreme cold and hot air masses respectively, conditions are moderated in the middle Rockies.

Parts of three major continental river systems originate in Glacier National Park: the Columbia (via the Flathead River); the Saskatchewan (via the St. Mary's River) and the Missouri-Mississippi (via the Milk River). The biogeography of stoneflies in the Flathead River system, therefore, is influenced by historical stocks contained in each basin. Pleistocene ice apparently persisted in Glacier National Park well into recent

times, therefore harboring endemic species as well as gradually
opening up migrational pathways from three different directions.
However, the present species composition of Flathead stoneflies
is ca. 95 percent Rocky Mountain; only a few Pacific coastal or
Canadian Shield forms are present.

Baumann et al. (1979) reported 151 Plecoptera species in the
Rocky Mountains from the Yukon to New Mexico. Sixty-two percent
(94 species) of the Rocky Mountain fauna are known to occur in
the Flathead River system. Eleven additional species are
reported from adjacent rivers (within one county distant). Thus,
we believe 105 species or 70 percent of the Rocky Mountain fauna
actually exist in the Flathead. Only four of these species
apparently are eurythermal enough to spatially overlap more than
three stream orders in the Flathead River continuum, which
indicates the degree to which stoneflies have specialized in this
river system. K. W. Stewart presently is documenting taxonomy
and distribution of stoneflies in Alaska and he believes no more
than ca. 35-40 species exist in the entire state (per. comm.).
Likewise, Ward (1982) collected only 33 species of stoneflies
over an altitudinal gradient of nearly 2000 m in the southern
Rocky Mountains. Thus, it appears that Plecoptera species
richness within the Rocky Mountains maximizes in the Flathead
region.

Temperate Versus Tropical Latitudes

The temporal constancy of the tropics has long been held by
competition theorists as the explanation for higher biotic
diversity compared to that in more temperate latitudes. Recently
this has been challenged. Wolda (1978) showed that the range in
annual fluctuations of various environmental variables (e.g.,
rainfall, food resources) was at least as great as in selected
temperate regions; insect diversity may be a function of such
environmental fluctuations. Furthermore, Weins (1977) has argued
that, although patterns do exist in species inter-relationships
and community organization, "these patterns may be produced by
species-specific pre-adaptions to different resources such as
competition, predation, physical disturbance or re-current, but
predictable enviromental fluctuations." This implies that one
should look for periodic phenomena which may influence diversity
patterns, even in tropical situations.

Stout and Vandermeer (1975) concluded that species richness
of insects is higher in tropical streams. They compared
species-sample curves derived from the number of insect species
clinging to rocks in tropical versus temperate streams. They
found both altitudinal and seasonal trends in both data sets but
reasoned that competitive interactions largely explained observed
patterns. However, Illies (1969) found the Neotropical stonefly

fauna to be surprisingly depauperate (i.e., 68 species in South America versus 340 in Europe) even though rhithron streams can be found throughout the middle latitudes. Although low diversity in South America may be attributable to historical zoogeographical phenomena, Illies did find that the "normal" complement of feeding groups were represented in the Neotropical stonefly fauna of rhithron streams.

Light-trapping of Trichoptera by McElravy et al. (1981) near the Rio Chiriqui, Panama, over several years indicated that, although the caddisfly fauna was generally more diverse than most of the comparison sites in the Holarctic region, species richness was similar to patterns observed in streams in South Carolina and Pennsylvania. The tropical diversity pattern also followed a typically "seasonal" trend. All of this indicates to us: 1) that species diversity in tropical stream insects may well be largely influenced by patterns of environmental variability and periodic, unpredictable disturbance in the same manner as discussed above for more temperate regions, and 2) that diversity of all or most lotic forms may actually be lower in the tropics compared to optimal habitats in more northerly latitudes.

CONCLUSIONS

Most of the ideas presented are theoretical and based in part on largely untested concepts such as the Stream Continuum Concept (Vannote et al., 1980), Huston's (1977) generalized hypothesis of species diversity, the Thermal Equilibria Model of Vannote and Sweeney (1980) and Connell's (1978) Intermediate Disturbance Hypothesis. Yet, an interesting framework for studying stream insects emerges: predictable environmental variability (particularly temperature) may grossly determine the number of species present in a given stream segment, while biotic interactions, such as predation, may fine tune richness and actively promote evenness. Intermediate levels of disturbance, which introduce optimal environmental variability into stream systems, preclude the attainment of competitive equilibria, thus allowing coexistence of a greater total number of species and fewer dominant species.

This framework or hypothesis of insect diversity in streams is testable. Streams regulated by dams, particularly those with hypolimnial releases, are essentially experimentally manipulated systems. As we have previously suggested (Ward and Stanford, 1983b), with increasing distance downstream from a dam, energy flows will reset and correspondingly predictable diversity patterns should be manifest in the context of the river continuum. These patterns should be most clearly defined in the macrobenthos, which is often dominated by insects. The propensity of the system to retain predictability as variability

increases downstream from the constant conditions of the
tailwater segment represents an environmental gradient in which
predation and other biotic responses can be evaluated as the
system as a whole recovers spatially.

ACKNOWLEDGMENT

Dr. Stanford was supported in part by Grant No. R008223-01-0
from the U.S. Environmental Protection Agency during preparation
of this paper.

REFERENCES

Brown, A. V. and L. C. Fitzpatrick. 1978. Life history and
    population energetics of the dobson fly, Cordydalus
    cornutus. Ecology 59:1091-1108.
Colwell, R. K. 1974. Predictability, constancy, and contingency
    of periodic phenomena. Ecology 55: 1148-1153.
Connell, J. H. 1978. Diversity in tropical rain forests and
    coral reefs. Science 199: 1302-1310.
Hauer, F. R. and J. A. Stanford. 1981. Larval specialization
    and phenotypic variation in Arctopsyche grandis
    (Trichoptera: Hydropsychidae). Ecology 62:645-653.
Horwitz, R. J. 1978. Temporal variability patterns and the
    distributional patterns of stream fishes. Ecol. Monogr. 48:
    307-321.
Huston, M. 1979. A general hypothesis of species diversity.
    Amer. Nat. 113: 81-101.
Hynes, H. B. N. 1970. The ecology of stream insects. Ann. Rev.
    Entomol. 15:25-42.
Illies, J. 1969. Biogeography and ecology of neotropical
    freshwater insects, especially those from running waters.
    pp. 685-708. In: Fittkau, E. J. et al. (eds.), Biogeography
    and ecology in South America. Dr. W. Junk. The Hague.
Levin, S. A. and R. T. Paine. 1974. Disturbance, patch
    formation and community structure. Proc. Nat. Acad. Sci.
    71: 2744-2747.
Levins, R. 1968. Evolution in changing environments. Princeton
    Univ. Press, Princeton.
MacArthur, J. W. 1975. Environmental fluctuations and species
    diversity, pp. 74-80. In: Cody, M. L. and J. H. Diamond
    (eds.), Ecology and evolution of communities. Belknap
    Press, Cambridge, Mass.
McElravy, E. P., V. H. Resh, H. Wolda and O. S. Flint, Jr. 1981.
    Diversity, seasonality and annual variability of caddisflies
    from a "non-seasonal" tropical environment, pp. 149-156.
    In: Moretti, G. P. (ed.), Proc. Third Internat. Symp.

Trichoptera. Series Entomologica, Vol. 20. Dr. W. Junk. The Hague.

McLeod, M. J., D. J. Hornbach, S. I. Guttman, C. M. Way and A. J. Burky. 1981. Environmental heterogeneity, genetic polymorphism, and reproductive strategies. Amer. Nat. 118: 129-134.

Slobodkin, L. B. and K. H. L. Sanders. 1969. On the contribution of environmental predictability to species diversity, pp. 82-95. In: Diversity and stability in ecological systems. Brookhaven National Laboratory. Upton, New York.

Southwood, T. R. E. 1977. Habitat, the templet for ecological strategies? J. Anim. Ecol. 46: 337-366.

Stanford, J. A. and A. R. Gaufin. 1974. Hyporheic communities of two Montana rivers. Science 185: 700-702.

Stanford, J. A. and J. V. Ward. 1981. Preliminary interpretations of the distribution of Hydropsychidae in a regulated river, p. 323-328. In: Moretti, G. P. (ed.), Proc. Third Internat. Symp. Trichoptera. Series Entomologica, Vol. 20. Dr. W. Junk. The Hague.

Stearns, S. C. 1981. On measuring fluctuating environments: predictability, constancy, and contingency. Ecology 62: 185-199.

Stout, J. and J. Vandermeer. 1975. Comparison of species richness for stream-inhabiting insects in tropical and mid-latitude streams. Amer. Nat. 109: 263-280.

Sweeney, B. W. 1978. Bioenergetic and developmental response of a mayfly to thermal variation. Limnol. Oceanogr. 23: 461-477.

Thienemann, A. 1954. Ein drittes biozonotische Grundprinzip. Arch. Hydrobiol. 49: 421-422.

Vannote, R. L., G. W. Minshall, K. W. Cummins, J. R. Sedell, and C. E. Cushing. 1980. The river continuum concept. Can. J. Fish. Aquat. Sci. 37:130-137.

Vannote, R. L. and B. W. Sweeney. 1980. Geographic analysis of thermal equilibria, a conceptual model for evaluating the effect of natural and modified thermal regimes on aquatic insect communities. Am. Nat. 115: 667-695.

Ward, J. W. 1982. Altitudinal zonation of Plecoptera in a Rocky Mountain stream. Aq. Insects (in press).

Ward, J. W. and J. A. Stanford. 1979. Ecological factors controlling stream zoobenthos with emphasis on thermal modification of regulated streams, pp. 35-56. In: Ward, J. V. and J. A. Stanford (eds.), The ecology of regulated streams, Plenum Press, New York.

Ward, J. V. and J. A. Stanford. 1982. Thermal responses in the evolutionary ecology of aquatic insects. Ann. Rev. Entomol. 27: 91-117.

Ward, J. V. and J. A. Stanford. 1983a. The intermediate
    disturbance hypothesis: an explanation for biotic diversity
    patterns in lotic ecosystems. In: Fontaine, T. D. and S. M.
    Bartell (eds.), The dynamics of lotic ecosystems. Ann Arbor
    Science, Ann Arbor, Michigan.
Ward, J. V. and J. A. Stanford. 1983b. The serial discontinuity
    concept of lotic ecosystems. In: Fontaine, T. D. and S. M.
    Bartell (eds.), The dynamics of lotic ecosystems. Ann Arbor
    Science, Ann Arbor, Michigan.
Weins, J. A. 1977. On competition and variable environments.
    Amer. Sci. 65:590-597.
Wolda, H. 1978. Seasonal fluctuations in rainfall, food, and
    abundance of tropical insects. J. Anim. Ecol. 47:369-381.

APPLICATION OF ISLAND BIOGEOGRAPHIC THEORY TO STREAMS:

MACROINVERTEBRATE RECOLONIZATION OF THE TETON RIVER, IDAHO

G. Wayne Minshall, Douglas A. Andrews and
C. Yvonne Manuel-Faler

Department of Biology
Idaho State University
Pocatello, Idaho 83209

INTRODUCTION

The Teton River is located in southeastern Idaho and drains the west slope of the Rocky Mountains southwest of Yellowstone National Park (Fig. 1a). In the mid-1970's a dam was constructed near the lower end of the river. The dam stood 95 m high and, during its initial filling, retained a volume of 3.10×10^8 m^3. One June 5, 1976 the dam failed and released 70% of its volume within 2½ h. The resultant wall of water reached 23 m in height near the dam and further downstream, where the water left the canyon in which the dam was located and spread out across the flood plain, it was 15 m high. Discharges near our eventual study area jumped from 30 to 3500 m^3/s and velocities exceeded 12 m/s (Ray and Kjelstrom, 1978). About 9.4 km below the dam site, the Teton River splits into North and South Forks (Fig. 1a). The North Fork was severely damaged by the flood but the adjacent South Fork sustained much less structural upheaval. In order to facilitate channel restoration and bridge replacement, water was diverted from the North Fork in August and the streambed remained dry until mid-December.

Failure of the Teton Dam and subsequent dewatering of the channel in the North Fork provided an excellent opportunity to study long-term recolonization of the stream bottom and to examine the results in terms of Island Biogeographic theory. We studied recolonization by macroinvertebrates at four locations within the North Fork and at one site in the main river (Fig. 1b). Station 00 remained wetted the entire time; all the

279

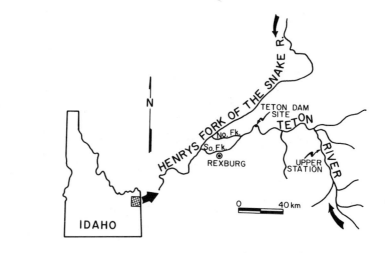

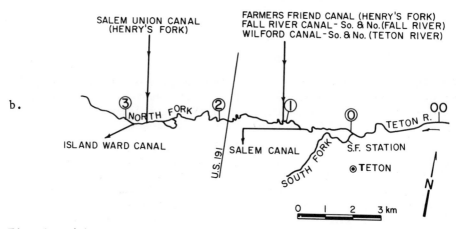

Fig. 1.   (a) Location of the Teton River in Eastern Idaho and
          (b) details of the study section including the location
          of the sampling station.

remainder were dry for several months. Irrigation withdrawal
above station 1 caused that site to dry up periodically even
after normal flow was restored to the North Fork. But irrigation
return flows above stations 2 and 3 kept them from a similar
fate. Sampling began 4 d after return of water to the North Fork
and continued through 1296 d.

     A number of investigators have examined colonization of
substrates by stream macroinvertebrates. In most cases these

have involved relatively short periods (days to weeks) and
restricted universes in which the initially uncolonized areas
were immediately adjacent to the source of colonizers (Waters,
1964; Dickson and Cairns, 1972; Nilsen and Larimore, 1973;
Ulfstrand et al., 1974; Stauffer et al., 1975; Townsend and
Hildrew, 1976; Khalaf and Tachet, 1977; Sheldon, 1977; Meir et
al., 1979; Trush, 1979; Shaw and Minshall, 1980). In only a few
cases have studies of colonization in streams exceeded a couple
of months and involved substantial areas (Crisp and Gledhill,
1970; Williams and Hynes, 1977; Gore, 1979, 1982). The present
study is of particular interest because it involves a period
(3½ y) and a distance (several km) much greater than previously
investigated.

The species equilibrium model of MacArthur and Wilson (1963,
1967) was derived to account for patterns of species diversity on
relatively homogenous habitat patches ("islands") separated by
dissimilar intervening areas. Originally the model was conceived
in terms of oceanic islands, although an analogy between true
geographic islands and continental habitat islands was
recognized. Subsequently, its applicability has been tested on
habitats which are insular in nature, ranging from individual
rocks (Osman, 1979) and flower blossoms (Siefert, 1975) to
mountain tops (Vuilleumier, 1970; Brown, 1971). In flowing water
systems, various aspects of the species equilibrium model have
been tested with respect to the colonization of artificial
substrates (Dickson and Cairns, 1972; Stauffer et al., 1975),
distribution among coastal rivers (Sepkoski and Rex, 1974), and
colonization of newly constructed stream channels (Williams and
Hynes, 1977; Gore, 1979, 1982). Groups of freshwater organisms
which have been examined in regard to the theory include diatoms
(Patrick, 1967), protozoans (Cairns et al., 1969; Cairns and
Ruthven, 1970), macroinvertebrates (Culver, 1970; Dickson and
Cairns, 1972; Stauffer et al., 1975; Stout and Vandermeer, 1975;
Williams and Hynes, 1977; Gore, 1979, 1982; Trush, 1979), snails
(Lassen, 1975), mussels (Sepkoski and Rex, 1974), and fish
(Barbour and Brown, 1974).

The attractiveness of the species equilibrium theory lies in
its ability to recognize the dynamic nature of ecological systems
and to predict changes in diversity (Osman, 1978). MacArthur and
Wilson (1963, 1967) suggested that changes in the number of
species in a community over time result from interaction between
an extrinsic immigration process that adds species and an
intrinsic extinction process in which species are eliminated by
ecological processes and chance. Basically, the model states
that the rate of establishment of new species will be high early
in colonization and will decrease geometrically with time;
extinction is expected to act conversely. Equilibrium
(stability) will be achieved when the immigration rate (I) is
equal to the extinction rate (E).

In terms of the theory, anything that affects immigration
and extinction (area, distance from source, etc.) will affect
equilibrium. Consequently, a number of predictions can be made
about immigration, extinction, and the resultant species
equilibrium including: (1) the more distant a habitat patch is
from the source of colonization (or the lower its I for any other
reason), the fewer the species that will be found there; (2)
reduction of the species pool of immigrants will reduce the
number of species; and (3) the smaller the size of the habitat
patch (or the greater its E for any other reason), the fewer the
species. Research on streams which supports these various
predictions (P_i) includes: P_1-Gore, 1979, 1982; P_2-Patrick,
1967; P_3-Patrick, 1967; Sepkoski and Rex, 1974; Hart, 1978;
Trush, 1979.

Applications of the model, it if can be shown to hold for
streams, include (1) determination of sample size (species-area
curves), (2) standardization of sample size (e.g., Stout and
Vandermeer, 1975) or frequency (see Waters, 1964 for description
of the problem), (3) estimation of immigration (i.e., drift)
rates (Minshall et al., 1982b), and (4) prediction or evaluation
of rate or extent of recovery from perturbation. The present
study provides general support for the MacArthur-Wilson model in
regard to both temporal and spatial patterns of immigration and
extinction, and uses the information to evaluate recovery of the
Teton River from catastrophic flood and other subsequent
disturbances of the stream.

RESULTS AND DISCUSSION

Abundance, Richness, and Diversity

Samples of macroinvertebrates were collected in triplicate
to a depth of about 12 cm with a 625-cm^2 Hess sampler (Waters and
Knapp, 1961) fitted with a 390-um nylon monofilament net. Total
numbers (Fig. 2) appeared to stabilize by the end of 439 d,
although considerable variation occurred over the remainder of
the study, probably due to seasonal effects. Published values of
the length of time required for the attainment of equilibrium
density range from 14-21 d (Khalaf and Tachet, 1977; Sheldon,
1977) to 32-64 d Shaw and Minshall (1980) or more (Williams and
Hynes, 1977; Gore, 1982). But, collectively the results of these
and the present study support the contention that rate of
recovery is related to the extent of the uncolonized area and its
proximity to a source of colonizers. The mainstream Teton
station responded most rapidly and supported the highest density
by the end of the year. Station 3 was colonized much later than

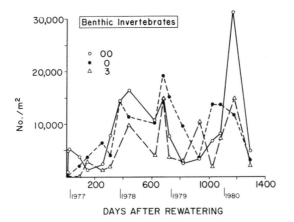

Fig. 2.   Total numbers of benthic invertebrates collected from
three of the Teton River study sites.

the other sites, at a time coinciding with the start of the
irrigation season, and probably was aided by recruitment from
irrigation return flows.   Rates of recovery at stations 0 and 3
correspond in a general way to their relative distance
downstream, but stations 1 and 2 were out of sequence and much
lower in abundance (see e.g., Fig. 10), indicating local
disturbances at both of those sites.   The ranking at the end .of
the study is the same as that at the end of the first year
although considerable variation occurred in the interim.   Much of
this temporal variation probably was due to seasonal fluctuations
in recruitment, mortality, immigration, and emergence which
obscured trends due to long-term recolonization (Osman, 1978).
Recovery at station 1 appears to have been held back by recurrent
drying up of the stream due to irrigation withdrawal.   The reason
for the depressed values at station 2 is unknown but could be
release of toxic substances (such as farm-related pesticides)
buried in the stream in the aftermath of the flood or to the fact
that, with station 1 dry part of the time, station 2 received
fewer immigrants.

The number of taxa present on a given date (= "colonization"; Osman, 1978) (Fig. 3) and Shannon-Weiner diversity (Fig.
4) showed patterns similar to that of total abundance.
Generally, values were inversely correlated with distance
downstream.   Initially, the downstream stations were colonized by
five or fewer taxa.   Data indicate there was a latent period
during which the number of taxa remained about the same (Table
1).   For station 0 this period lasted through day 16 whereas
at stations 1 and 2 it extended through days 148 and 310,

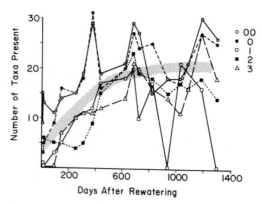

Fig. 3. Number of benthic invertebrate taxa present in the Teton
River on each sampling date. The shaded curve is a
general trend line fitted by eye.

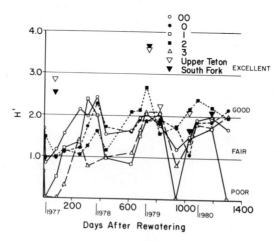

Fig. 4. Shannon-Weiner diversities (H') of the benthic
invertebrate fauna following the Teton flood.

Table 1. Changes in the number of invertebrate taxa present on individual sampling dates.

Station	Days before no. of taxa ≥5	Regression of no. of taxa on time		No. of taxa at equilibrium
		slope	r^2	
00	0	0.028	0.59	23
0	16 – 93	0.040	0.56	21
1	148 – 254	0.037	0.94	17
2	310 – 375	0.048	0.93	18
3	93 – 148	0.022	0.75	15

respectively. After this initial period, numbers of taxa increased at a linear rate that ranged from 0.022 to 0.049 taxa per day for about 300 d and then leveled off. This pattern is similar to that reported by Gore (1979) in that for any given date, numbers of taxa are generally greater at the upstream stations. it differs in that the slopes of the regression line during the period of increase in taxa are not correlated with distance downstream. H' values for the Teton River are much lower than those for unaffected streams in the area (e.g., Minshall, 1981), indicating that total recovery of this aspect of the community had not occurred even after 3½ y. This continued low diversity may reflect decreased habitat heterogeneity associated with the flood. The values are similar to those found on relatively homogeneous artificial substrates (Dickson and Cairns, 1972; Stauffer et al., 1975,; Meier et al., 1979).

Cumulative number of taxa over time (Fig. 5) showed a curvilinear relationship. The asymptote was reached around 700 d, which is about 300 d later than that for total abundance, richness, and species diversity. Also, the values on any date generally correspond to respective distances below the source of colonizers (except at station 1). This supports the predictions that colonization rates will vary with distance from the "mainland" and that more-distant islands will support fewer species than those nearer the source of colonization (MacArthur and Wilson, 1963, 1967). However, in streams the "distance effect" may break down once maximum densities have been reached, since with time the source of colonization will progressively shift downstream and the distances between "mainland" and "islands" will decrease (ultimately to almost 0) (Gore, 1982).

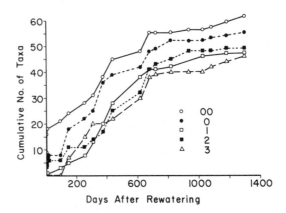

Fig. 5. Cumulative changes in species richness of benthic
         invertebrates in the Teton River.

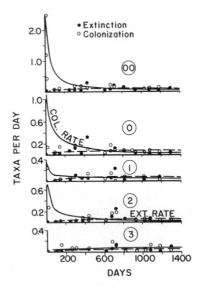

Fig. 6. Colonization rates by new taxa (immigration) and
         extinction rates for benthic invertebrates in the Teton
         River. Rates were calculated as described by Williams
         and Hynes (1977).

Colonization rates for new taxa (= "immigration"; Osman, 1978) and extinction rates for the Teton River, expressed as number of taxa per day (Fig. 6), also are in agreement with theoretical considerations (MacArthur and Wilson, 1967; Simberloff, 1969). Immigration rates at most of the Teton River sites did not level off until around day 220 or later, at least 100 d later than estimated by Williams and Hynes (1977) and Gore (1979) for newly constructed stream channels. This difference probably is due to the larger size of the initially uncolonized reach in the Teton River (10's of km vs approx. 400-750 m).

Equilibrium is the point where the curves for immigration and extinction cross and about where the number of species in an area will be stabilized (MacArthur and Wilson, 1967) (Fig. 6) although a slight additional decline may occur due to biological interactive effects (Simberloff, 1969). In the Teton River this equilibrium occurred between 600 and 650 d at all sites except station 3. The intercept of the two curves moved to the left with increased distance from the source of colonization as expected (MacArthur and Wilson, 1967; Simberloff, 1969). Even though equilibrium appears to have been attained by ca 625 d, the number of taxa present (Fig. 3) did not fully stabilize. Dickson and Cairns (1972) found a similar situation for macroinvertebrates colonizing an artificial substrate and attributed it to sampling variability. Although this certainly is possible in the case of the Teton River data as well, it is just as likely due to well-known seasonal variations in life cycles and annual variations in carrying capacity (see e.g., Minshall, 1981). The situation at station 3 suggests continual extinction and reinvasion of species over extended periods and failure of the community to stabilize.

Extinction rates at equilibrium are equivalent to turnover rates (Simberloff and Wilson, 1969; Brown and Kodric-Brown, 1977). Turnover rates for the Teton River were roughly 0.02 taxa/d and did not vary appreciably with distance from the source. Turnover rate for stream invertebrates was 1.35 taxa/d at 42 d for floating artificial substrates (Dickson and Cairns, 1972) and about 0.10 at 109 d for a newly formed stream channel (Williams and Hynes, 1977), indicating that the values for the Teton River are quite low. The continued low immigration and extinction rates at all sites following establishment of equilibrium, suggests that stream habitats fit the MacArthur-Wilson model even though they are inhabited mainly by seasonally recurring temporary inhabitants. This conclusion indicates that the lack of equilibrium found by Williams and Hynes (1977) at 373 d probably was due to incomplete recovery of the stream rather than to the inapplicability of the model.

In general, our data differ from the findings of other investigators concerning the length of time until colonization is completed (e.g., Townsend and Hildrew, 1976; Shaw and Minshall, 1980). Those studies indicate recolonization times of around 60 d or less, whereas in the Teton River in excess of a year was required. However, in those cases where much shorter recovery times were involved, the denuded areas were much smaller and were situated much closer to the source of colonists. Colonization in the Teton River is consistent with the magnitude of the disturbance and suggest that communities in large streams have a relatively high degree of resilience. Also, when taken collectively, the existing studies on colonization in streams further support the prediction that rate of colonization or recovery will vary directly with distance from the source of colonizers.

Functional Feeding Group Abundance

The Teton River invertebrates were placed into functional feeding groups (Cummins, 1973), based on our experience with the fauna in other Idaho streams and published records (Merritt and Cummins, 1978). In general, grazers predominated during the early stages of recolonization and were replaced by collectors after 4 to 6 months (Fig. 7). But at station 3, where no colonization occurred for the first five months, collectors always accounted for the majority of invertebrates. Shredders were never abundant. The number of grazers was loosely associated with algal standing crop (Fig. 8). The lack of grazers at station 3 during the early months was associated with low algal biomass, but their subsequent build-up there did not keep pace with algal growth. In a similar manner, the number of collectors tended to increase with the buildup of detritus (Fig. 9) (although the relationship was even less clear than that between grazers and chlorophyll). On the other hand, there was no comparable relationship between the number of potential prey and predators (Fig. 10). The long lag in the establishment of the predators is a striking aspect of the trophic structure data. Gore (1982) reported similar trends during the 500 d of colonization covered by his study. Collectors and grazers arrived early (with increase in detrital accumulations), and predators arrived very late in the process. The lack of a really close association between functional feeding group abundance and standing crops of potential foods, as in the present study, is not uncommon (e.g., Hawkins and Sedell, 1981; Minshall et al., 1982a) and may be due to a number of factors including high sample variance and the fact that other variables besides food are affecting the community.

The method of Heatwold and Levins (1972) was used to examine "return to equilibrium trophic structure" in the Teton River

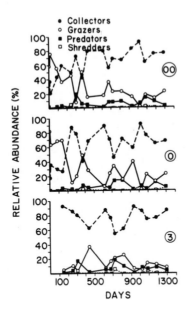

Fig. 7. Relative abundance (%) of benthic invertebrate
         functional feeding groups in the Teton River.

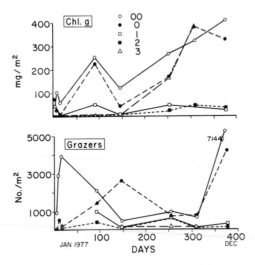

Fig. 8. Chlorophyll a content of the periphyton and abundance of
         benthic invertebrate grazers in the Teton River.

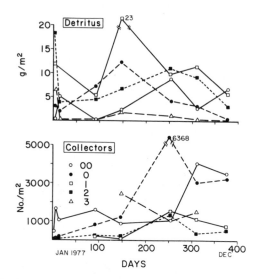

Fig. 9. Total detritus standing crops (g ash-free dry mass) and
         abundance of benthic invertebrate collectors in the
         Teton River.

during recolonization but, in order to avoid some of the
criticisms of Simberloff (1976), we used data on the abundances
of each functional feeding group rather than on number of taxa.
This approach involves calculation of the sum-of-the-squared
deviations from equilibrium for each census date, with the
results expressed as "departure from equilibrium" (D_t). Since a
pre-disturbance estimate of equilibrium trophic structure was not
available, we used mean values from five sets of samples obtained
during the study period from a site just above the former
backwaters of the dam (Fig. 1a, "Upper station"). The pattern
(Fig. 11) was similar in all cases, with a period of rapid
recovery followed by a much slower long-term return toward
equilibrium. In general, the D_t values decreased by 50% in the
first 100 d. Degree of recovery followed a pattern similar to
that found for richness and total abundance, with station 00
showing the greatest recovery, followed by stations 0 and 3.
These estimates of departure from trophic equilibrium are meant
only to indicate the rate of return of trophic conditions in the
community to pre-defaunation conditions and to suggest that in
these terms recolonization had not yet reached completion by the
end of the study. We do not intend to infer, as Heathwole and
Levins (1972) did, that the process of recovery is deterministic.
It may, in fact, be strictly stochastic as argued by Simberloff
(1976, 1978).

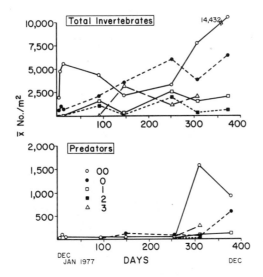

Fig. 10.   Abundance of all benthic invertebrates and of the
           predator component in the Teton River.

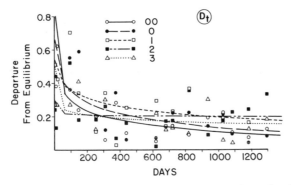

Fig. 11.   Departure from trophic equilibrium by the benthic
           invertebrate community following the Teton flood.

Drift

        Colonization of the North Fork was similar to that of
offshore islands since transport of immigrants by currents from a
source area (mainland) is involved in both cases. Water-borne
transport of invertebrates in streams is known as drift.
Virtually all of the early colonization in the North Fork was due
to drift, since dewatering destroyed all resident populations
(verified by sampling to a depth of about 50 cm; although drying

Table 2. Density of total macroinvertebrates in benthos (n/m^2) and drift (n/m^3). Most of the missing data are due to interference by ice.

Day	Date	Station 00 Benthos	Drift Day	Drift Night	Station 0 Benthos	Drift Day	Drift Night	Station 2 Benthos	Drift Day	Drift Night
9	12-26-77	4869	5.40	--	672	13.80	--	32	0.03	--
93	3-20-77	3797	0.97	--	2065	20.45	--	598	0.75	--
148	5-14-77	1290	--	--	3719	2.15	2.86	202	1.53	20.80
254	8-28-77	2067	0.42	1.70	6368	0.28	1.55	1776	0.75	0.58
310	10-23-77	8044	19.49	3.32	3915	--	--	267	1.01	0.99
376	12-27-77	14432	2.84	--	8452	0.58	--	392	--	--
439	2-28-78	17148	3.12	1.14	11855	2.54	2.09	5157	1.79	0.85
621	8-29-78	10673	0.92	10.27	9497	0.58	1.43	1538	6.66	1.09
681	10-28-78	14632	15.50	44.40	19118	6.43	2.10	3727	0.58	2.76
820	3-27-79	2981	4.12	0.81	7978	1.98	0.56	1605	0.14	0.53
937	7-10-79	4897	3.31	8.20	3374	5.54	34.70	2021	1.00	17.90
1027	10-10-79	6217	3.62	0.00	13630	0.57	1.58	1735	0.37	0.90
1182	3-13-80	31780	1.54	--	24298	0.65	2.06	1758	3.12	1.83

out of the stream extended well beyond that), initial reentry of
water was outside the reproductive period of most inhabitants,
and there was not time enough for upstream migration from the
Snake River. The data on recolonization (Figs. 2, 3; Table 2)
indicate that potential colonists (especially Baetis and
Chironomidae) can move over relatively long distances (km) in a
reasonably short time (days). These findings provide a new
dimension to the capacity of invertebrates to drift, since most
previous workers (e.g., see Townsend and Hildrew, 1976) have
emphasized the short-distance (2-100 m) nature of this
phenomenon. The data on drift (Table 2) also indicate that it is
not density dependent, as found by Diamond (1967), since drift
densities at the end of the study (when standing crops were high)
were not appreciably different from those at the beginning (when
standing crops were low). Further, correlations between drift
densities and benthic standing crops were not statistically
significant for any station.

Establishment of Community Equilibrium (Recovery)

A particularly vexing aspect of this study is that there was
little or no agreement between the various measures as to when
recovery of the benthic community from the Teton Dam failure
occurred (Table 3). Estimates vary from a year to over $3\frac{1}{2}$ y
depending on the measure selected. A. L. Sheldon (pers. comm.)
has pointed out that these recovery times are very sensitive to
changes in the procedures used to estimate them. Species
equilibrium is the only one of the indices used that is defined
by the intersection of two lines and, therefore, an unequivocal
point. However, the procedure used here of choosing the point
where a major inflection in a curve first occurs should give
reasonably comparable results and should provide a conservative
estimate of the time required for 90% of the asymptotic level to
be reached.

A common procedure in evaluating effects of, or recovery
from, pollution is to examine abundance, richness, or a
combination to the two. However, it is clear that these do not
cover all aspects of community organization and may indicate
"recovery" well before it has occurred. Equilibration in the
number of taxa per collection (richness) lagged behind that of
total numbers. This was especially evident at stations 1 and 2.
Gore (1979) found a similar response for a newly constructed
section of a Wyoming stream, where the lag was 20-30 d. He noted
that a delay in maximum diversity compared with density has been
observed in other colonizing communities by MacArthur and Wilson
(1967). MacArthur and Wilson suggest that this lag period
represents a time of dynamic adjustment during which a return
toward the dominance structure of undisturbed areas is made. As
noted elsewhere, species equilibrium has the greatest body of

Table 3. Estimates of time (in days) required for various
measures of the invertebrate community to show recovery
from the Teton flood.

Colonization Rate	$\geq$200
Total Numbers	~375
Number of Taxa per collection (Richness)	~440
Shannon-Weiner Diversity	~440
Species Equilibrium (Immigr. = Extinc.)	~625
Cumulative Number of Taxa	~700
Trophic Equilibrium (D_t = 0)	>1295

theory and empirical evidence to support it as a measure of
"recovery". But it suffers from the fact that abundance and its
partitioning among species are not taken into account. Trophic
equilibrium, as calculated here, has not been used anywhere else
and requires further evaluation. However, it and species
equilibrium provide more conservative estimates of recovery times
than conventional measures and deserve serious consideration for
use in resource management and pollution assessment. Whichever
measure is used it is clear that the Teton flood and resulting
channel restoration activities resulted in a severe perturbation
of the macroinvertebrate community that required over a year for
recovery.

ACKNOWLEDGMENTS

The research on which this paper is based was supported by a
contract from the U.S. Soil Conservation Service. We thank Dale
Bruns, Louis Farley, Dave Grant, Patricia Kolbet, Bob Seibell,
Terry Skinner, Tim Tickner, and especially Ron Pace for their
help with various aspects of the data collection and processing.
J. A. Gore, A. L. Sheldon, Judy N. Minshall and an unidentified
reviewer read the manuscript and offered a number of helpful
comments.

REFERENCES

Barbour, C. D. and J. H. Brown. 1974. Fish species diversity in
    lakes. Amer. Nat. 108:473-489.
Brown, J. 1971. Mammals on mountaintops: nonequilibrium
    insular biogeography. Amer. Nat. 105:467-478.
Brown, J. H. and A. Kodrick-Brown. 1977. Turnover rates in
    insular biogeography: effect of immigration on extinction.
    Ecology 58:445-449.
Cairns, J., Jr., M. L. Dahlberg, K. L. Dickson, N. Smith and W.
    T. Waller. 1969. The relationship of fresh water protozoan
    communities to the MacArthur-Wilson equilibrium model.
    Amer. Nat. 103:439-454.
Cairns, J., Jr., and J. A. Ruthven. 1970. Artificial
    microhabitat size and the number of colonizing protozoan
    species. Trans. Amer. Microsc. Soc. 89:100-109.
Crisp, D. T. and T. Gledhill. 1970. A quantitative description
    of the recovery of the bottom fauna in a muddy reach of a
    mill stream in southern England after draining and dredging.
    Arch. Hydrobiol. 67:502-541.
Culver, D. C. 1970. Analysis of simple cave communities. I.
    Caves as islands. Evolution 24:463-474.
Cummins, K. W. 1973. Trophic relations of aquatic insects.
    Ann. Rev. Ent. 18:183-206.
Dickson, K. L. and J. Cairns, Jr. 1972. The relationship of
    fresh-water macroinvertebrate communities collected by
    floating artificial substrates to the MacArthur-Wilson
    equilibrium model. Amer. Midl. Nat. 88:68-75.
Dimond, J. B. 1967. Evidence that drift of stream benthos is
    density related. Ecology 48:855-857.
Gore, J. A. 1979. Patterns of initial benthic recolonization of
    a reclaimed coal strip-mined river channel. Can. J. Zool.
    57:2429-2439.
Gore, J. A. 1982. Benthic invertebrate colonization: source
    distance effects on community composition. Hydrobiologia,
    in press.
Hawkins, C. P. and J. R. Sedell. 1981. Longitudinal and
    seasonal changes in functional organization of
    macroinvertebrate communities in four Oregon streams.
    Ecology 62:387-397.
Hart, D. D. 1978. Diversity in stream insects: regulation by
    rock size and microspatial complexity. Verh. Internat.
    Verein. Limnol. 20:1376-1381.
Heatwole, H. and R. Levins. 1972. Trophic structure stability
    and faunal change during recolonization. Ecology
    53:531-534.
Khalaf, G. and H. Tachet. 1977. La dynamique de colonisation
    des substrats artificiels par les macroinvertebres d'un
    cours d'eau. Anals. Limnol. 13:169-190.

Lassen, H. H. 1975. The diversity of freshwater snails in view of the equilibrium theory of island biogeography. Oecologia 19:1-8.

MacArthur, R. H. and E. O. Wilson. 1963. An equilibrium theory of insular zoogeography. Evolution 17:373-387.

MacArthur, R. H. and E. O. Wilson. 1967. The theory of island biogeography. Princeton. 203 pp.

Meier, P. G., D. L. Penrose and L. Polak. 1979. The rate of colonization by macro-invertebrates on artificial substrate samples. Freshwat. Biol. 9:381-392.

Merritt, R. W. and K. W. Cummins. 1978. An introduction to the aquatic insects of North America. Kendall/Hunt Publ. Co., Dubuque. 441 pp.

Minshall, G. W. 1981. Structure and temporal variations of the benthic macroinvertebrate community inhabiting Mink Creek, Idaho, U.S.A., a 3rd order Rocky Mountain stream. J. Freshwat. Ecol. 1:13-26.

Minshall, G. W., J. T. Brock, and T. W. LaPoint. 1982a. Characterization and dynamics of benthic organic matter and invertebrate functional feeding group relationships in the upper Salmon River, Idaho (U.S.A.). Int. Rev. ges. Hydrobiol. 67:793-820.

Minshall, G. W., R. C. Petersen, and C. F. Nimz. 1982b. Species richness in streams of different size from the same drainage basin. Unpublished Typescript. Idaho State University, Pocatello.

Molles, M. C., Jr. 1978. Fish species diversity on model and natural reef patches: experimental insular biogeography. Ecol. Monogr. 48:289-305.

Nilsen, H. C. and R. W. Larimore. 1973. Establishment of invertebrate communities on log substrates in the Kaskaskia River, Illinois. Ecology 54:366-374.

Osman, R. W. 1978. The influence of seasonality and stability on the species equilibrium. Ecology 59:383-399.

Patrick, R. 1967. The effect of invasion rate, species pool, and size of area on the structure of the diatom community. Proc. Nat. Acad. Sci. 58:1335-1342.

Ray, H. A. and L. C. Kjelstrom. 1978. The flood in southeastern Idaho from the Teton Dam failure of June 5, 1976. U.S. Geol. Surv. Open-File Report 77-765. 48 pp.

Seifert, R. P. 1975. Clumps of Heliconia inflorescences as ecological islands. Ecology 56:1416-1422.

Sepkoski, J. J., Jr. and M. A. Rex. 1974. Distribution of freshwater mussels: coastal rivers as biogeographic islands. Syst. Zool. 23:165-188.

Shaw, D. W. and G. W. Minshall. 1980. Colonization of an introduced substrate by stream macroinvertebrates. Oikos 34:259-271.

Sheldon, A. L. 1977. Colonization curves: application to
    stream insects on semi-natural substrates. Oikos
    29:256-261.
Simberloff, D. S. 1969. Experimental zoogeography of islands:
    a model for insular colonization. Ecology 50:296-314.
Simberloff, D. 1976. Trophic structure determination and
    equilibrium in an arthropod community. Ecology 57:395-398.
Simberloff, D. 1978. Using island biogeographic distributions
    to determine if colonization is stochastic. Amer. Nat.
    112:713-726.
Simberloff, D. S. and E. O. Wilson. 1969. Experimental
    zoogeography of islands: the colonization of empty islands.
    Ecology 50:278-289.
Stauffer, J. R., H. A. Beiles, J. W. Cox, K. L. Dickson, and D.
    E. Simonet. 1975. Colonization of macrobenthic communities
    on artificial substrates. Rev. de Biol. 10:49-61.
Stout, J. and J. Vandermeer. 1975. Comparison of species
    richness for stream-inhabiting insects in tropical and
    mid-latitude streams. Amer. Nat. 109:263-280.
Trush, W. J., Jr. 1979. The effects of area and surface
    complexity on the structure and formation of stream benthic
    communities. Unpubl. M.S. thesis. Virginia Polytechnic
    Institute, Blacksburg. 149 pp.
Townsend, C. R. and A. G. Hildrew. 1976. Field experiments on
    the drifting, colonization, and continuous redistribution of
    stream benthos. J. Anim. Ecol. 45:759-772.
Vuilleumier, F. 1970. Insular biogeography in continental
    regions. I. The northern Andes of South America. Amer.
    Nat. 104:373-388.
Waters, T. F. 1964. Recolonization of dunuded stream bottom
    areas by drift. Trans. Amer. Fish Soc. 93:311-315.
Waters, T. F. and R. J. Knapp. 1961. An improved stream bottom
    fauna sampler. Trans. Amer. Fish Soc. 90:225-226.
Williams, D. D. and H. B. N. Hynes. 1977. Benthic community
    development in a new stream. Can. J. Zool. 55:1071-1076.

ORGANIC MATTER BUDGETS FOR STREAM ECOSYSTEMS:

PROBLEMS IN THEIR EVALUATION

K.W. Cummins[1], J.R. Sedell[2], F.J. Swanson[2], G.W. Minshall[3], S.G. Fisher[4], C.E. Cushing[5], R.C. Petersen[6], and R.L. Vannote[7]

1. Dept. of Fisheries and Wildlife, Oregon State Univ., Corvallis, Oregon 97331; 2. U.S. Forest Service, Corvallis, Oregon 97331; 3. Dept. of Biology, Idaho State Univ., Pocatello, Idaho 83209; 4. Dept. of Zoology, Arizona State Univ., Tempe, Arizona 85281; 5. Ecosystem Dept., Battelle Pacific N.W. Lab., Richland, Washington 99352; 6. Dept. of Limnology, Univ. Lund, Lund 3-220003, Sweden; 7. Stroud Water Research Center, Avondale, Pennsylvania 19311

INTRODUCTION

Since the pioneering work at Hubbard Brook (Fisher and Likens, 1972, 1973; Bormann et al., 1969, 1974; Bormann and Likens, 1979), there has been ever increasing interest in watershed budgets, both for total organic matter, usually expressed as carbon (Wetzel et al., 1972), and various ions (Fisher and Likens, 1973; Johnson and Swank, 1973; Swank and Douglass, 1975; Fisher, 1977; Webster and Patten, 1979; Fahey, 1979; Mulholland and Kuenzler, 1979; Gurtz et al., 1980; Mulholand, 1981). The primary interest in stream dynamics within a budget context has been in the rate of loss of organic matter from the land as well as storage and biological conversion of organic matter in the stream. Impetus for most studies has come from the realization that energetics of small streams (generally orders 1 to 3 [Strahler, 1957]) are heavily dependent on organic nutritional resources of terrestrial origin (Ross, 1963; Hynes, 1963; Cummins, 1974; Hynes, 1975). New insights into the structure and function of running water ecosystems and terrestrial-aquatic linkages (Waring, 1980) are based on the changing terrestrial dependence with increasing channel size (Cummins, 1975, 1977; Vannote et al., 1980; Minshall et al.,

1983), varying stream-side vegetation (Minshall, 1978), and the
dynamics of input, storage or processing, and output of organic
matter (Vannote et al., 1980; Minshall et al., 1983; Elwood et
al., 1982; Newbold et al., 1982a,b).

The role of organic matter in running waters has been
documented and discussed to the point that streams and rivers are
no longer viewed primarily as open export systems of terrestrial
products, but rather as sites of production and processing of
organic material (Hynes, 1970; Whitton, 1975). This new image
has come largely from annual energy budget estimates for streams,
(e.g., Odum, 1957a; Teal, 1957; Nelson and Scott, 1962; Tilly,
1968; Hall, 1972; Fisher and Likens, 1973; Sedell et al., 1974;
Mann, 1975). Stream systems, from small headwaters to large
rivers, import, produce, process, and store organic matter
(Vannote et al., 1980). The processing and resultant partial
release of organic and inorganic nutrients from one stream reach
to the next has been characterized as processing along a
continuum (Vannote et al., 1980) or spiraling (Webster et al.,
1975; O'Neill et al., 1975; Webster and Patten, 1979; Newbold et
al., 1982). Only part of this material is exported downstream to
the next order or laterally to the upper bank or flood plain
without significant alteration. Complex, highly specialized
biological communities reside in running waters and not only
alter the quantities but also the quality of organic material and
inorganic nutrients exported or stored relative to that imported.
An example is the alteration of the size distribution of
particulate organic matter (POM i.e., detritus defined here as
all particles > 0.45 m plus microbial biomass; Cummins, 1974),
by aggregation and disaggregation, that is exported from a reach
relative to that imported. Fisher and Likens (1973) provided the
initial conceptual basis for examining different lotic ecosystems
using the classical two-dimensional P/R plot (Photosynthesis/
Respiration [Odum, 1957a, b]) with a third axis representing the
system's import-export balance. All ecosystems, whether they
accumulate materials, have net export, or are at steady state,
can be located in this three dimensional space. In this mode a
steady state system has its import equal to export and P/R = 1.
To maintain a steady state when gross photosynthesis (P) and
community respiration (R) are not equal, the system must either
import or export energy. The mode also allows that P may equal R
in non-steady state systems.

To evaluate such a model, all inputs (detritus and
photosynthesis) outputs (transport and respiration), and changes
in storage must be measured independently. Because this is such
a laborious task, one or more parameters are routinely obtained
by difference, and in no case has detrital storage been
adequately examined, especially over greater than annual time

periods. By assuming a steady state (i.e., inputs equal outputs
and storage pools experience no net change), the annual energy
budget can be balanced by attributing the difference between
total energy input and output of organic matter to a single
measured parameter or flux. By definition, in steady state
systems, inputs which are not exported from storage compartments
must be utilized or transformed. Because nutrient (e.g. organic
matter) storage and turnover are considered key characteristics
of stream ecosystems, an important ecological question is whether
their assumed steady state is meaningful on an annual basis for
the purpose of constructing such material balance budgets.
Certainly none of the parameters are time invariant. Lake and
terrestrial ecologists usually can place these systems within an
historical perspective and view the present state of the system
as a result of the past. For example, sediments in lakes and
annual rings in trees provide a record of past events. Stream
ecosystem history is not recorded as neatly, therefore, the
history of past events has not been adequately considered in
short term studies. Stream ecosystem structure and function are
very much dependent on recent flood history, long term variation
in runoff, and dynamics of riparian vegetation. The storage and
export values commonly measured reflect past and present annual
runoff, flood size and frequency patterns, vegetation, and
erosional conditions in the watershed as influenced by both
natural and man-induced variation. If stream organic budgets are
not placed in their historical context, their usefulness for
comparison with other sites and ecosystem types, as well as the
significance of relationships among photosynthesis, respiration,
and input/export ratio is subject to serious question.

In preparing and evaluating material budgets for running
water ecosystems, it is essential to consider patterns of
movement, processing, and storage in both space and time (Fig.
1). Typically, annual budget estimates have been based on
sampled input and output from a stream segment or small watershed
(e.g., Hall, 1972; Fisher and Likens, 1973; Sedell et al., 1974;
Minshall, 1978). There are fundamental differences between
segment and watershed budgets and the validity of comparisons
must be carefully examined. Regardless of budget type, it is
necessary to recognize the importance of episodic events, such as
floods (channel), and fire (watershed) of greater than annual
frequency of recurrence. In addition, if the emphasis of an
investigation is biological (e.g., processing efficiency, i.e.,
fraction of organic inputs annually coverted to CO_2), the

biological response time must be carefully evaluated relative to
the period of inputs and their availability, and the source of
material collected at output (Meyer and Likens, 1979).

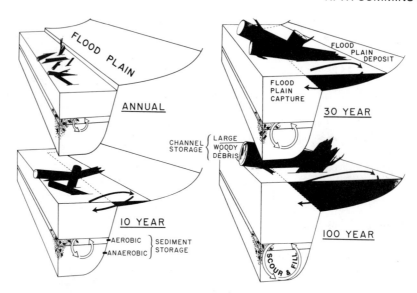

Fig. 1.   Diagramatic representation of spatial and temporal
          relationships in the retention and displacement of
          particulate organic matter (detritus) in streams. The
          general pattern of recurrence of floods of various
          severities (annual, 10-, 30- and 100-year) is depicted
          together with: 1) depth of scour and fill = sediment
          storage; 2) extent of flood plain inundation = deposit
          onto, and capture from, the flood plain; and 3) size of
          channel debris dislodged or moved = channel storage of
          large woody debris. For example, a 100-year flood
          scours deeper into the sediments, extends further onto
          the flood plain, and moves larger debris jams than
          floods of lesser magnitude. Depending upon the nature
          and depth of the sediments, there may be an anaerobic
          zone where, because decomposition rates are greatly
          reduced, periods of organic storage are extended,
          unless disturbed by flood scouring.

TEMPORAL AND SPATIAL PROBLEMS IN ESTIMATING AND INTERPRETING
STREAM ORGANIC MATTER BUDGETS

        Ecologists commonly assume that certain natural systems are
at steady state, in other words, material storage does not
exhibit net changes over a study period (Morisawa, 1968; Fisher
and Likens, 1973). The appropriateness of a steady state model
in stream ecology or geomorphology depends largely on the time

span and the area considered (Schumm and Lichty, 1965).
Generally, the smaller the stream area and shorter the time
period, the more applicable should be the steady state
assumption. Some geomorphologists (Trimble, 1975; Schumm et al.,
1975; Maddock, 1976) argue that a drainage basin can not be
considered as being at steady state at any particular point in
its history. Leopold and Maddock (1953) proposed that a tendency
toward steady state existed for certain physical properties of
streams, and that adjustments between changing streamflow
variables could be defined by power function equations. They
considered stream channel morphology to be in "quasi-equilibrium"
because the substantial scatter about regressed relationships
made it uncertain as to whether steady state had been attained.
To these geomorphologists, the dynamic equilibrium of a stream
channel involves adjustments to the history of water and sediment
discharge (Megahan and Nolan, 1976) over the previous 5-10 years
or more and does not represent a steady state condition at any
one particular year of water discharge. Because channel and
streamflow conditions are constantly changing, the equilibrium is
described in a statistical rather than absolute sense (Maddock
1976).

     However, organic matter budgets are commonly viewed as
absolute, not statistical, although during any given year in
which a budget is developed, steady state is probably not a valid
assumption (e.g., Welton, 1980). During any single year,
increase or decrease in storage dominates a budget depending on
water discharge and POM input.

     The assumption of steady state of organic matter storage in
a stream is further complicated by variation in storage
characteristics over a greater than annual time scale (e.g.,
Welton, 1980). Major storms may alter the volume of material in
storage to such an extent that return to pre-storm channel
characteristics occurs only over a period of years to decades.
Furthermore, channel storage capacity of forest streams may vary
over the history of the adjacent forest, because periods of tree
mortality leading to inputs of large woody debris increase both
the total standing crop and the channel capacity for storage of
fine organic detritus. Therefore, even if a stream experiences
no net change in storage for a year, that year may not be
"typical" of long-term conditions which are constantly changing
in response to previous storms and the history of the riparian
zone.

Discharge History of Streams

     The importance of flooding to stream organic matter budget
studies over annual and longer cycles has been largely ignored.
For example, the recurrence interval of peak flows for the study

period is not standard information in stream ecology research papers. As much as 80% of the particulate organic matter (POM > 0.5 μm) exported by a stream over a year can be discharged during one or two storms (Hobbie and Likens, 1973; Bormann and Likens, 1979; Crisp and Robson, 1979. Because the origin of this material, whether it be within sediments, channel bed or banks, or the flood plain, and its residence time in the stream system are unknown, the biological significance of such events to the community has yet to be determined. High water capture of a boardwalk placed in the riparian zone along a reach of Augusta Creek, Michigan provides a specific example. Exclusion from an organic budget of the lumber which collected at the downstream weir during the storm seems logical, but what about the many other pieces of organic matter with similar histories from the same location? Clearly, both the source of the inputs and time scale in which they enter the channel and are transported are important if the emphasis is on stream biology.

Few published studies (e.g. Dawson, 1980) have placed the field sampling period in the context of either the annual flood cycle or longer term discharge patterns. It is clearly important to know the relationship of the study period to major flood or drought events. At a minimum, material or energy budget measurements should be placed in general perspective of flood return frequencies of the study year and several preceding years. Figures 2A and B present recurrence frequencies (Leopold et al., 1968; Morisawa, 1968) of peak discharge, maximum temperature, and annual degree days for the North Santiam River, near Detroit, Oregon. Such long term flow-temperature perspectives allow comparison of a study year to overall averages and conditions prior to sampling. This issue is further complicated because the history of sediment movement into and through mountain stream systems has probably been in large part keyed to major storms. In forested areas of the H. J. Andrews Experimental Forest, Oregon, for example, significant debris avalanches, a major supplier of sediment to channels, and debris torrents which flush steep channels, have been triggered by storms with a 5 year and greater return period (Swanson and Dyrness, 1975). For example, many of the mass movement events occurring over a 30 year sample record were triggered by a December 1964 storm that took place within the period but had a return interval greater than 100 years.

The preceding discussion points up the difficulty in accounting for infrequent episodic events that strongly influence sediment transport and channel morphology. When a major event having a return period greater than the sample period occurs during a study, it dominates the record. If such an event does not occur within the time of the study, a major part of the long-term system behavior is missed. Further, the

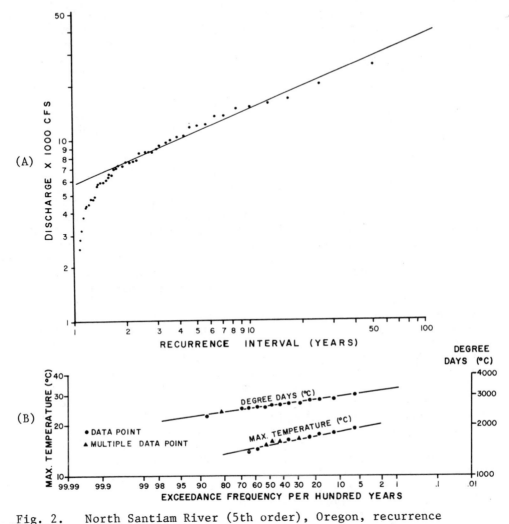

Fig. 2.   North Santiam River (5th order), Oregon, recurrence
           intervals for maximum yearly floods (A) and temperature
           (B) based on a 20 year record. Degree days calculated
           as. mean daily temperature times number of days having
           that mean and summed over the water year (October 1 –

           September 30). Recurrence interval $(T_R) = \dfrac{1 + n}{m}$,

           where n = number of years of record and m = rank of
           discharge or temperature for each year. (The
           reciprocal of T_R is the probability of occurrence of a

           discharge or temperature of that magnitude in any given
           year.)

interdependence of material transfer during successive peak flow
events poses problems in framing a particular study period within
a longer term perspective. Return periods of stream flow are
calculated assuming independence of successive events. However,
the amount of organic and inorganic material carried by a given
discharge depends on a variety of factors, such as magnitudes and
timing of preceding events (Paustian and Bestcha, 1981; Bilby and
Likens, 1981). Consequently, return periods calculated for a
sequence of peak flows do not necessarily reflect the relative
magnitude of those events in terms of organic matter transport.

The record of runoff for the McKenzie River at McKenzie
Bridge, Oregon, offers an indication of historical variations in
precipitation, runoff, and peak annual flow in the area of a
major stream ecology research site (Fig. 3). The runoff record
is plotted as cumulative departure from the mean runoff for the
62 year period (Fig. 3) to show historic trends in runoff. There
is no evidence that annual precipitation totals in the Northwest
are serially correlated (Dowdy and Matalas, 1969). The negative
slope of cumulated values from 1928 to 1945 reflects a period of
lower than average runoff while a positive slope between 1947 and
1958 indicates above average runoff. Runoff during the dry
period, which was 14% less than mean conditions, also exhibited
average peak annual discharges of only 69% of peak flows during
the wet period. Occurrence of major floods is not restricted to
generally wet periods. The flood of December 1964, the largest
well-documented historic flood of a regional scale in the Pacific
Northwest and California, occurred during both a year, and during
a two decade period, of average runoff. This long term runoff
record suggests that during periods of dry years flushing of the
stream can be considerably reduced. Although there are no
organic matter budgets for these early wetter and drier periods,
it is likely that accumulation of refractory organic materials in
the stream was appreciably greater during dry years and there may
have been an annual net loss from the watershed during the wet
periods, which involved both higher total annual and peak flows.

Two annual organic budgets constructed for a small watershed
(WS10) in the H. J. Andrews Experimental Forest, Oregon are an
example of the importance of runoff history; the budgets include
an extremely dry year, with 167 cm precipitation in water year
1973 (Oct. - Sept.) and a wet year with 303 cm in 1974 (Franklin
et al., 1981). Precipitation is typically seasonal, totaling
approximately 240 cm and falling mostly as rain between October
and March. Water year 1972 had 3 of the 10 highest storm flows
recorded in the 24-year records of runoff for the Experimental
Forest. Thus, these budgets were constructed at a time when
storage of organic materials probably was relatively low.
Wet-year (1974) organic inputs by litterfall and lateral

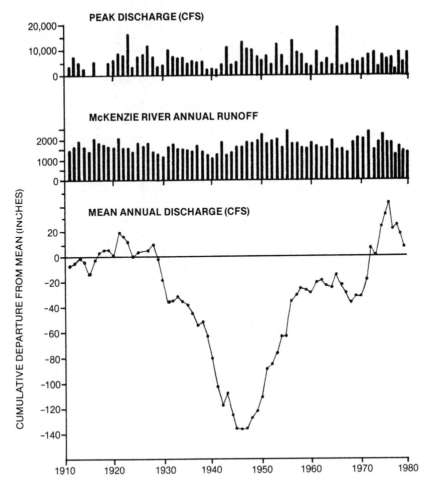

Fig. 3.   Runoff history of the McKenzie River, Oregon, showing
peak discharge, mean annual discharge, and cumulative
departure from the mean (Dalrymple, 1965) of the entire
62 year round record. (That is, the record for each
year is subtracted from the mean, keeping track of
sign, and these differences cumulated and plotted
through time.)

(surface) movement exceeded dry-year (1973) inputs by 5% and 65%
respectively. Clearly, inputs change significantly with runoff.
The dissolved organic matter (DOM< 0.45 μm) loss was 246% higher
in 1974 because solution loss is related to runoff. Even though
the annual runoff for 1974 was the second highest in the 63-year
record for the upper McKenzie drainage, the peak discharge had a

one year return period, and on that basis, POM export could be considered average or below. Therefore, the various elements of an organic budget for streams are extremely responsive to the magnitude and frequency of storm events. Although inputs (except landslides) may be relatively less variable, storage, export, and biological processes must all be evaluated in relation to runoff and return frequency if valid comparisons within and between running water systems are to be made. The discussion above has emphasized the effects of peak flows, but it is also important to consider the role of other episodic events, such as high winds without appreciable precipitation, which can result in major pulses of POM input without significant export.

Organic Export Rating Curves

Although rating curves (suspended load vs discharge) for inorganic particulate losses from a watershed have proven useful for calculating sediment transport, relationships between stream discharge and organic matter concentration are approximations at best (Fisher and Likens, 1973; Bormann et al., 1974; Brinson, 1976; Bilby and Likens, 1979; Bormann and Likens, 1979; Schlesinger and Melack, 1981). Even the best regressions of particulate organic concentration vs water discharge explain only about 50% of the POM concentration from discharge data (Bormann et al., 1974). When corrections for differences in the percent organic matter relative to total particulates at various flow rates are included, errors in calculating POM losses from a watershed are even greater. Studies of inorganic and organic components of transport have revealed different patterns for each of the rising and falling limbs of a peak flow hydrograph (Paustian and Bestcha, 1981).

Rating curves for coarse particulate organic matter (CPOM, > 1mm) are particularly poor. Fisher and Likens (1973) obtained an R^2 of 0.05 for their rating curve and rejected its use on that basis. Sedell and co-workers (Oregon State University, unpublished data) obtained a CPOM rating curve for a first order Oregon stream (WS10) with an $R^2 = 0.14$, based on samples collected over a full year by passing the entire discharge through an 80 ml net. These results indicate the high degree of uncertainty in using rating curves for calculating CPOM export.

A rating curve for fine particulate organic material (FPOM, <1mm >0.45 μm) also was developed for the Oregon stream Watershed 10 (Fig. 4). High FPOM export at moderate flows of 18-23 L/s (Fig. 4) represents the first storm in the fall, indicating the importance of timing as well as magnitude of discharge. Various curve forms produce poor fits to the data in

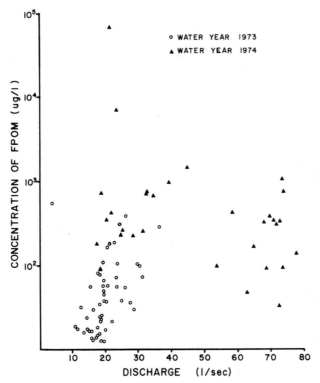

Fig. 4.   Fine particulate organic matter (FPOM) vs discharge
relationship for Watershed 10, Oregon, over two water
years. The high value of FPOM at low discharge levels
represent the first storm of the fall-winter rainy
season.

Figure 4, particularly at high flows when the greatest percentage
of annual POM transport occurs. Thus, peak discharge, the
interval between storms, and storm sequence are all important
factors in determining the total export of organic matter from a
watershed during a storm event and a myriad of other factors such
as time of year, amount of litterfall, decomposition rates of
POM, etc.

    In an effort to obtain a better fit of peak discharge to
total organic export, 30 storms creating discharges > 20 L/s in
Watershed 10 were examined (Table 1). Linear regressions of
total export against peak discharge, interval between storms, and
interval between storms times peak discharge were calculated.
All yielded similar results for this area of variable
precipitation, but in mesic areas with fairly even precipitation
regimes, the combination of interval between storms, magnitude of

Table 1. Various linear regression analyses of particulate organic matter (POM) export (> 75 μm) recorded for Watershed 10, Oregon, during all storms producing discharge peaks >20 liters sec^{-1} in water years 1973 and 1974.

X (independent variable)	Y (dependent variable)	N	R^2	Slope	Intercept	S.E. of Y	S.E. of Slope	S.E. Intercept
Peak discharge (1 sec^{-1})	Kg POM export over duration of flood	28	0.57	3.35	0.01	0.37	0.34	0.54
Peak discharge times day interval between floods	same	27	0.51	4.87	0.14	0.39	0.13	0.30
Day interval between floods	same	27	0.43	5.98	1.51	0.43	1.20	0.89

the previous storm, and peak discharge may be useful for
estimation of total POM transport as well as inputs to a reach.

In general, then, the relationship between discharge and the
export of organic matter from small stream watersheds is
non-linear, time dependent, and site specific (i.e., dependent on
retention characteristics). Such serial correlation
relationships are likely to be represented best by hysteresis
curves (Paustian and Bestcha, 1979; Whitfield and Schreier, 1981)
and sampling needs to be continuous rather than by conventional
grab-samples (Dawson, 1980; Whitfield and Schreier, 1981).

Relationship Between Organic Matter Export and Decomposition

Loss of organic material from a stream bed can result from
both export (downstream and to the upper bank or flood plain) and
decomposition. Decomposition rates for assumed steady state
systems have been calculated by Fisher and Likens (1973) and
Borman et al. (1974) from a modified equation by Olson (1963):
$$X = I/ (k+K_1), \text{ where,}$$

X = size of organic pool in the stream bed at steady state,
in kg

I = input rate in kg of organic matter yr^{-1}

k = fractional loss rate of X due to decomposition yr^{-1}

k_1 = fractional loss rate of X due to export yr^{-1}

For a forested watershed (WS6) in the Hubbard Brook
Experimental Forest (New Hampshire) a dynamic equilibrium was
assumed and five years of export data were averaged to derive a
mean decomposition value (k) of 0.43 per year (Bormann et al.,
1974). An energy budget can be calculated for 1965-1966 using
export and storage data from Bormann et al. (1974), litter input
data from Gosz et al. (1972), and blow-in data from Fisher and
Likens (1973), $k = 0.46$ and $k_1 = 0.021$. For year the 1966-1967,
which included several large storms, $k = 0.40$ and $k_1 = 0.088$.
These calculations of k and k_1 values over the five year period
show that even in a year with a large storm only 8.8% of the
estimated particulate inputs is exported as POM and in the year
of smallest discharge events it is only 2.1%. Assuming no net
change in storage, the resultant POM decomposition ranged from 82
to 96% of inputs. These calculations portray stream ecosystems
as: 1) retentive and affording ample opportunity for biological
processing of detrital inputs and 2) highly variable annual
transport systems, with export varying more than 400% while
decomposition rates varied only 14% over the 5 year period of

study. Therefore, it is probable that in a wet year with no
large storms, the DOM fraction would be emphasized; in a year
having a 10-20 year storm, POM losses would dominate a budget.

In the modified Olson (1963) equation, decomposition of
detritus is independent of temperature and dependent upon the
size of the storage pool. Decomposition of detritus in streams,
however, has been shown to be strongly related to temperature
(e.g., Bolling et al., 1975a; Suberkropp et al., 1975) and,
therefore, k is a function of temperature. For example, 90-95%
decomposition of leaf litter, composed of species which are
processed at medium to fast rates, takes about 1000 degree days
(Petersen and Cummins, 1974). The annual accumulation of degree
days, and variations between years are extremely important in
determining annual losses due to microbial respiration. A
maximum year-to-year variation of approximately 1000 degree days
was observed over a 20 year period in the North Santiam River
(Fig. 2B). This represents a potential annual fluctuation of 30
to 40% in decomposition losses attributable to temperature alone,
which would influence the time of depletion of higher quality
inputs as well as conditioning and use of lower quality litter.
Therefore, even if storage were adequately measured and related
to storm events within a given annual cycle and between years,
this budget approach cannot be used independent of temperature.

Sampling schemes used to determine parameter values X, I,
and k_1 in Olson's equation must deal with scales in time
(turnover frequency) and space (storage pool dimensions)
appropriate to characterize each parameter. Failure to do so
will lead to spurious decomposition rate (k) estimates. For
example, Olson's equation is dependent on definition of storage
pool size under circumstances commonly encountered in budget
calculations. That is, for two streams with identical I and k_1,
but with twice as much POM storage (X) in one, apparent
fractional loss from storage due to decomposition (k) would be
57% lower in the stream with lower standing crop (using data for
I, k_1, and X from Bormann et al., 1974). Therefore, for a given
stream, k will vary with the definition of X. If the storage
term is considered to include logs, deep sediment, and bank
storage, then input and export terms must take into account
events which alter these components. Ideally each storage
compartment should be characterized in terms of its particular X,
I, k, and k_1 values, so that decomposition rate is more
realistically a function of temperature, substrate quality, and
oxygen environment and not obtained by difference. Thus,
although the Olson (1963) model is a starting point, it must be
significantly modified, and conditions carefully specified in
order to be appropriate.

Detrital (POM) Storage

       As stated previously, the assumptions of annual steady state
in stream budget work has been perpetuated by failure to
adequately measure storage and the input-output balance is
misleading and primarily a function of physical retention or
release rather than biological activity. In addition, the
importance of, and interaction between, POM storage pools change
with increasing stream size, from headwater tributaries to large
rivers. For example, channel woody debris is particularly
significant in small streams lacking sufficient hydraulic force
to move large jams while flood plain deposition of POM is more
significant for large rivers.

       Storage in and associated with stream channels is
apportioned among several sites (compartments) which are
identifiable spatially and on the basis of turnover rates (Fig.
1). Although the compartments intergrade, three are sufficiently
discrete to make their distinction useful: 1) in the channel
sediments; 2) in the channel on or above the sediments –
primarily in pools and in coarse woody debris jams; and, 3) on
the flood plain or upper bank (Fig. 1). Rates of POM processing
are controlled by such factors as dissolved oxygen concentration;
size and degradability (quality) of the organic matter;
temperature; and extent, frequency, and persistence of wetting
(Merritt and Lawson, 1979). The major detritus processing sites
are the aerobic sediment layer and exposed portions of POM
accumulations in debris jams. The processing rates are much
slower in the deeper, anaerobic sediment layers, and large woody
debris accumulations have processing times much greater than an
annual cycle. On the flood plain (upper bank), processing is
slower during cold or dry periods than in the stream (Fig. 1).
Large wood is processed more rapidly out of the stream channel
during warm periods because of significantly higher fungal and
invertebrate activity.

       The organic storage of Bear Brook (Fisher and Likens, 1973)
was assumed to be at steady state, with a detritus reservoir of
about 80% of the annual input or output of energy. A standing
crop less than the annual organic flux supports the assumption of
steady state. However, detrital storage would be particularly
important in budget calculations when it is large in relation to
the other terms of the material balance equations. The storage
component can be large when 1) channel conditions are conducive
to retaining a large amount of organic material, and/or 2) a high
proportion of the total sediment yield consists of bedload
including organics. The potential for greater channel sediment
storage increases with decreasing stream order, particularly in
forested streams of mountainous areas. These streams are
characterized by high erosion and litter production rates, and

slope profile irregularities, but have channel obstructions such as logs and boulders for which there is usually insufficient hyraulic force for movement. Since slope irregularities, and obstructions create a stairstep effect, deposition of organics behind these obstructions must be considered in addition to physical processes such as scour and fill of the channel bottom.

Sediment storage. Detritus in sediment is subject to aerobic or anaerobic processing (Fig. 1). Depth of the aerobic zone is dependent upon the size and heterogeneity of sediment particle sizes, rate of microbial oxygen consumption, channel gradient, average flow conditions, and seasonal and annual flood patterns. In the cases in which total standing crop of particulate organics (excluding large woody debris) has been measured, sampling was generally in the aerobic portion of the sediment profile subjected to annual scour and fill (Fig. 1). POM is generally assumed to be available annually for biological processing to CO_2 and conversion to dissolved organic matter

(DOM) and living biomass. Provided the organics and associated biota remain aerobic, at least for significant periods during the warmer portion of the annual cycle, processing can be accomplished within a year (Petersen and Cummins, 1974; Boling et al., 1975a,b). POM that remains buried in anaerobic zones is processed at significantly slower rates than that which remains aerobic over the annual cycle, as evidenced by intact, sulfide-blackened leaves excavated from deep sediments in the spring. On the other hand, since invertebrates have been recovered to depths of 45 cm or more into the substrates in streams with loose, heterogeneous beds (Coleman and Hynes, 1970; Hynes, 1974; Hynes et al., 1976), the depth of this aerobic processing zone needs to be carefully examined for each stream.

Concentration of CPOM and FPOM stored in the beds of two old-growth forest streams in Oregon was greatest in the sediments below the 10 cm depth normally sampled (Figs. 5 and 6). Studies will scour chains in such streams indicate that the depth of scour and fill (Fig. 1) for a 10-year storm is 15 to 20 cm (Moring and Lantz, 1975). Thus, storms could mobilize from storage two to four times the particulate detritus recovered in surface sediments by usual sampling techniques. Similar to flood scour, POM may be depleted from deep sediment storage or buried under new deposits through scour and fill that frequently occurs following forest clearcutting (Fig. 6) or channelization.

Channel debris. Until recently (Froehlich, 1973; Keller and Swanson, 1979; Bilby and Likens, 1980; Bilby, 1981) large woody debris has been ignored in stream budgets. The assumptions, for budget purposes, that the amount of large woody debris in channels was small and its processing rates insignificant were

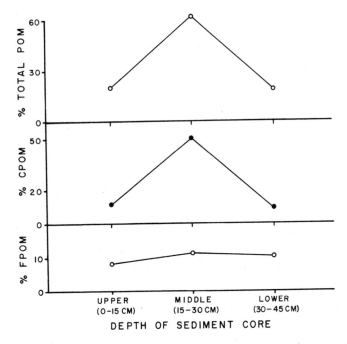

Fig. 5.   Changes in relative % storage of total (POM; >75 μm),
           coarse (CPOM; >1 mm) and fine particulate organic
           matter (FPOM: <1 mm >75 μm) at three riffle sediment
           depths. Based on ash free dry weight of seven,
           approximately 20 cm diameter, frozen cores in Flynn
           Creek, Oregon, August 5, 1975. Total POM ranged from
           about 450 gm^{-2} to 1600 g m^{-2}, CV's as % ranged
           from 20-50. Unpublished data, Hess and Brown, Dept. of
           Forest Engineering, Oregon State University.

untested. It is obvious that the assumption of insignificant
quantities of debris is not valid, particularly in debris-laden
headwater channels of the Pacific Northwest (Fig. 7A). The
amount of woody debris in first and second order Oregon streams
has been measured at 8-25 kg/m^2 organic dry weight (Froehlich,
1973; Keller and Swanson, 1979; Anderson et al., 1978; Anderson
and Sedell, 1979; Triska and Cromack, 1979). Assuming a maximum
100+ year processing time to convert woody debris to finer
particulates, the annual contribution of large wood-derived FPOM
to the mean aerobic sediment standing crop would be about 5 to
10% based on data available for Oregon and Michigan streams
(Sedell and Cummins, unpublished data). Furthermore, a leaching
rate (assuming gradual conversion of cellulose and lingin to more

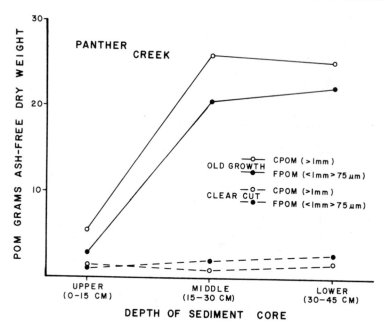

Fig. 6.   Comparison of coarse (CPOM; >1 mm) and fine (FPOM; <
           1 mm >75 μm) particulate organic matter in riffle
           sediments of Panther Creek, Oregon, a 2nd order
           tributary of the Alsea River. Old-growth Douglas fir
           (120 year) and clear cut (two years after cutting) are
           compared. Samples were three to five frozen cores,
           approximately 15 cm in diameter, taken in each section
           September 9, 1973. CV's and % ranged from 30-50.
           Unpublished data, Sedell (Dept. of Fisheries and
           Wildlife) and Brown (Dept. of Forest Engineering),
           Oregon State University.

labile substrates) of only 0.1% per year from large debris would
be sufficient to account for the entire annual measured DOM
export from Watershed 10 (Oregon). Processing of large wood is
apparently continuously concentrated in the outer millimenter or
so (Aumen, Oregon State University, unpublished data). Input of
large woody debris, which varies greatly from year to year, can
not be assessed by litterfall or bank traps. For example, in a
100 m reach (about 1 m wide) of Watershed 2 in the H. J. Andrews
Experimental Forest the blow-down of several tree tops doubled
the standing crop of organic debris in a single storm. All the

large (> 10 cm diameter) woody debris was measured in a 260 m reach of a third-order section of Augusta Creek, Michigan, which appeared to have a low concentration of wood (Fig. 7B). Even in this stream, the measured loading of woody debris--approximately 1.2 kg/m^2 organic dry weight, was about 15 times the mean annual detrius standing crop of the aerobic sediments, excluding the large wood measured by conventional traps. Assuming a 30 year processing time for large wood (hardwoods are processed more rapidly than wood of conifers; data from experimental streams yield estimates of 10 years for medium [approximately 10 cm] and smaller wood) the breakdown of this CPOM would yield the equivalent of about 14% of the FPOM standing crop annually. Because the amount of large woody debris in the particular study reach was probably below the average for streams in the area due to land use practices, and processing rate is probably greater, this estimate is conservative.

Another important feature of large woody debris is its role in retention of finer CPOM and FPOM and DOM (Bilby and Likens, 1980). Retention of FPOM in debris jams and associated sediments may result in its storage for periods of more than a year and the FPOM may require several years for processing. Whether the channel is in a period of loading of large debris, and whether large woody debris is retaining or releasing finer particulates, are critical assessments in compiling a budget.

Flood plain. Food plains are areas of river valley bottom inundated when bankfull channel capacity is periodically exceeded (Maddock, 1976). During overbank flow the flood plain or upper bank can serve as both a source and a sink for POM (Fig. 1). The organic content of flood plain and upperbank deposits has not been considered in budget determinations for lotic systems, nor has the movement of organic matter from the flood plain back into the channel. the rates of rise and fall of water level, peak discharge, channel and flood plain geometry, and flood plain vegetation all interact to determine the POM dynamics of a flood plain during overbank flow. Exposed and buried living and dead vegetation can be used to evaluate the balance between deposition and erosion on the flood plain, as well as the extent of inundation from floods of various magnitudes. Sigafoos (1964) reconstructed the depositional history of about 75 cm of sediment in the flood plain of the Potomac River which resulted from a mix of deposition and erosion over a 30-year period - the net effect being deposition.

Detritus exchange between the channel and flood plain is likely to involve important qualitative changes with respect to

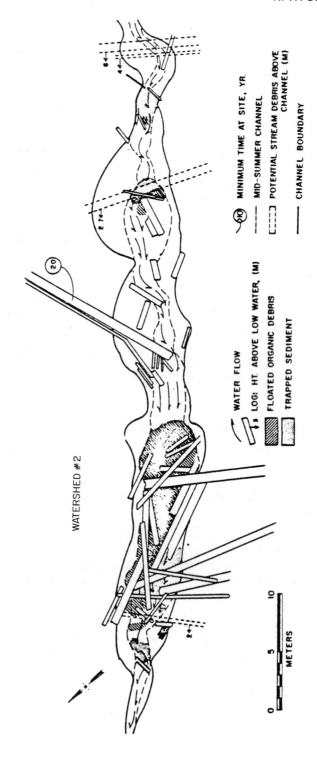

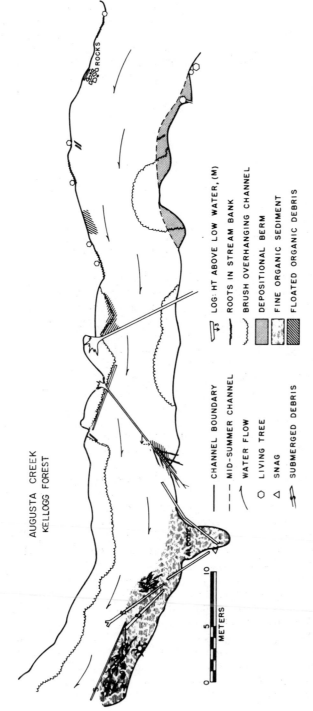

Fig. 7. Detailed reach maps showing wood debris in Watershed 2 (H. J. Andrews Experimental Forest) in the Cascades of Oregon (A) and Augusta Creek (Kellogg Forest, Michigan State University) in southwestern Michigan (B).

particle size, biochemistry, microbiology, and degree of conditioning in the terrestrial system (Merritt and Lawson, 1979; Triska and Cromack, 1979). Furthermore, as stated above, when temperatures are lower on the flood plain than in the stream and/or conditions are dry, biological processing of particulate detritus can be slower than in aerobic zones of the stream (Merritt and Lawson, 1979). Consequently, leaf litter deposited on the flood plain in the autumn and captured during spring high water, enters the stream in a less processed condition than litter that remained in the aerobic stream sediments over the same period.

Quantitative and qualitative data are required on the export from and import to the channel to and from the flood plain over annual and longer periods. However, typical litterfall and ground surface movement trap techniques are not adequate to characterize flood plain dynamics and other methods (e.g., some sort of mark and recapture procedure) would be required for assessment of organic matter input and output from flood plains.

Changes in POM storage along river continua. Input, storage, biological processing, and export of organic matter vary through a drainage basin from headwater streams to larger rivers. In the River Continuum concept described by Vannote et al. (1980), it was suggested that the size distribution of POM changes with increasing stream order such that both transport and surface sediment storage are generally characterized by decreasing amounts of CPOM as the influence of the riparian vegetation decreases with increasing channel width. This trend of decreasing CPOM to FPOM ratio in transport and storage (Sedell et al., 1978; Naiman and Sedell, 1979a, 1979b; Cummins et al., 1982), which more adequately characterizes the latter, is modified by tributaries and the CPOM generated from macro-algae (Minshall et al., 1983; Wallace et al., 1982). The changes in relative importance of certain size classes, which reflect different input, physical retention, and biological processing along the river continuum, necessitate modifications in the sampling methods applied to organic budget studies in streams and rivers of various size. For example, channel storage in debris jams would be most significant in first through about third order headwater streams, while upper bank, off channel pools, along banks and on point bars, and flood plain sites become more significant with increasing river size. The importance of deep sediment storage in organic budget calculations in a particular water year and the action of scour and fill along the river continuum are presently unknown. The problem of comparison among streams of different sizes is compounded because budgets for headwater streams typically approached on a watershed basis, differ significantly from analysis of individual reaches which is more feasible for larger rivers.

COMPARISON OF WATERSHED AND REACH (SEGMENT) DERIVED ORGANIC
BUDGETS

Theoretical Considerations

The components of watershed and reach organic budgets are
summarized in Figure 8. The watershed approach covering all
channels in a basin is appropriate for small first-, to about
third-order watersheds. For stream orders greater than about
three, however, a "representative" reach, or river segment,
usually is selected, preferably with few tributaries. An initial
step in either case is the determination of a water balance.
When the entire watershed is included, ground water inputs,
overland flow, and precipitation directly into channels should

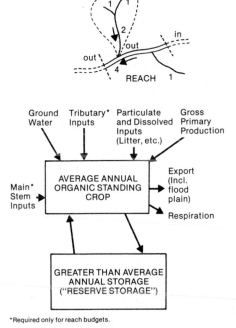

*Required only for reach budgets.

Fig. 8.   A comparison of components of watershed (A) and reach
           or segment (B) organic budgets. Examples of stream
           orders (numbers) and input and output measurement
           points are shown in the top portion of the figure.

equal mainstem output plus evaporation and channel related transpiration. The usual method to determine POM and DOM fluxes has been based on relationships between organic matter concentrations in grab samples of input and output at different rates of water discharge. Total organic fluxes are then calculated on the basis of transport rating and flow duration curves (e.g., Fisher and Likens, 1973). Storage is frequently the most poorly documented element of budgets (Fig. 8), because both spatial and temporal sampling problems may lead to significant over- or underestimates of this parameter for a given annual budget.

Budget comparisons. Data from seven study sites across the United States provide examples of broad trends and problems in comparing organic matter budgets for different streams (Tables 2-5). Tables 2 and 3 summarize background information on the study streams and methods used to estimate the organic budgets. The sites differ significantly in watershed area, stream slope, and other geomorphic characteristics. Differences in riparian vegetation represent important sources of variation among streams of a given order in different basins. For example, first-order streams in arid regions are bordered by sparse vegetation, so particulate inputs from surrounding areas are lower than in-stream primary production (Table 4). In Deep Creek, POM generated from aquatic macrophyte breakdown is the major component of the fraction. In contrast, a coastal coniferous forest stream may flow through a 70 meter high forest and receive low light inputs and abundant litterfall from canopy and understory vegetation of riparian plants (Campbell and Franklin, 1979). This variation in the degree of coupling between terrestrial and aquatic components varies predictabilty along the river continuum, even within a particular vegetation biome (Vannote et al., 1980).

Although obvious differences exist among the organic budgets of the lotic ecosystems compared (Tables 4 and 5), POM constituted a major component (25 to 71%) of the total organic matter input. Except for Deep Creek and Fort River, the amounts of POM imported, expressed on an areal basis, are similar. Of the approximately 2300 to 3200 kcal m^{-2} yr^{-1} POM delivered to WS10, Augusta Creek, and Bear Brook, 7 to 37% was exported, the remainder being converted to DOM, biomass, respired to CO_2 and stored. POM exports significantly exceeded imports only in Deep Creek and White Clay Creek. Values of gross primary production for the systems cluster into two groups at approximately 10 to 90 kcal m^{-2} yr^{-1} (0.2 to 2.5% of total inputs of their respective budgets). As expected, the small, heavily shaded streams had low primary production. In all but Deep Creek and Fort River,

Table 2. Selected geomorphic, hyrologic, meteorologic, physico-chemical and ecological data for the stream systems for which data are presented in Tables 3 and 4.

Parameters	Bear Brook	Watershed 10	Rattlesnake Creek	Augusta Creek	Deep Creek	White Clay Creek	Fort River
Stream Order	2	1	1	1	2	3	4
Biome	Deciduous (New Hampshire)	Coniferous (Oregon)	Arid (Washington)	Deciduous (Michigan)	Arid (Idaho)	Deciduous (Pennsylvania)	Deciduous (Massachusetts)
Geology	Stillimanite gneiss	Basalt	Basalt	Glacial till	Lacustrine sediments	Mica schist gneiss	Metamorphic Paleozoic rock
Watershed area Km^2 (Acres)	1.3 (321)	0.1 (25)	350.0 (86,485)	3.3 (815)	447.0 (110,453)	7.2 (1,779)	105.0 (25,946)
Dominant Watershed Activity	protected	protected	protected	grazing	grazing	farming, grazing	farming, grazing
Drainage, density (length/basis area)	1.31	5.00	0.01	1.10	0.09	2.79	--
Stream gradient (%)	14	47	2	0.2	0.2	1.0	0.2
Base flow discharge M^3/sec (cfs)	0.002 (0.07)	0.003 (0.01)	0.011 (0.40)	0.013 (0.45)	0.09 (0.12)	0.068 (2.40)	0.28 (10)
Annual precipitation (cm)	123	245	17	152	41	107	110
Maximum Temp (°C)	18	16	22	20	27	22	27.5
Annual cumulated degree days	3285	3165	5547	3256	6695	4398	--
Total alkalinity (mg/l)		10	127	185	233	52	--
NO$_3$ (mg/l)		<0.1	0.3	3.3	0.2	2.8	0.2
Total P (µg/l)		80	0.15	30	40	--	25
Water (detritus year(s) for which budget is presented (Table 4)	1968-1970	1972-1973, 1973-1974	1969-1970	1972-1973	1970-1972	1972-1974	1972-1973

Table 3. Methods used for determining organic budgets for six North American streams.

Stream System	INPUTS			OUTPUTS			
	Particulate Organic Matter (POM)	Dissolved Organic Matter (DOM)	Net Primary Production (Net PP)	POM	DOM	Respiration	Storage
Bear Brook	Litter fall (irregular intervals) and lateral transport traps (weekly intervals, branches >1cm diameter not included); non-winter drift and bottle samples, irregular intervals	Throughfall ground water and stream grab samples, irregular intervals (DOC <0.45 μm)	Light and dark bottle (mosses), three determinations	CPOM pending basin, FPOM (<1mm) bottle samples, irregular intervals	Stream grab samples, irregular intervals	Microbial respiration by difference; mosses dark bottles (3 determinations); macroconsumer respiration (calculated)	None detected-steady state standing crop in ponding basin
Watershed 10	Litter fall and lateral transport traps, monthly intervals (>0.45 μm)	Throughfall, storm event, groundwater seep and stream grab samples, monthly intervals, (DOC <0.45 μm)	Circulating chambers in summer months	Continuous whole stream seiving with 80 μm net. Irregular grabs	Same as inputs	Gilson respirometer of detritus quality and particle size, monthly intervals	By difference of the POM inputs to outputs
Rattlesnake Creek	Litter fall traps, Irregular intervals and weekly bottle samples (> 0.5 μm)	Weekly grab samples (< 0.5 μm)	Upstream-downstream pH	Weekly bottle samples (< 0.5 μm) plus loss of standing crop of insects and water cress flushed out by floods	Weekly grab samples (< 0.5 μm)	Upstream-downstream pH	Not measured

| | INPUTS | | | | OUTPUTS | | | |
Stream System	Particulate Organic Matter (POM)	Dissolved Organic Matter (DOM)	Net Primary Production (Net PP)	POM	DOM	Respiration	Storage
Augusta Creek	Mean monthly detritus standing crop for the detritus year	Difference (magnitude checked against biweekly DOC grab samples)	Circulating chambers, monthly intervals	By difference (magnitude checked against standing crop changes less respiration)	Weekly grab samples (< 0.5 μm)	Gilson respirometer of detritus by particle size and circulating chambers, monthly intervals	Not measured (assumed to be different than standing crop)
Deep Creek	Litter fall and blow in (irregular intervals) and drift (monthly)	Monthly grab samples (< 0.5 μm)	Upstream-downstream O_2 or pH	Drift (monthly)	Monthly grab samples (< 0.5 μm)	Upstream-downstream O_2 or pH	Coring and various measures of deposition and accrual
White Clay Creek	Litter fall and lateral transport traps (weekly intervals for 2 yrs.) POM standing crop, random transects,[2] equivalent to 1m FPOM 0.1m[2] core samples	Weekly grab samples (DOC < 0.45 μm)	Circulating chambers, 2-5 sample runs per week	CPOM (7 mm) wier on stream orders 2 and 3 before and after storm events in fall and winter, irregular spring and summer samples. FPOM (< 7mm > 50 μm) 24 hour pump samples, 2-5 per week	Weekly grab samples (DOC < 45 μm)	Circulating chambers 2-5 sample runs per week	Not measured
Fort River	Litterfall traps (biweekly) and drift (every 3 days: CPOM 1mm mesh net at timed intervals; FPOM grab samples)	Grab samples every 3 days (< 0.1 μm)	Diel O_2 curves (23 measurements over 18 months)	Drift (same as Inputs)	Same as Inputs	Diel O_2 curves[2]	None detected - steady state standing crop

Table 4. Energy budgets for 7 North American Streams (POM = particulate organic matter, > 0.45 m particle diameter; CPOM = coarse particulate organic matter, $> 1-2$ mm; FPOM = fine particulate organic matter, $< 1-2$ mm > 0.45 μm; DOM = dissolved organic matter, < 0.45 μm; GPP = gross primary production). All values kcal $m^{-2}yr^{-1}$ calculated by conversion of organic matter ash free dry wt using Cummins and Wuycheck (1971).

Parameters	Bear Brook	Watershed 10	Rattlesnake Creek[1]	Augusta Creek	Deep Creek	White Clay Creek	Fort River
Order	2	1	1	1	2	3	4
Budget Type	Reach (1700 m)	Watershed	Watershed	Watershed	Reach (Station 3)	Reach	Reach (1700 m)
INPUTS							
POM	3226	2633	2728	2317	142758	3030	36000
CPOM	(2733)	(2369)	(1215)	(355)	(2490)	(2153)	(2250)
FPOM	(493)	(263)	(1513)	(1962)	(140268)	(878)	(33750)
DOM	2807	979	3699	1974	149085	4117	100125
GPP	10	90	5000[3]	38	14245	3782	2700
SUBTOTAL	6033	3701	11427	4328.5	306088	10929	138825
OUTPUTS							
EXPORT							
POM	1201	176	2726	858	169099	3286	31500
CPOM[4]	(927)	(116)	(1213)	---	(2469)	(1894)	(900)
FPOM	(274)	(60)	(1513)	---	(166639)	(1392)	(30600)
DOM	(2797)	(1476)	(3201)	(1022)	(149085)	(4117)	(101700)
Respiration	2035	1053	3611[3]	2449	12124	3031	5625
SUBTOTAL	6033	2704	9538	4329	330308	10434	138825

Parameters	Bear Brook	Watershed 10	Rattlesnake Creek[1]	Augusta Creek	Deep Creek	White Clay Creek	Fort River
Storage	0	806	0	0	0	0	0
TOTAL	6033	3510	9538	4329	330308	10434	138825
Difference in Input-Output Balance[2]	0	191	1889	0	-24220	495	0

[1] Cushing and Wolf, 1982
[2] No difference reported in some because values obtained by difference.
[3] Does not include aerial respiration of emergent vegetation.
[4] Vascular hydrophyles plus benthic organic matter.

Table 5. Organic budget components, expressed as percent, for 7 North American streams.

Parameters	Bear Brook	Watershed 10	Rattlesnake Creek[1]	Augusta Creek	Deep Creek	White Clay Creek	Fort River
Order	2	1	1	1	2	3	4
Budget Type	Reach (1700 m)	Watershed	Watershed	Watershed	Reach (Station 3)	Reach	Reach (1700 m)
INPUTS							
POM	53.5	71.1	23.9	53.5	46.6	27.7	25.9
CPOM	(45.3)	(64.0)	(10.6)	(8.2)	(0.8)	(19.7)	(1.6)
FPOM	(8.2)	(7.0)	(13.2)	(45.3)	(45.8)	(8.0)	(24.3)
DOM	46.3	26.4	32.4	45.6	48.7	37.7	72.1
GPP	0.2	2.5	43.7	0.9	4.7	34.6	2.0
TOTAL	100.0	100.0	100.0	100.0	100.0	100.0	100.0
OUTPUTS							
Exports	66.2	26.6	62.1	43.4	96.3	71.0	95.9
POM	19.9	6.9	28.6	19.8	51.1	31.5	22.6
CPOM	(15.4)	(4.6)	(12.9)	---	(0.7)	(18.2)	(0.6)
FPOM	(4.5)	(2.3)	(15.9)	---	(50.4)	(13.3)	(22.0)
DOM	(46.3)	(19.6)	(33.6)	(23.6)	(45.1)	(39.5)	(73.3)
Respiration	38.8	41.6	37.9	56.6	3.7	29.0	4.1
Storage	0	31.8	0	0	0	0	0
TOTAL	100.0	100.0	100.0	100.0	100.0	100.0	100.0

community respiration was a significant portion (29 to 56.6%) of the organic budget output.

Many of the differences in budget characteristics are artifacts of system delimitation, thus in reaches of larger streams such as Deep Creek and Fort River, fluvial import of organic matter assumes greater importance relative to inputs measured on an area basis such as litterfall. Respiration, per unit area exhibits a twelve-fold range in the various systems when reported on a kcal/m^2 basis, and an eleven-fold range on the basis of percent of total output which is not areally based. The heavy macrophytic producing streams (Fort River and Deep Creek) have the highest absolute rates of community respiration yet the lowest percent output by respiration. The remaining output is by fluvial export which is naturally high in these rivers.

The hydrologic data shown in Table 2 should be interpreted in the context of the long term patterns in each watershed. Because the U.S.G.S. water year covers the period of October 1 through September 30, stream discharge records correspond well with the "detritus year", conveniently defined in the temperate zone as extending from one period of leaf fall to the next. Peak 24-hour flows for the budget years compared (Tables 4 and 5) are in the range of 1 to 5-year flood recurrence intervals of annual maximum daily discharge, based on 8-24 year records. Although similar flood years allow for more legitimate comparisons, the absolute organic budget values given for each site are not necessarily representative of the long term.

The ratio of export to import (E/I) and primary production to community respiration (P/R) for the seven streams are compared in Table 6. First-order streams (except Rattlesnake Springs) and second-order Bear Brook had P/R ratios less than one. Watershed 10, Augusta Creek, and Bear Brook are all heavily shaded systems with significant allochthonous inputs, and measured organic matter export was two-thirds or less of imports in these streams. In Rattlesnake Springs, Deep Creek, and Fort River systems with P/R > 1, and E/I near to or > 1, the majority of POM particulate export was derived from extensive beds of aquatic macrophytes. The seven study streams' relationships shown in Table 6 follow the prediction of Fisher and Likens (1973) that in systems having P/R < 1 (heterotrophic) imports exceed exports, when P/R > 1 exports exceed imports, and they balance when P/R = 1 (see Fort River, Table 4). However, the ratios given in Table 6 are only approximate, because of problems in accurately measuring detrital export and derivation of major budget items by difference, which essentially makes them a function of export.

Table 6. Comparison of primary production/community respiration (P/R) and organic matter export/organic matter import (E/I) ratios for the seven flowing water systems (United States) compared in Tables 2-5)

Ecosystem Parameters	Watershed 10 (Oregon)	Augusta Creek (Michigan)	Rattlesnake Springs (Washington)	Bear Brook (New Hampshire)	Deep Creek (Idaho)	White Clay Creek (Pennsylvania)	Fort River Reach (Massachusetts)
Stream Order	1	1	1	2	2	3	4
E/I	0.26	0.44	0.83	0.66	1.09	1.04	0.98
P/R	0.09[1]	0.50	1.38	0.01[1]	1.18	1.25[1]	0.96[1]

[1]Minimal estimates calculated assuming gross primary production equals twice net.

Comparison of reach and watershed methods of budget calculations. It would be useful to be able to compare hydrologically and geomorphically diverse stream ecosystems in terms of over-all biologic functions with some index of efficiency of organic matter processing. The desired index must be applicable to both reach and watershed stream budgets and should be insensitive to such arbitrary features as ecosystem size. Fisher and Likens (1973) suggested that Respiration per Total Input (R/I), termed ecosystem efficiency, may be an index of general utility in this regard.

For stream ecosystems studied on a reach basis, for example Bear Brook, inputs such as litter and primary production, as well as respiratory output, occur on an areal basis. If the size of the study reach were increased, these parameters would increase proportionately (on a whole system basis). However mainstream input (transport from upstream) is constant, regardless of the study reach length below the input site. Therefore, to arbitrarily increase reach length is to automatically increase efficiency since R increases while mainstem input declines in importance on a unit areas basis. Thus, ecosystem efficiency, as originally defined by Fisher and Likens (1973), is of no utility in comparing diverse stream ecosystems.

For whole watershed stream studies, however, mainstem and tributary inputs are irrelevant since all channels are designated as part of the system. Inputs of organic matter in flowing water occur only as groundwater or, in some cases, overland flow directly from the terrestrial system. To increase system size by considering all channels in a watershed does not automatically alter the ratio of R to total input, since all fluxes can be calculated on an areal basis. Groundwater input to stream channels is a function of watershed area, and if drainage density remains constant as system size is increased, so does areal groundwater input (baring groundwater losses). Therefore, watershed budget studies can be compared legitimately, regardless of size.

We suggest two efficiency indices of general utility:

Ecosystem Efficiency (EE) = $R/(P+L+G)$
Retention Efficiency (RE) = $(R \pm \Delta S)/(P+L+G)$
where R = ecosystem respiration; P = gross primary production; L = litter input; G = groundwater input; and S = organic matter storage.

All units can be expressed as kilocalories $\cdot$ m^{-2} $\cdot$ yr^{-1} (or as grams organic matter $\cdot$ m^{-2} $\cdot$ yr^{-1}).

Ecosystem efficiency (Fisher and Likens, 1973) indicates the extent to which all organic matter inputs are respired by the system per unit time (e.g., annually). Retention efficiency indicates the extent to which all inputs are either respired or stored and is the preferred general expression for handling non steady-state systems. At steady state, the expressions are equivalent, although the difference between fast processing and low retention needs to be carefully distinguished.

Bear Brook data can be converted to a watershed basis and compared to Bear Brook reach data (Table 7). Since all channels are now included, ecosystem area increases 2.4 fold as do all areal fluxes. Tributary and mainstem inputs become zero and DOM input declines per m^{-2}. Ecosystem efficiency (EE) for Bear Brook as a watershed is then 0.61 as compared to 0.37 for Bear Brook as a reach ecosystem. The latter value is unique to the arbitrarily defined reach studied and has no general applicability.

Ecosystem efficiency for the other apparently steady-state watershed stream, Augusta Creek, is 0.56. Rattlesnake Creek and Watershed 10 are not at steady state and must be described with Retention Efficiencies. RE's for these two systems are 0.22 and 0.49 respectively compared to 0.56 and 0.61 for Augusta Creek and Bear Brook.

These data suggest that for a variety of watershed streams (except Rattlesnake Creek), more than half of all inputs are retained and/or processed to CO^2 on an annual basis. If we extend the model unaltered to the watersheds of large rivers, we would predict that less than half of all organic matter inputs to streams enter oceans. Realized retention efficiencies of large rivers may in fact be greater than those of small watershed streams. For example, litter inputs decline per unit of channel area in wider streams and primary production may not increase compensatorily. Large rivers and small to mid-sized streams with high drainage density are quite difficult to treat directly as watershed streams and have consequently been studied as stream reach ecosystems (e.g., Deep Creek, White Clay Creek, Fort River). As previously stated, budget data for reaches cannot be used to compute EE and RE due to the problem of system size. How then might stream reaches be compared? Stream metabolism index (SMI; Fisher, 1977) can be used to compare stream reaches and is defined as:

$$SMI = (R \pm \Delta S)/(P+L+T+M+G) - (Q_f)(M)$$

Where T = organic matter input via tributaries; M = Organic Matter input at mainstem upstream site; Q_f = discharge at mainstem output site/discharge at mainstem input site. Other variables are as previously defined.

Table 7. Comparison of reach and whole system budgets for Bear Brook, New Hampshire.

	Reach		Watershed	
	(Segment)[1]		(All Channels)[2]	
Length	1704m		5174m	
Area	5877m^2		14,029m^2	
	Kg stream^{-1}	g m^{-2}	Kg stream^{-1}	g m^{-2}
Litter	3260		7791	
Lateral transport	547		1162	
Litter fall	43		103	
	(3850)	(655)	(9556)	(681)
CPOM	640	109	0	
FPOM	155	26	0	
DOM, surface	1580	269	0	
subsurface	1800	306	3380	
	(4175)	(710)	(3380)	(241)
P_G	13	2	31	2
Total Input	8038	1368	12,967	924
CPOM	1370	233	1370	98
FPOM	330	56	330	24
DOM	3380	575	3380	241
	(5080)	(864)	(5080)	(362)
Respiration	2958	503	7887	562
Total Ouput	8038	1368	12,967	924
R/Input	.37		.61	
R/Particulate	.37		.61	
SMI	1.0		.82	

[1] Fisher and Likens, 1973.
[2] Unpublished but calculated from data in Fisher and Likens, 1973.

As with EE and RE, SMI measures the efficiency of
respiration or storage of some organic input. The denominator of
the expression represents all inputs that enter the reach from
all sources less mainstem input (fluvial transport) corrected for
accrual of water through the reach. That is R ± ∆S is judged
relative to excess inputs which would otherwise cause organic
matter concentration to increase across the system. An increase
in organic matter concentration in transport across the system is
termed loading in this context, thus if SMI = 1.0, output water
would have the same organic matter concentration as mainstem
input water, and the system would not load. If SMI < 1.0, organic
matter concentration would increase across the system and if SMI >
1.0, concentration would decline. The specified performance
criterion used here (zero loading) is admittedly arbitrary - the
system may merely "hold its own" and prevent concentration
increases linearly.

For watershed stream systems, SMI reduces as follows:

$$SMI = (R \pm \Delta S) / P+L+O+O+G - (Q_f)(M)$$

and, since M is analogous to G;
$$SMI = (R \pm \Delta S) / P+L+G) - (1.0)(G)$$
$$SMI = (R \pm \Delta S) / P+L$$

Thus any watershed stream which respires or stores organic
matter equivalent to primary production and litter inputs will
have SMI = 1.0 and will export water with an organic matter
concentration equivalent to the DOM in groundwater. If SMI = 1.0
all the way to the sea, organic matter concentration in water
entering estuaries will be equivalent to that in groundwater
entering headwater streams. In fact, SMI may vary widely from
headwater to estuaries, being > 1.0 in some regions (e.g., below
sites of organic enrichment) and < 1.0 in others.

Reach SMI's were computed for Bear Brook and Fort River
systems (Table 7). Bear Brook had an SMI of 1.0 while Fort River
SMI = 0.66. If Fort River is assumed to be a steady-state
system, respiration is only 66% of that required to prevent
loading. As estimated, Bear Brook does not load because all
"excess" inputs are respired.

Watershed stream SMI's for WS10 and Bear Brook are 0.62 and
0.82 respectively, although WS10 exhibits an almost two-fold
range in two consecutive years (Table 7). Both systems load, in
that SMI < 1.0. Loading in watershed streams occurs as
groundwater containing only DOM rapidly picks up more DOM (Kaplan
et al., 1980) and POM via litter inputs. Thus, we might expect
that loading is the general rule in headwater systems as a
consequence of the SMI as defined.

While all three efficiencies described here are legitimate indices for comparing different streams on a watershed basis, all are sensitive to discharge and accompanying fluvial transport of organic matter. In the two years' of budget data available for WS10 in Oregon, inputs were low in the relatively dry water year 1973, ΔS was high, and RE was 0.73 (Table 7). In wet water year 1974, input was higher, yet respiration remained constant and RE dropped to 0.34. Retention efficiency was thus generally an inverse function of discharge which in turn increased total input and ΔS was negative. Yet in both years, ΔS was positive. During years with unusually high discharges we can envisage a negative ΔS that exceeds respiration. Under those conditions, RE would be negative. In steady-state systems increased discharge lowers RE simply by increasing organic matter input at (presumably) constant R. SMI behaves similarly in both watershed and reach systems even though the absolute values are different. At relatively constant R, increased discharge lowers SMI, thus the system loads to a greater extent during wet than during dry years. For WS10, SMI's for dry and wet years are 0.84 and 0.46, respectively.

In summary, we see that while the efficiency indices proposed here are conceptually sound and can be applied to all watershed stream ecosystems, only SMI can be applied to reach systems. To date, no satisfactory analogue of RE suitable for use on reach systems, has been devised, and thus reach and watershed stream systems cannot be compared in this regard. More importantly perhaps, all efficiency indices are highly sensitive to fluctuations in discharge and decomposition rate of organic matter. Because we have shown that discharge exhibits great year to year variation, efficiency values of whatever type, are of little utility when based on data from a single year. Not only is discharge a critical variable shaping seasonal and even diel patterns in efficiency, but several other factors also greatly influence efficiency on a short-term basis. Temperature, cumulative degree days, insolation, litter input, primary production, and the heterotrophic-autotrophic status of communities, among other variables, shown much greater variation of diel or seasonal periods than from year to year. All influence efficiency. Thus the annual organic budget at best represents a temporal compromise which is too short for revealing stream function on the time scale of geomorphic change and too long for elucidating biologic control of critical processes.

Whole Basin Budgets by Stream Order

As discussed above, quantification of organic budgets has most frequently involved inputs and outputs from entire small (first order) watersheds (e.g., Fisher and Likens, 1973; Sedell et al., 1974) and occasionally from discrete reaches (e.g.,

Fisher, 1977). The River Continuum Concept visualizes the stream-river as a continuum of nutrient turnover processes and population assemblages which are predictably adapted to the most probable physical state of a river system along its drainage network (Vannote et al., 1980).

Comparisons of the four different biome stream-river systems studied by the River Continuum Group were made by estimating annual organic carbon budgets. Seasonal terrestrial input, primary production, respiration, storage, and transport data from the 16 sample sites were calculated by stream order using basin geomorphic and hydrologic characteristics (Table 8 and 9). The annual carbon budget gives a first approximation of absolute carbon fluxes for all channels of a given order in each basin. By comparing the inputs (litter, gross primary production, and [non-flood event] transport to order n from n-1) to outputs (respiration and transport) and the difference between them to POM storage (except large wood), each basin was evaluated with respect to changes in carbon storage as to general aggrading (storing organic carbon) or degrading (exporting) condition. Storage divided by the excess of inputs or outputs provides an estimate of the number of years of storage for the given output rate (Table 9) - that is, given the current loss/accumulation rate, the number of years required to remove/accumulate the present storage pool. About 60% of the 23 (by order) systems showed a pattern of aggradation with outputs less than inputs.

Headwater Oregon streams (orders 1 and 2) and all Michigan streams (except first-order) were characterized by large amounts of storage (Table 9). All other stream orders in all four biomes appeared to be quite active - either rapidly degrading or aggrading with estimated periods for accrual or loss of existing storage of less than two years (< 0.1 to 1.7). Only one of the 23 order-systems (Oregon fifth order) was evaluated as being close to equilibrium (inputs-outputs = 5.6 Ton C). In all but one case (Idaho second order), the difference between inputs and outputs was about an order of magnitude (11 cases) or less (11 cases) than storage. At the basin level, all systems except Oregon (due to large sixth and seventh order export) appear to be aggrading. Storage in Michigan streams larger than order 1 was dominated by massive amounts of FPOM, representing between about 2 and 13 years accrual at measured input-output rates.

Inclusion of large wood (generally > 10 cm diameters) in the whole basin estimates changes the input-output balance considerably. Excluding wood-dominated Oregon streams, all basins were estimated to be aggrading with enough storage to account for between 1 and 14 years of accrual at existing input-output rates. By contrast, about 60 years would be required to remove coarse wood in the Oregon basin.

Table 8. Basin characteristics used in calculating order-
specific budgets. Stream length-watershed basin
relationships as: Watershed area = (Stream Length)^B.

OREGON

Order	No. of Streams	Mean Watershed Area (Km^2)	Mean Length (Km)	Mean Stream Width (m)	Total Channel Area (Km^2)
1	346[a]	0.2	0.96	0.6	0.20
2	87	0.2	1.1	1.8	0.17
3	22[a]	5.2	2.8	3.0	0.18
4	8	0.8	3.0	7.5	0.18
5	3[a]	16.5	6.6	12.0	0.24
6	2	178.9	15.3	26.0	0.80
7	1[a]	484.6	27.9	40.0	1.10
Total	469	484.6	590.9	--	2.87

IDAHO

Order	No. of Streams	Mean Watershed Area (Km^2)	Mean Length (Km)	Mean Stream Width (m)	Total Channel Area (Km^2)
1	852	0.2	0.5	0.3	0.13
2	252[a]	0.8	0.8	0.5	0.04
3	63	7.4	2.3	3.2	0.29
4	16[a]	70.6	5.0	6.0	0.26
5	2[a]	738.1	35.0	30.0	1.80
6	1[a]	1238.3	56.0	46.0	0.97
7	--	--	--	--	--
Total	1186	1238.3	728.8	--	3.49

MICHIGAN

Order	No. of Streams	Mean Watershed Area (Km^2)	Mean Length (Km)	Mean Stream Width (m)	Total Channel Area (Km^2)
1	80[a]	0.4	0.6	1.5	0.07
2	13[a]	3.4	26.4	6.2	2.13
3	8[a]	3.6	23.1	8.0	1.11
4	2	103.3	148.4	27.8	8.25
5	1[a]	777.3	451.9	47.5	21.50
6	--	--	--	--	--
7	--	--	--	--	--
Total	104	777.3	1275.3	--	33.06

(Contiued)

Table 8.   Continued

PENNSYLVANIA

Order	No. of Streams	Mean Watershed Area (Km^2)	Mean Length (Km)	Mean Stream Width (m)	Total Channel Area (Km^2)
1	59[a]	0.3	0.5	1.5	0.04
2	17[a]	0.8	2.5	3.0	0.13
3	6[a]	1.0	5.3	6.2	0.16
4	3[a]	24.5	9.1	11.8	0.32
5	1	37.5	17.7	17.4	0.31
6	--	--	--	--	--
7	--	--	--	--	--
Total	86	37.5	139.7	--	0.96

[a] Stream order for which a sampling site was included in the Continuum study.
[b] Upper Salmon River from headwaters to Yankee Fork.

Table 9. Watershed budgets by stream order for the basins in each of the regional biomes. All values in metric tons carbon per year for all channels of a given stream order in a basin (i.e., Ton C · order n^{-1} · yr^{-1}).

	OREGON	IDAHO	MICHIGAN	PENNSYLVANIA
ORDER 1				
Inputs				
Litter	73.6	3.3	22.7	21.9
Gross Primary Production	3.6	11.2	2.3	2.7
Total	76.6	14.5	25.0	24.6
Outputs				
Transport	39.5	459.6	180.6	38.0
Respiration	8.1	15.2	6.0	5.7
Total	47.6	474.8	186.6	43.7
Storage[a]	142.0	12.7	21.1	8.1
Input-Output[b]	29.0	-445.1	-161.6	-19.1
Years of Storage[c]	4.9	0.1	0.1	0.4
ORDER 2				
Inputs				
Litter	4.8	1.4	471.5	14.8
Gross Primary Production	39.5	459.6	180.6	38.0
Transport[d]	62.1	1.0	319.5	71.2
Total	106.4	464.4	971.6	211.4
Outputs				
Transport	24.1	220.0	27.7	18.8
Respiration	7.6	1.9	473.3	18.5
Total	31.7	221.9	504.0	37.3
Storage[a]	138.9	3.9	949.4	12.1
Input-Output[b]	74.7	242.5	476.6	174.1
	(A)	(A)	(A)	(A)
Years of Storage[c]	1.9	0.1	2.0	0.1

(Continued)

Table 9. Continued.

	OREGON	IDAHO	MICHIGAN	PENNSYLVANIA
ORDER 3				
Inputs				
Litter	5.7	13.7	174.3	87.7
Gross Primary				
Production	7.0	26.0	233.1	72.2
Transport[d]	24.1	220.0	27.7	18.8
Total	96.8	259.7	435.1	178.7
Outputs				
Transport	21.1	531.3	96.9	50.5
Respiration	8.7	22.0	274.2	19.3
Total	29.8	553.3	371.1	69.8
Storage[a]	21.1	24.0	535.0	20.2
Input-Output[b]	67.0	-293.6	64.0	108.9
	(A)	(D)	(A)	(A)
Years of Storage[c]	0.3	0.1	8.4	0.2
ORDER 4				
Inputs				
Litter	65.7	12.3	288.9	175.4
Gross Primary				
Production	8.5	43.4	2498.1	60.0
Transport[d]	21.1	531.3	96.9	50.5
Total	95.3	587.0	3183.9	285.9
Outputs				
Transport[d]	2.8	131.1	466.0	269.4
Respiration	8.7	26.8	2473.4	60.4
Total	11.5	157.9	2939.4	329.8
Storage[a]	14.1	17.7	3852.8	27.9
Input-Output[b]	83.8	429.1	244.5	-43.9
	(A)	(A)	(A)	(D)
Years of Storage[c]	0.2	0.04	15.8	0.6

	OREGON	IDAHO	MICHIGAN	PENNSYLVANIA
ORDER 5				
Inputs				
Litter	87.6	11.9	850.9	115.1
Gross Primary Production	13.2	376.7	8869.6	74.0
Transport[d]	2.8	131.1	466.0	269.4
Total	103.6	519.7	10186.5	458.5
Outputs				
Transport	86.5	619.9	1245.5	91.6
Respiration	11.5	289.8	7580.9	79.6
Total	98.0	909.7	8826.4	171.2
Storage[a]	9.6	290.7	9718.0	15.0
Input-Output[b]	5.6	-390.0	1360.1	287.3
	(A)	(D)	(A)	(A)
Years of Storage[c]	1.7	0.7	7.1	0.1
ORDER 6				
Inputs				
Litter	135.1	14.8		
Gross Primary Production	52.0	101.2		
Transport[d]	86.5	619.9		
Total	273.6	735.9		
Outputs				
Transport	726.4	67.8		
Respiration	43.0	71.3		
Total	769.4	139.1		
Storage[a]	63.6	102.3		
Input-Output[b]	-495.8	596.8		
		(A)		
Years of Storage[c]	0.1	0.2		

(Continued)

Table 9. Continued.

	OREGON	IDAHO	MICHIGAN	PENNSYLVANIA
ORDER 7				
Inputs				
Litter	120.5			
Gross Primary Production	82.9			
Transport[d]	726.4			
Total	929.8			
Outputs				
Transport	1162.3			
Respiration	65.3			
Total	1227.6			
Storage[a]	42.9			
Input–Output[b]	−297.8			
Years of Storage[c]	0.1			
BASIN TOTAL				
Storage[a]	432.2	451.3	15076.3	83.2
Input	1725.7	2466.0	14650.6	1067.8
Output	3308.3	2337.0	12175.2	680.1
Input Output[b]	−533.8	139.7	1983.6	507.3
	(D)	(A)	(A)	(A)
Years of Storage[c]	0.8	3.5	6.1	0.2
Coarse Wood Storage[e]	32350	1537	5309	583
Years of Storage (including wood)	61.4	14.2	10.3	1.3

[a] Storage not including coarse wood (approximately > 10 cm diameter; also fine wood < 10 cm > 2 cm probably underestimated).

[b] If output > inputs (I–O = negative value) the system was degrading (D) the year of measurement and generally storage should be small. If output < input (I–O = positive value) the system was aggrading (A) and storage should be large. Balanced inputs and outputs indicated equilibrium (E) for the sample year.

[c] When inputs (I) exceed outputs (O) (I-O) = positive) the years of storage entry indicates length of time that would be required to accumulate the observed storage at the calculated aggradation rate. When outputs exceed inputs (I-O) = negative) the years of storage entry indicates length of time that would be required to remove observed storage at the calculated degradation rate.

[d] This assumes that all transport measured in stream order n-1 is input for stream order n.

[e] Coarse wood approximately >10 cm volumes estimated by separate inventories. Annual coarse wood export from basin assumed = 0.

ECOSYSTEM COMPARISONS

Despite the problems in calculating stream budget-derived efficiencies, lotic researchers (e.g., Fisher and Likens, 1973; Bormann et al., 1974; Bormann and Likens, 1979) have joined system ecologists in making efficiency calculations (Reichle, 1975; Reichle et al., 1975; Webster et al., 1975; O'Neill et al., 1975). The importance of allochthonous inputs and the unidirectional flow of running waters make comparisons of streams with terrestrial and lentic systems difficult, although steep sloping terrestrial systems also exhibit unidirectional "flow". Streams are not mere conduits which export terrestrial products from watersheds, but, as with other systems, are physically retentive and biologically active. Organic matter and inorganic nutrients are partially cycled within a given reach of stream or river with some portion being released to the reach below or retained in storage (Elwood et al., 1982; Minshall et al., 1983; Newbold et al., 1982). If the terrestrial community through which a stream flows accumulates organic matter over long periods of time, the stream should behave in a generally similar fashion, although specific processing rates and storage capacities would differ.

If comparisons of lotic with terrestrial ecosystems are made on the basis of net ecosystem production (NEP; Reichle, 1975; Reichle et al., 1975), particulate and dissolved organic matter inputs (I) must be included in the stream calculations. The comparative relationships would be (Batzli, 1974):

Terrestrial NEP = GPP - R_E

Stream (aquatic) NEP = $[GPP + I] - [R_E + E] = \Delta S$,

where GPP = gross primary production, R_E = ecosystem respiration, E = organic matter exports, and ΔS = change in storage pools Based on measurements made on WS10, net ecosystem production can be calculated as:

$$NEP = (90 + 3612) - (1053 + 1651) = 988 \; Kcal \; m^{-2} \; yr^{-1}$$

(mean of actual measurement of ΔS = 806 Kcal m^{-2} yr^{-1})

If ecosystem productivity is calculated as NEP divided by GPP + I (Reichle, 1975; Reichle et al., 1975), the value is 0.27 for WS10, higher than for various terrestrial communities (0.05 – 0.24 [Reichle et al., 1975]) and in the range for lakes (0.30 – 0.53, calculated from Wissmar, unpublished data, University of Washington). However, given the uncertainties of budget measurements in streams, particularly temporal and spatial variations in storage, the calculation of such ecosystem parameters will require very extensive data sets. Also, the range of values through the drainage net (Vannote et al., 1980) would predictably cover at least as wide a range as terrestrial systems, particularly since stream system differences are compounded by varying degrees of terrestrial (riparian) influence. Nevertheless, when adequate data permit it, within and between ecosystem comparisons of material balance budgets should provide useful insights into system function.

DISCUSSION

    As pointed out initially, a primary objective of stream organic budget assessments has been the determination of ecosystem functional properties, which would allow comparisons within and between biomes. Because of the temporal and spatial problems inherent in material balance budgets for organic matter in streams, the results can be ambiguous and misleading. For example, budgets constructed primarily from export data are more a feature of physical retention characteristics than biological function. Therefore, present budget-derived perceptions of stream ecosystem efficiency are dependent on storage and export phenomena which tend to obscure the significance of biological processes. Since organic matter budgets for streams have been used in the calculation in efficiencies employed in ecosystem comparisons, the conclusions resulting from such comparisons must be evaluated carefully.

    The basic problem with existing organic budgets is that the steady state, or dynamic equilibrium, that is the range over which the parameters fluctuate, is not time invariant (Botkin and Sobel, 1975) for running water systems. Depending upon the time frame of reference, flood events of different magnitude can be

variously considered as perturbations that displace streams from
stable trajectories or as fundamental ecosystem features and,
therefore, within the boundary conditions (i.e., variance) of the
trajectory. Major components of lotic ecosystems, such as
organic storage, are largely dependent on the timing and
magnitude of discharge events, and periodic adjustments in the
size of the organic storage pool are features of the system, not
perturbations (e.g., Dawson, 1980). Since any definition of a
steady state or dynamic equilibrium condition for stream
ecosystems is highly time dependent, so are the boundary
conditions for "normal stability." Therefore, the concepts of a
system's resistance to, and resilience after, a change (Webster
et al., 1975, this volume), for example in storage, must be so
narrowly defined for a given running water system that their
general usefulness may be significantly reduced.

As we have shown, existing material balance budgets for
streams reflect physical transport and storage characteristics to
a much greater extent than the metabolic properties of the stream
ecosystem. The large majority of materials are exported during
flows which occur for less than ten percent of the time. Since
flows vary from year to year, time dependent budgets are not
useful unless they have been constructed over a long period
(possibly ten years or more). While organic budgets have been
useful in showing the retentive capacity of streams to be fairly
efficient, the separation of the metabolic properties of the
ecosystem from the physical hydraulic ones has yet to be done.
Flood transported organic particulate matter is seldom quantified
nor is the rate of decomposition of the material estimated. For
example, the origin of major exports from a watershed (or basin)
may be from outside the channel or in other deep storage pools
that are characterized by minimal biological activity. Also, the
passage of the material may be so rapid as to preclude
significant biological response.

A flow-dependent analysis is available (e.g., Sedell et al.,
1978) which might prove useful in characterizing organic
transport between reaches and allow for the interpretation of the
variability between geomorphic setting and hydrologic regime. An
additional advantage is that one runoff season would be
sufficient if representative flows are monitored. Since the
sediment and water moving through a stream channel are the
primary independent variables influencing channel morphology,
quantitative relations have been established between water and
sediment and all aspects of channel morphology such as dimension,
shape, gradient, and pattern have been related to stream
discharge. The ability of water to do work, or stream energy,
combined with a measure of reach retention might serve as a
suitable common denominator for comparing differences in organic

transport and storage characteristics between streams (Minshall et al., 1983).

Geomorphologists and hydrologists (Leopold et al., 1968; Pfankuch, 1975) have shown a relationship between stream energy and the size of suspended and bedload inorganic material transported by streams. Stream power, defined in terms of discharge, percent channel slope, and density (or mass) of water per unit channel width (Leopold and Langbein, 1962; Langbein and Leopold, 1964; Leopold et al., 1968), seems to be a logical unit for comparing streams of different sizes. However, the power expression needs to be scaled with a roughness term to account for a stream's retention characteristics prior to plotting against biological parameters (e.g., primary production). Some empirically derived scaling system for retention is badly needed (see Minshall et al., 1983 for possible approaches).

Further evaluation of ecosystem properties will come with improved understanding of storage dynamics. Input-output dynamics of the storage pools in aquatic ecosystems need to be examined in terms of 1) the frequency and magnitude of events which reset the quantity and composition of material in storage, 2) the relationship between the source of exported material, its movement through a system and the stream biological response time (for example, if material that was rarely or ever in the channel is moved through and out of a watershed at a rate that does not allow for significant biological activity (Meyer and Likens, 1979), its inclusion in biologically oriented budget calculations is not warranted and 3) the more continuous, low level leakage in and out of a storage compartment between major turnover events, and the internal processing dynamics of each storage pool should be evaluated in relation to temperature, substrate quality, conditioning (microbial colonization and metabolism of the organic substrates), and oxygen environment.

CONCLUSIONS

Many of the problems with organic matter budgets discussed in this paper suggest that the field of stream ecology will not benefit from the development of budgets for numerous systems based on short-term (1-2 yr) records and in which some key parameters are determined by difference. Small-scale research programs should focus on comparisons of selected processes between diverse stream ecosystems rather than attempt to determine whole system budgets. What is needed is the determination of total stream ecosystem budgets at a few selected sites where existing long-term data sets can be continued and augmented by improved monitoring of storage dynamics.

These studies should view the system on scales appropriate to system behavior. For example, log dynamics must be viewed on the time-scale of decades and centuries, but this does not mean that the studies themselves must be of this duration. Using dendrochronologic and mapping methods, it is possible to reconstruct histories of debris inputs to, and catastrophic export from, streams for periods up to more than a centruy (Sigafoos, 1964; Swanson et al., 1976). Improved understanding of storage dynamics will arise from other types of studies not commonly a part of budget development. Such studies include long-term monitoring of decomposition by repeated sampling of marked organic materials placed on the stream bed and in the sediment. There is also a particular need for study of the effects of flood on the storage and biota in streams. Stream ecologists need to be opportunistic and prepared to respond to infrequent episodic events.

Floods, large debris in streams, and major changes in storage are elements of fluvial geomorphology. It is geomorphic processes and features which establish the physical template on which stream biology is developed and maintained. Future progress in stream ecosystem analysis will be dependent in large measure on the ability of stream ecologists to incorporate understanding of physical and historical processes into models of lotic ecosystem behavior and functioning.

ACKNOWLEDGMENTS

This paper was first written in 1976. Most of the ideas were not new at that time and certainly stemmed from each author's interactions with other groups and projects as well as their own research. Since then, the paper has gone through many versions and revisions and was widely circulated. Others have integrated their concepts with ours and presented elsewhere some aspects of the problems treated here. However, we believe now, as we did in 1976, that the paper integrates a variety of perspectives addressing problem areas in stream ecosystem energy budget work and constitutes a useful contribution.

Original data collection and concept development reported in this paper were supported by the following research grants: River Continuum Project (NSF BMS-75-07333 and DEB-7811671); U.S. Department of Energy contracts DE AT06-79-EV1004 and AT(45-1)-1830; Coniferous Biome (NSF 7602656); Desert Biome (NSF GB-15886). River Continuum Contribution No. 22; Technical Paper No. 0000 Oregon Agricultural Experiment Station.

We wish to thank S. V. Gregory, J. Van Sickle, J. L. Meyer, L. Kaplan, J. D. Hall and R. J. Naiman for their reviews of the manuscript.

REFERENCES

Anderson, N.H., J.R. Sedell, L.M. Roberts and F.J. Triska. 1978.
The role of aquatic invertebrates in processing wood debris
in coniferous forest streams. Am. Midl. Nat. 100:64-82.

Anderson, N.H. and J.R. Sedell. 1979. Detritus processing by
macroinvertebrates in stream ecosystems. Ann. Rev. Ent.
24:351-357.

Batzli, G.O. 1974. Production, assimilation and accumulation of
organic matter in ecosystems. J. Theor. Biol. 45:205-217.

Bilby, R.E. 1981. Role of organic debris dams in regulating the
export of dissolved and particulate matter from a forested
watershed. Ecology 62:1234-1243.

Bilby, R.E. and G.E. likens. 1979. Effects of hydrologic
fluctuations on the transport of fine particulate organic
carbon in a small stream. Limnol. Oceanogr. 24:69-75.

Bilby, R.E. and G.E. Likens. 1980. Importance of organic debris
dams in the structure and function of stream ecosystems.
Ecology 61:1107-1113.

Boling, R.H., E.D. Goodman, J.O. Zimmer, K.W. Cummins, R.C.
Petersen, J.A. Van Sickle, and S.R. Reice. 1975a. Toward a
model of detritus processing in a woodland stream. Ecology.
56:141-151.

Boling, R.H., R.C. Petersen, and K.W. Cummins. 1975b. Ecosystem
modeling for small woodland streams, pp. 183-204. In: B.C.
Patten (ed.), Systems analysis and simulation in ecology.
Vol. 3 Academic Press, NY. 601 pp.

Bormann, F.H., G.E. Likens, and J.S. Eaton. 1969. Biotic
regulation of particulate and solution losses from a forest
ecosystem. BioScience 19:600-610.

Bormann, F.H., G.E. Likens, T.G. Siccama, R.S. Pierce, and J.S.
Eaton. 1974. The export of nutrients and recovery of
stable conditions following deforestation at Hubbard Brook.
Ecol. Monogr. 44:255-277.

Bormann, F.H. and G.E. Likens. 1979. Pattern and process in a
terrestrial ecosystem. Springer-Verlag, N.Y. 253 pp.

Botkin, D.B. and M.J. Sobel. 1975. Stability in time-varying
ecosystems. Amer. Nat. 109:625-646.

Brinson, M.M. 1976. Organic matter losses from four watersheds
in the humid tropics. Limnol. Oceanogr. 21:572-582.

Campbell, A.G. and J.F. Franklin. 1979. Riparian vegetation in
Oregon's western Cascade Mountains: composition, biomass
and autumn phenology. U.S. Int. Biol. Prog. Conif. Forest
Biome. Bull. 14, 90 pp.

Coleman, M.J. and H.B.N. Hynes. 1970. The vertical distribution
of the invertebrate fauna in the bed of a stream. Limnol.
Oceanogr. 15:31-40.

Crisp, D.T. and S. Robson. 1979. Some effects of discharge on
the transport of animals and peat in a north Pennine
headstream. J. Appl. Ecol. 16:721-736.

Cummins, K.W. 1974. Stream ecosystem structure and function. BioScience 24:631-641.

Cummins, K.W. 1975. The ecology of running waters; theory and practice. Proc. Sandusky River Basin Symp. Int. Joint Commission on Great Lakes, p. 277-293.

Cummins, K.W. 1977. From headwater streams to rivers. Am. Biol. Teacher 39:305-312.

Cummins, K.W. and J.C. Wuycheck. 1971. Caloric equivalents for investigations in ecological energetics. Mitt. Int. Verein. Limnol. 18:1-158.

Cummins, K.W., M.J. Klug, G.M. Ward, G.L. Spengler, R.W. Speaker, R.W. Ovink, D.C. Mahan and R.C. Petersen. 1981. Trends in particulate organic matter fluxes, community processes and macroinvertebrate functional groups along a Great Lakes drainage basin river continuum. Verh. Int. Verein. Limnol. 21:841-849.

Cushing, C.E. and E.G. Wolf. 1982. Organic energy budget of Rattlesnake Springs, Washington. Amer. Midl. Nat. 107:404-407.

Dawson, F.H. 1980. The origin, composition, and downstream transport of plant material in a small chalk stream. Freshwat. Biol. 10:419-435.

Dowdy, D.R. and N.C. Matalas. 1969. Analysis of variance, covariance, and time series, pp. 8-78 - 8-90. In: V.T. Chow (ed.), Handbook of applied hydrology. McGraw-Hill Co., New York.

Elwood, J.W., J.D. Newbold, R.V. O'Neill and W. Van Winkle. 1983. Resource spiraling: an operational paradigm for analyzing lotic ecosystems. In: T.D. Fontaine and S.M. Bartell (eds.), The dynamics of lotic ecosystems. Ann Arbor Science, Ann Arbor, Michigan (in press).

Fahey, T.J. 1979. Changes in nutrient content of snow water during outflow from a Rocky Mountain coniferous forest. Oikos 32:422-428.

Fisher, S.G. 1977. Organic matter processing by a stream-segment ecosystem: Fort River, Massachusetts, U.S.A. Int. Rev. ges. Hydrobiol. 62:701-727.

Fisher, S.G. and G.E. Likens. 1972. Stream ecosystem: organic energy budget. Bioscience 22:33-35.

Fisher, S.G. and G.E. Likens. 1973. Energy flow in Bear Brook, New Hampshire: an integrative approach to stream ecosystem metabolism. Ecol. Monogr. 43:421-439.

Froehlich, H.A. 1973. Natural and man-caused slash in headwater streams. Loggers Handbook. 638 pp.

Gosz, J.R., G.E. Likens, and F.H. Bormann. 1972. Nutrient content of litter-fall on the Hubbard Brook Experimental Forest, New Hampshire. Ecology. 53:769-784.

Gurtz, M.E., J.R. Webster, and J.B. Wallace. 1980. Seston dynamics in southern Appalachian streams: effects of clearcutting. Can. J. Fish. Aquat. Sci. 37:624-631.

Hall, C.A.S. 1972. Migration and metabolism in a temperate stream ecosystem. Ecology 53:585-604.

Hobbie, J.E. and G.E. Likens. 1973. The output of phosphorous, dissolved organic carbon, and fine particulate carbon from Hubbard Brook watersheds. Limnol. Oceanogr. 18:734-742.

Hynes, H.B.N. 1963. Imported organic matter and secondary productivity in streams. Proc. 14 Int. Congr. Zool. 3:324-329.

Hynes, H.B.N. 1970. The ecology of running waters. Univ. Toronto Press. 555 pp.

Hynes, H.B.N. 1974. Further studies on the distribution of stream animals in the substratum. Limnol. Oceanogr. 19:92-99.

Hynes, H.B.N. 1975. The stream and its valley. Verh. Int. Verein. Limnol. 19:1-15.

Hynes, H.B.N., D.D. Williams, and N.E. Williams. 1976. Distribution of the benthos within the substratum of a welsh mountain stream. Oikos 27:307-310.

Johnson, P.L. and W.T. Swank. 1973. Studies of cation budgets in the southern Appalachians on four experimental watersheds with contrasting vegetation. Ecology 54:70-80.

Keller, E.A. and F.J. Swanson. 1979. Effects of large organic material on channel form and fluvial processes. Earch Surface Processes 4:361-380.

Langbein, W.B. and L.B. Leopold. 1964. Quasi-equilibrium states in channel morphology. Am. J. Sci. 262:782-794.

Leopold, L.B., M.G. Wolman, and J.R. Miller. 1968. Fluvial processes in geomorphology. W.H. Freeman, San Francisco. 622 pp.

Leopold, L.B. and T. Maddock, Jr. 1953. The hydraulic geometry of stream channels and some physiographic implications. U.S. Geol. Surv. Prof. Paper 252. 57 pp.

Leopold, L.B. and W.B. Langbein. 1962. The concept of entropy in landscape evolution. U.S. Geol. Surv. Prof. Paper 500A. 20 pp.

Maddock, T., Jr. 1976. A primer on flood plain dynamics. J. Soil Wat. Cons. 31:44-47.

Mann, K.H. 1975. Patterns of energy flow, pp. 248-263. In: B.A. Whitton (ed.), River ecology. Blackwell Sci. Publ., Oxford. 694 pp.

Megahan, W.F. and R.A. Nowlin. 1976. Sediment storage in channels draining small forested watersheds in the mountains of central Idaho, pp. 4-115 - 4-126. In: Proc. Third Federal Inter-Agency Sedimentation Conference. Denver, Colorado. PB-245.

Merritt, R.W. and D.L. Lawson. 1979. Leaf litter processing in
    flood plain and stream communities, pp. 93-105. In:
    R.R. Johnson and F.J. McCormick (eds.), Strategies for
    protection and management of flood plain wetlands and other
    riparian ecosystems. Proc. Symp. Forest Serv. U.S.D.A. Gen.
    Techn. Rept. WO-12.
Meyer, J.L. and G.E. Likens. 1979. Transport and transformation
    of phosphorous in a forest stream ecosystem. Ecology
    60:1255-1269.
Minshall, G.W. 1978. Autotrophy in stream ecosystems.
    BioScience 28:767-771.
Minshall, G.W., R.C. Petersen, K.W. Cummins, T.L. Bott, J.R.
    Sedell, C.E. Cushing, and R.L. Vannote. 1983. Interbiome
    comparisons of stream ecosystem dynamics. Ecol. Monogr.
    53:1-25.
Moring, J.R. and R.L. Lantz. 1975. The Alsea Watershed Study:
    Effects of logging on the aquatic resources of three
    headwater streams of the Alsea River, Oregon. Fishery
    Research Report No. 9, Oregon Dept. of Fish and Wildlife,
    Corvallis, Oregon.
Morisawa, M. 1968. Streams: their dynamics and morphology.
    McGraw-Hill, N.Y. 175 p.
Mulholland, P.R. 1981. Organic carbon flow in a swamp-stream
    ecosystem. Ecol. Monogr. 51:307-322.
Mulholland, P.R. and E.J. Kuenzler. 1979. Organic carbon export
    from upland and forested wetland watersheds. Limnol.
    Oceanogr. 24:960-966.
Naiman, R.J. and J.R. Sedell. 1979a. Characterization of
    particulate organic matter transported by some Cascade
    Mountain streams. J. Fish. Res. Bd. Can. 36:17-31.
Naiman, R.J. and J.R. Sedell. 1979b. Benthic organic matter as
    a function of stream order in Oregon. Arch. Hydrobiol.
    87:404-422.
Nelson, D.J. and D.C. Scott. 1962. Role of detritus in the
    productivity of a rock outcrop community of a Piedmont
    stream. Limnol. Oceanogr. 7:396-413.
Newbold, J.D., P.I. Mulholland, J.W. Elwood, and R.V. O'Neill.
    1982a. Organic carbon spiralling in stream ecosystems.
    Oikos 38:266-272.
Newbold, J.D., R.V. O'Neill, J.W. Elwood, and W. Van Winkle.
    1982b. Nutrient spiralling in streams: implications for
    nutrient limitation and invertebrate activity. Amer. Natur.
    120:628-652.
Odum, H.T. 1957a. Trophic structure and productivity of Silver
    Springs, Florida. Ecol. Monogr. 27:55-112.
Odum, H.T. 1957b. Primary production measurements in eleven
    Florida springs and a marine turtle-grass community.
    Limnol. Oceanogr. 2:85-97.
Olson, J.S. 1963. Energy storage and the balance of producers
    and decomposers in ecological systems. Ecology 44:322-331.

O'Neill, R.V., W.F. Harris, B.S. Ausmus, and D.E. Reichle. 1975.
    A theoretical basis for ecosystem analysis with partcular
    reference to element cycling, pp. 28-40. In: F.G. Howell,
    H.B. Gentry, and M.H. Smith (eds.), Mineral cycling in
    southwestern ecosystems. E.R.D.A. Symposium Series,
    Washington, D.C. (Conf. 74-0531).

Paustian, S.J. and R.L. Bestcha. 1979. The suspended sediment
    regime of an Oregon Coast Range stream. Wat. Res. Bull.
    15:144-154.

Petersen, R.C. and K.W. Cummins. 1974. Leaf processing in a
    woodland stream. Freshwat. Biol. 4:343-368.

Pfankuch, D.J. 1975. Stream reach inventory and channel
    stability evaluation. Publ. U.S. Dept. of Agric. Forest
    Ser., Northern Region (RI-75-002). 26 pp.

Reichle, D.E. 1975. Advances in ecosystem analysis. BioScience
    25:257-264.

Reichle, D.E., R.V. O'Neill, and W.F. Harris. 1975. Principles
    of energy and material exchange in ecosystems, pp. 27-43.
    In: W.H. van Dobben and R.H. Lowe-Connell (eds.), Unifying
    concepts in ecology. Dr. W. Junk, The Hague. 320 pp.

Ross, H.H. 1963. Stream communities and terrestrial biomes.
    Arch. Hydrobiol. 59:235-242.

Schlesinger, W.H. and J.M. Melack. 1981. Transport of organic
    carbon in the world's rivers. Tellus 33:172-187.

Schumm, S.A. and R.W. Lichty. 1965. Time, space, and causality
    in geomorphology. Am. J. Sci. 263:110-119.

Schumm, S.A., M.P. Mosley, and G.L. Zimpfer. 1975. Unsteady
    state denudation. Science 191:871.

Sedell, J.R., F.J. Triska, J.D. Hall, and N.H. Anderson. 1974.
    Sources and fates of organic inputs in coniferous forest
    streams, pp. 57-69. In: R.H. Waring and R.L. Edmonds
    (eds.), Integrated Research in the Coniferous Forest Biome,
    Bull. 5. Coniferous Forest Biome, U.S. I.B.P. 78 pp.

Sedell, J.R., R.J. Naiman, K.W. Cummins, G.W. Minshall, and R.L.
    Vannote. 1978. Transport of particulate organic matter in
    streams as a function of physical processes. Verh. Int.
    Verein. 20:1366-1375.

Sigafoos, R.S. 1964. Botanical evidence of floods and
    flood-plain deposition. U.S. Geol. Surv. Prof. Paper 485
    (a):1-35.

Strahler, A.N. 1957. Quantitative analysis of watershed
    geomorphology. Am. Geophys. Union. Trans. 38:913-920.

Suberkropp, K.F., M.J. Klug, and K.W. Cummins. 1975. Community
    processing of leaf litter in woodland streams. Verh. Int.
    Verein. Limnol. 19:1653-1658.

Swank, W.T. and J.E. Douglas. 1975. Nutrient flux in
    undisturbed and manipulated forest ecosystems in the
    southern Appalachian mountains. Publ. Assoc. Internat. Sci.
    Hydrol. Symp. Tokyo (Dec. 1975) 177:445-456.

Swanson, F.J. and C.T. Dyrness. 1975. Impact of clearcutting
    and road construction on soil erosion by landslides in the
    western Cascade Range, Oregon. Geology 3:393-396.

Swanson, F.J. and G.W. Lienkaemper. 1978. Physical consequences
    of large organic debris in Pacific Northwest streams. USDA
    Forest Serv. Gen. Techn. Rept. PNW-69. 12 pp.

Teal, J.M. 1957. Community metabolism in a temperate cold
    spring. Ecol. Mongr. 27:283-302.

Tilly, L.J. 1968. Structure and dynamics of Cone Spring. Ecol.
    Monogr. 38:169-197.

Trimble, S.W. 1975. Denudation studies: Can we assume stream
    steady state? Science 188:1207-1208.

Triska, F.J. and K. Cromack. 1979. The role of wood debris in
    forests and streams, pp. 171-198. In: R.H. Waring (ed.),
    Forests: fresh perspectives from ecosystem analysis.
    Oregon State Univ. Press, Corvallis. 198 pp.

Vannote, R.L., G.W. Minshall, K.W. Cummins, J.R. Sedell and C.E.
    Cushing. 1980. The river continuum concept. Can. J. Fish.
    Aquat. Sci. 37:130-137.

Wallace, J.B., D.H. Ross, and J.L. Meyer. 1982. Seston and
    dissolved organic matter dynamics in a southern Appalachian
    stream. Ecology 62:824-838.

Waring, R.H. (ed.). 1980. Forests: fresh perspectives form
    ecosystem analysis. Oregon State Univ. Press. 198 pp.

Webster, J.R. and B.C. Patten. 1979. Effects of watershed
    perturbation on stream potassium and calcium dynamics.
    Ecol. Monogr. 19:51-72.

Webster, J.R., J.B. Waide, B.C. Patten. 1975. Nutrient
    recycling and the stability of ecosystems, pp. 1-27. In:
    F.G. Howell, J.B. Gentry, and M.H. Smith (eds.), Mineral
    cycling in southeastern ecosystems. E.R.D.A. Symposium
    Series, Washington, D.C. (Conf. 74-9531).

Welton, J.S. 1980. Dynamics of sediment and organic detritus in
    a small chalk stream. Arch. Hydrobiol. 90:162-181.

Wetzel, R.G., P.H. Rich, M.C. Miller, and H.L. Allen. 1972.
    Metabolism of dissolved and particulate detrital carbon in a
    temperate hardwater lake. Mem. Ist. Ital. Idrobiol. 29
    (Suppl.):185-243.

Whitfield, P.H. and H. Schreier. 1981. Hysteresis in
    relationships between discharge and water chemistry in the
    Fraser River basin, British Columbia. Limnol. Oceanogr.
    26:1179-1182.

Whitton, B.A. (ed.). 1975. River ecology. Blackwell Sci.
    Publ., Oxford. 694 pp.

Wolman, M..G. and J.P. Miller. 1960. Magnitude and frequency of
    forces in geomorphic processes. J. Geol. 68:54-74.

STABILITY OF STREAM ECOSYSTEMS

J.R. Webster[1], M. E. Gurtz[2], J. J. Hains[3],
J. L. Meyer[4], W. T. Swank[5], J. B. Waide[6], and
J. B. Wallace[7]

1. Department of Biology, Virginia Polytechnic
Institute and State University, Blacksburg, Virginia
24061;   2. Division of Biology, Kansas State
University, Manhattan, Kansas 66506;   3. Department of
Zoology, Clemson University, Clemson, South Carolina
29631;   4. Department of Zoology, University of
Georgia, Athens, Georgia 30602;   5. Coweeta Hydrologic
Laboratory, Otto, North Carolina 28763;   6. U.S. Army
Corps of Engineers, Waterways Experiment Station,
Vicksburg, Mississippi 39180;   7. Department of
Entomology, University of Georgia, Athens, Georgia
30602

INTRODUCTION

The ability of ecosystems to recover from external
disturbances, that is, their stability, is a fundamental property
of these systems. Quantification of the ability for various
ecosystems to recover and understanding of the mechanisms behind
stability are currently areas of major ecological research. In
this paper we present an overview of how the stability concept
has been used in ecology and a more specific discussion of the
application of these ideas to stream ecosystems. This is
followed by a case study in which we have been observing the
stability of small streams in response to watershed logging and
comparing stream stability to stability of the adjacent forest
ecosystem.

THE CONCEPT OF ECOLOGICAL STABILITY

Discussions of ecological stability began with and have been primarily focused around a relationship between stability and system complexity (e.g., Odum, 1953). MacArthur (1955) formalized this idea by providing a measure of pathway complexity, and Margalef (1957, 1963) suggested using species diversity as a measure of stability. The diversity-stability relationship has been supported with a variety of lines of evidence (e.g., Elton, 1958; Dunbar, 1960; Pimentel, 1961), but experimental studies have generally failed to demonstrate an obligate relationship between diversity and stability (e.g., Watt, 1964; Patten and Witkamp, 1967; Hairston et al., 1968; Hurd et al., 1971; Murdoch et al., 1972; Larson, 1974; Goodman, 1975), though they also have not invalidated such a relationship (McNaughton, 1977). Lack of support for a causal relationship between diversity and stability also has come from theoretical (model) studies, most of which suggested that the more complex the model, the less likely it was to be stable (e.g., Gardner and Ashby, 1970; May, 1971a, 1971b, 1972; Smith, 1972; Hubbell, 1973). However, model studies also have suggested that there are situations where increasing the complexity of ecological models may lead to greater probability of stability (e.g., Smith, 1972; Roberts, 1974; Webber, 1974; DeAngelis, 1975; Waide and Webster, 1975). Also Van Voris et al. (1980) demonstrated experimentally a correlation between functional diversity and stability.

Discussions of ecological stability have been hampered by diverse and poorly stated definitions. Two concepts predominate (e.g., Margalef, 1968). First is the notion of constancy or lack of variability. Ecosystems (or communities) which show little temporal change--for example, show year-to-year constancy in population sizes and community composition--are considered stable; whereas ecosystems with large temporal variability, such as tundra ecosystems, are considered unstable. The other notion of ecological stability involves the ability of an ecosystem to respond to disturbance, that is, its ability to exhibit secondary succession (Connell and Slatyer, 1977; McIntosh, 1980). An ecosystem is considered stable if its response to a disturbance is small and its return to its original state is relatively rapid. An ecosystem is unstable if it is greatly changed by disturbance, returns slowly to its original state, or if it never recovers to the original state.

These two notions of stability are related since the constancy or lack of constancy shown by an ecosystem is a reflection of its ability to respond to the almost continuous disturbances to which it is subjected (Lewontin, 1969). Ecosystems which show the greatest constancy should be those with the greatest ability to respond to disturbance. However, it may also be argued that the ability to dampen small "normal" disturbances may not contribute to the ability to respond to large disturbances. An ecosystem which experiences only small normal environmental variability may appear stable when in fact it is poorly adapted to respond to an unusual disturbance. Ecosystems with highly fluctuating environments become "desensitized" to exogenous perturbations (Watt, 1968; Copeland, 1970; Peterman, 1980).

Attempts to quantify ecological stability have been based mostly on engineering concepts of stability (e.g., Lewontin, 1969; May, 1973). Stability in the sense of Liapunov (or neighborhood stability) concerns the ability of a system to return to steady state if disturbed an infinitesimal distance from steady state. It has been argued that this stability concept is not applicable to ecosystems (Lewontin, 1969; Preston, 1969; Holling, 1973, 1974; Botkin and Sobel, 1975). However, it also has been argued that by their existence ecosystems demonstrate stability in the sense of Liapunov (Webster et al., 1975). This stability then becomes a starting point for more relevant discussions which allow comparisons of the relative stability of various types of ecosystems. Several authors have suggested that relative stability can be divided into various components (Patten and Witkamp, 1967; Child and Shugart, 1972; Waide et al., 1974; Hurd and Wolf, 1974). Two aspects of disturbance response seem most important. One involves the ability of the ecosystem to exhibit minimal response to disturbance. The other aspect relates to the rapidity with which an ecosystem responds once it has been displaced by a disturbance. These two aspects of relative stability have been given a variety of names (cf. Orians, 1974; Cairns and Dickson, 1977; Connell and Slatyer, 1977). We will use the terms resistance and resilience respectively (Webster et al., 1975).

Only a few workers have quantified resistance and resilience of ecosystems or ecosystem models and they used several techniques (e.g., Patten and Witkamp, 1967; Webster, 1975; Webster et al., 1975; Stauffer et al., 1978; Swank and Waide, 1980). While no clearly applicable method exists for evaluating relative ecosystem stability, these studies indicate several important points. Ecosystem resistance is related to a large, slow ecosystem component which damps or buffers disturbances (Golley, 1974; O'Neill et al., 1975; O'Neill and Reichle, 1980). This component of the ecosystem, which is usually soil organic matter, sediment, or detritus, provides an inertia which makes it difficult to displace the system from steady state. Resilience is related to rapid turnover of ecosystem components. Ecosystems in which the autotrophs are phytoplankton are much more resilient than ecosystems dominated by trees. Generally, the two aspects of relative stability are inversely correlated (Webster et al.,

1975). However, there are exceptions. The arctic tundra is an
ecosystem with relatively slow turnover compared to other
terrestrial ecosystems, perhaps because of the harsh environment
and short growing season, and therefore has relatively low
resilience. This ecosystem also has a low accumulation of
biomass and thus low resistance. On the other hand, Watson and
Loucks (1979) suggested that Lake Wingra, as represented by their
models, has both large, slow components which confer resistance
and rapid turnover pools which provide resilience.

Because most studies of ecological stability have been based
on traditional engineering stability concepts, there has been
little consideration of the type or magnitude of disturbance.
Such considerations are obviously needed in ecosystem studies.

APPLICATION OF ECOSYSTEM STABILITY CONCEPTS TO STREAMS

In 1960, Margalef noted that running waters had failed to
attract the interest of general ecologists. This has continued
to be especially true with respect to studies of stream ecosystem
stability. Further, the few attempts to apply traditional
approaches of ecosystem stability (diversity, stability,
maturity) and more recent stability concepts (relative stability)
to streams have often led to contradictions.

Traditional Stability Concepts

Traditional indicators of ecosystem stability (diversity,
maturity) show considerable disagreement when applied to streams
(Table 1). Species diversity or number of stream fish species
has been found to generally increase with stream order (Kuehne,
1962; Harrell et al., 1967; Sheldon, 1968; Platts, 1979), though
streams larger than 6th-order were not included in any of these
studies. Whiteside and McNatt (1972) found that fish species
diversity increased through the first four stream orders but
decreased in the 5th-order stream. Harrel and Dorris (1968)
found a downstream increase in macroinvertebrate diversity
through 5th-order and a decrease in the number of macroinvertebrate species in the highest order (6th) reach. Headwater
streams usually are considered to have moderate or low biotic
diversity, mid-order reaches have higher diversity, and in very
large streams diversity is fairly low (Vannote et al., 1980).
According to Margalef (1963), stability and maturity should
follow this same trend. However, Margalef (1960) suggested that
there is a general downstream trend from low to high maturity
(and thus low to high diversity). We might speculate that the
low observed diversity in large rivers is the result of several
factors. First, large rivers have not been widely studied.
Second, organisms of large rivers (e.g., oligochaetes and

Table 1. Indicators of stream ecosystem stability. ***, high;
          **, intermediate; *, low.

	Low Order Stream (headwater)	Middle Order Stream	High Order Stream	Reference
Biotic Diversity	**	***	*	Vannote et al., 1980
Functional Group Diversity	**	**	*	Cummins, 1975
Maturity	*	**	***	Margalef, 1960
Geologic Age	*	**	***	Cummins, 1975
Exploit Other Subsystems	*	**	***	Hall, 1972
Pigment Ratio	**	*	—	Motten and Hall, 1972
P:R and/or P:B Ratio	**	*	—	Motten and Hall, 1972
Detritus Base	***	*	**	Fisher and Likens, 1973
Stored Detritus	***	**	*	Naiman and Sedell, 1979

chironomids) are seldom identified to species and diversity
estimates are based on higher taxonomic categories. Finally, and
perhaps most importantly, most studies of large rivers have been
done on polluted ecosystems, since few large pristine rivers
still exist. As documented by Wilhm (1972) and many others,
pollution decreases stream diversity and may cancel normal trends
in diversity (Tramer and Rogers, 1973). In a pristine condition,
large rivers were characterized by large floodplains and
backwater areas, many woody snags, and often very clear water
with light reaching the bottom (Bartram, 1791; Bates, 1863;

Bakeless, 1961). Because of the diversity of habitats in such a
river, one might expect to find a highly diverse community.

Headwater streams commonly are regarded as immature (e.g.,
Margalef, 1960; Hall, 1972; Fisher and Likens, 1973). Survival
strategies of benthic stream communities in general are typical
of pioneer communities: small body size, short generation time,
high reproductive rate, and resting stages resistant to
unfavorable conditions (Patrick, 1970). Fisher (this volume)
stated that in typical streams of mesic areas which experience
frequent flooding, succession is truncated, and these streams are
dominated by pioneer type species. Nelson and Scott (1962)
suggested that river communities are in a frequent disclimax
condition as a consequence of periodic high water.

Geologically, headwater streams are younger than the streams
into which they drain (e.g., Leopold et al., 1964). However, the
most primitive representations of most orders of aquatic insects
are found in headwater streams (Cummins, 1975). This is not
surprising if, as it is currently believed, insects evolved
terrestrially and invaded freshwater from land, since headwater
streams represent the maximum interface between land and water
(Vannote et al., 1980). Also, one might argue that while a
specific stream is younger than the stream it drains into, the
headwater stream ecosystem predates the large river type of
ecosystem. In an evolutionary sense, then, the headwater type of
stream ecosystem is more mature.

A number of other factors do not support Margalef's (1960)
suggestion of a downstream increase in maturity. Margalef
(1960, 1963) suggested that immature ecosystems should be
exploited by mature ecosystems, and, as Hall (1972) noted, small
streams export energy to large streams. However, physical
factors (e.g., gravity) are so important, it is questionable
whether this criterion should be applied. Headwater streams
receive large amounts of energy from forests, but we probably
wouldn't consider a stream more mature than a forest.

Motten and Hall (1972) found that two other factors, pigment
ratio and P/R (production/respiration) ratio, indicated greater
maturity at their upstream station, and suggested that in small
streams, internal organizing mechanisms were so overridden by
external factors that analysis of developmental strategies is
spurious.

Finally, Odum (1969) suggested that a detritus base is
typical of mature ecosystems. Fisher and likens (1973) pointed
out that, on this basis, headwater streams are very mature
ecosystems. Middle-order streams where autotrophic production is
highest (Vannote et al., 1980) would be the least mature.

Further, if organic detritus can substitute for living biomass as suggested by Fisher and Likens (1973), and we bring in Margalef's (1963) idea that biomass is the "keeper of organization" in ecosystems, then maturity clearly decreases downstream. Naiman and Sedell (1979) documented a downstream decrease in the benthic standing crop of detritus in streams and pointed out that the large accumulation of detritus, primarily logs, in headwater streams contributes to the resistance of these ecosystems. They further suggested that as stream order increases, the pathway for obtaining stability changes from one of resistance to resilience.

In general, there is no weight of evidence supporting a downstream increase in stream ecosystem stability. In fact, if maturity, geologic age, and exploitation are ruled out of Table 1 as inappropriate or reversed, as discussed above, then the weight of evidence would seem to support a greater stability in the headwaters. Probably the best indicator of stream ecosystem maturity and perhaps stability would be estimates of energy utilization efficiency (Fisher and Likens, 1973). However, estimates of small stream efficiency are few (e.g., Fisher and Likens, 1973; Webster and Patten, 1979; Mulholland, 1981), and estimates for larger streams are almost non-existent (Fisher, 1977) (But see also Cummins et al., this volume).

Assimilative Capacity

Assimilative capacity is another way of viewing stream ecosystem stability, especially concerning its response to man-induced stress or pollution. Assimilative capacity has been widely used--in a narrow sense related to the response to a decrease in dissolved oxygen due to organic waste and more broadly as the ability of a stream to purify itself when subject to inputs of various types of wastes (Velz, 1976). Cairns (1977a, b) said that assimilative capacity should be defined as the ability of an aquatic ecosystem to assimilate a substance without degrading or damaging the system's ability to maintain its community structure and functional characteristics.

There is considerable debate as to whether streams have any assimilative capacity, that is, whether there is a threshold below which pollution does not change the ecosystem (Woodwell, 1975). Campbell (1981) stated that addition of a very small quantity of a pollutant to a stream may result in an extremely small change--perhaps the death of a few bacteria--but because the stream has responded, it can not be said to have an assimilative capacity. On the other hand, Cairns (personal communication) argues that the ability of the stream to respond to disturbance without change in the basic functional capabilities of the stream ecosystem is evidence of its assimilative capacity even if this response involves the loss of

a few organisms or even a few species. Perhaps we should speak
of assimilative capacity as the bounds within which a stream
ecosystem has the ability to return (in time or distance) to its
original state at least in terms of its function. This region
has been referred to as a domain of attraction (Holling, 1973;
Peterman, 1980).

In general, the bounds of the domain of attraction of a
stream appear to be large. Because of the ability of rivers to
cleanse themselves of impurities (Cairns and Dickson, 1977),
rivers are one of the main recipients of anthropogenic wastes.
Two major factors contribute to this form of stream ecosystem
stability. First is the ability of the stream to biologically
transform the pollution. The second, but possibly more
important, factor is streamflow. In contrast to other types of
ecosystems into which wastes might be dumped, discharges into
streams "are immediately carried away from the doorstep of the
discharger" (Cairns and Dickson, 1977). For this latter reason,
streams have a much greater assimilative capacity than any other
type of ecosystems. Cline et al. (1979), for example, found that
stream runoff and spates ameliorated short-term construction
impacts on a Colorado stream resulting in higher stability than
had been predicted.

Relative Stability

As discussed above, a number of authors have suggested that
we look for relative measures of the ability of ecosystems to
minimize response to disturbance. Based on its rapid turnover
and small autotrophic component, O'Neill et al. (1975) ranked a
small headwater spring as the most sensitive to environmental
changes of the various ecosystems they studied. In a later
comparison of relative stability, O'Neill (1976) found that a
model of this spring ecosystem had the next to smallest stability
of the six ecosystems he examined. Webster et al. (1975) ranked
their stream model lowest in both resistance and resilience among
their eight ecosystem models; however, it should be noted that
the stream model was unique among the models in their study
because it lacked nutrient cycling. Based partly on these same
eight models, but using different techniques to estimate
resistance and resilience, Webster (1975) suggested that though
streams have low resistance to disturbance, they have very high
resilience. This high resilience has been borne out by studies
showing rapid recovery of disturbed streams relative to
terrestrial ecosystems (e.g., Webster and Patten, 1979). Factors
which contribute to the high resilience of stream ecosystems
include the short life cycles of the organisms in the community,
rapid recolonization mechanisms (drift, flying adults),
recolonization from the hyporheic zone, annual renewal of
allochthonous inputs (Fisher and Likens, 1973), and, probably

most important, the continual flushing by the current (Webster and Patten, 1979).

Unidirectional flow of water in streams is the primary factor which has made it difficult to get a theoretical grasp on stream ecosystems (Hutchinson, 1963). Margalef (1968) suggested that stream ecosystems should be conceptualized as processes rather than as stable organizations. O'Neill et al. (1979) concluded that the spatial dimension, which is necessary when modeling streams, can significantly alter conclusions about the dynamic behavior of the system in its response to perturbation. Current is an important mechanism of stream stability because it removes or at least displaces effects of disturbance; however, from the standpoint of resistance, current is a force of instability as it continually removes accumulated biomass and detritus, especially during storms. This instability is countered by instream mechanisms which act to retain materials within the stream (Vannote et al., 1980; Minshall et al., 1983), such as debris dams (Bormann et al., 1969; Fisher and Likens, 1973; Sedell et al., 1978; Bilby and Likens, 1980), filter feeders (Wallace et al., 1977), and rapid nutrient uptake (Ball and Hooper, 1963; Elwood and Nelson, 1972; Meyer, 1979; Newbold et al., 1981).

AN EXPERIMENTAL STUDY OF STREAM ECOSYSTEM STABILITY: BIG HURRICANE BRANCH

In 1974 we began a study to evaluate the stability of a stream ecosystem in response to commercial logging (Monk et al., 1977). A primary objective was to compare the stability of the stream with the stability of the forest. Defining stability as the ability of an ecosystem to maintain primary and secondary production following disturbance, we predicted prior to this study that the stream would be less resistant but more resilient than the forest. The forest was cut in 1977 and in this paper we present an overview of some of our observations on the response of the stream.

Site Description

This study was conducted on Watershed (WS) 7 at Coweeta Hydrolic Laboratory in the southern Appalachian Mountains of North Carolina. Prior to logging in 1977, the only experimental manipulation to this watershed since the U.S. Forest Service began management in 1924 was a woodland grazing experiment. Six cattle grazed the watershed from 1941 through 1949 but had little long-term impact on the vegetation (Johnson, 1952; Williams, 1954). Twenty-five years later there were no discernible effects of grazing on water chemistry or stream flow (Swank and Douglass,

1977). Pre-clearcut vegetation on the watershed was separated
into four main types. A pine-hardwood association occurred at
higher elevations and along ridges. An oak-hickory association
at intermediate elevations occurred in two distinct types based
on dominant tree species, litter depth and decay rate, and soil
moisture and temperature: mesic on east-facing slopes and xeric
on west-facing slopes. At lower elevations along streams, the
vegetation was a typical cove hardwood association.

Watershed 7 is drained by a second-order stream, Big
Hurricane Branch. The substrate of this stream varies from
sections of steep exposed bedrock to short sandy reaches of low
gradient and infrequent small pools. Prior to logging, dissolved
nutrient levels in Big Hurricane Branch were quite low (Table 2).

Hugh White Creek, which drains Coweeta Watershed 14, was
used as a reference stream for biological research on Big
Hurricane Branch. Vegetation on WS 14 is similar to that on WS 7
prior to logging, although WS 14 has a northwestern aspect and WS
7 has a southern aspect. Hugh White Creek was selected as a
reference stream because of its similarity in size and discharge
to Big Hurricane Branch. Comparisons of physical and chemical
characteristics of the two streams and their watersheds are given
in Table 2.

During April–June, 1976, three roads with a total length of
3 km were built on WS 7 for logging access. Two of the roads
crossed the main stream. Approximately 5% of the watershed area
was disturbed by road building. Logging began in January 1977
and was completed in June. Site preparation--i.e., clearfelling
trees that remained after logging--was completed in October 1977.
A mobile cable system was used for most logging; however, tractor
skidding was used on more gentle slopes (about 8.9 ha). A total
of 15.9 ha was not logged (but was site prepared) due to
insufficient volume of marketable timber. Mineral soil was
exposed on less than 10% of the total watershed area. Following
logging, most of the logging debris which fell in or over the
stream was removed from the main channel. Neither debris dams
that existed prior to logging nor debris in tributary channels
were removed.

Abiotic Factors

Streamflow. Both Big Hurricane Branch and Hugh White Creek
are equipped with sharp-crested V-notch weirs for continuous
stream flow measurement (Fig. 1). Hugh White Creek has had a
slightly higher annual discharge than Big Hurricane Branch (Table
2), and prior to logging WS 7 there was a consistent linear
relationship between annual discharge in the two streams.
However, due to decreased summer transpiration loss following

Table 2. Physical and chemical characteristics of Hugh White
          Creek (WS 14) and Big Hurricane Branch (WS 7) prior to
          logging WS 7.

	WS 14 Hugh White Creek	WS 7 Big Hurricane Branch
Watershed area (ha)	61.1	59.5
Maximum elevation of WS (m)	996	1060
Minimum elevation of WS (m)	708	724
Main stream channel length (m)	1077	1225
Mean bank-full width[a] (cm)	406	256
Mean mid-stream depth[a] (cm)	6.4	10.5
Main channel gradient ($m \cdot m^{-1}$)	0.16	0.19
Mean annual discharge[b] ($l \cdot sec^{-1}$)	19.5	17.7
Maximum discharge during study period ($l \cdot sec^{-1}$)	–	1281.2
Mean annual elemental concentrations[c] ($mg\ l^{-1}$)		
NO_3-N	0.004	0.002
NH_4-N	0.004	0.004
PO_4-P	0.002	0.002
Cl	0.540	0.699
SO_4-S	0.362	0.475
K	0.350	0.492
Na	0.739	0.946
Ca	0.460	0.846
Mg	0.280	0.372
pH	6.61	6.82

[a] Measurements taken every 5 m for the first 500 m above the weir pond.
[b] Based on 40 and 26 years of record, respectively. Data from Coweeta Hydrological Laboratory.
[c] Data from Swank and Douglass, 1977.

logging (e.g., Swift et al., 1975; Bormann and Likens, 1979), annual discharge in Big Hurricane Branch has increased relative to Hugh White Creek (Fig. 2). The monthly distribution of increase occurs primarily in summer but continues through fall, and early winter, a timing pattern similar to many other cutting experiments conducted at Coweeta (Douglass and Swank, 1975).

Rainfall has had a significant effect on the response of Big Hurricane Branch to logging operations. In May 1976, when logging roads were under construction, there were two large storms, with precipitation amounts of 17 and 21 cm, the second having a recurrence interval of 100 years. Some of the effects of these storms are described below. The effects of a severe drought during 1980 have not yet been fully evaluated. Streamflows during this period reached previously unrecorded lows.

Stream temperatures. As has been found in other studies (e.g., Swift and Messer, 1971), summer water temperatures increased in Big Hurricane Branch following logging (Fig. 3). After removal of the canopy which shaded the stream, water temperatures during summer months were several degrees higher than temperatures predicted from a pre-logging regression against water temperatures in a stream draining an adjacent, forested watershed. The most recent data suggest that with canopy regrowth, water temperatures are decreasing.

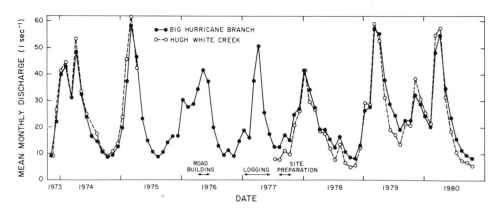

Fig. 1.   Mean monthly discharge in Big Hurricane Branch (WS 7) and Hugh White Creek (WS 14) during the study period. Periods of disturbance on WS 7 are also indicated.

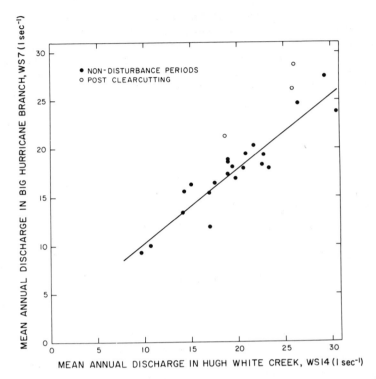

Fig. 2. Comparison of mean annual discharge in Big Hurricane Branch before and after logging with mean annual discharge from an undisturbed watershed, WS 14, drained by Hugh White Creek. The regression line fits the points for non-disturbance periods.

Dissolved nutrient levels. Small but detectable changes in concentrations of dissolved ions occurred in Big Hurricane Branch following logging. During and following road-building (1976), but prior to logging, increases were found in concentrations of Ca, SO_4-S, K, Cl, Mg, Na, and SiO_2. However, NO_3-N, NH_4-N, and PO_4-P showed no response during the following year (1977). In the next year (1978), the most conspicuous change was a large increase in the concentration of dissolved NO_3-N, beginning in August 1977. The concentration of NO_3-N subsequently increased from pre-disturbance levels of 0.002 $mg \cdot 1^{-1}$ to 0.04 $mg \cdot 1^{-1}$ or higher. The concentration of PO_4-P remained near pre-disturbance levels.

Sediment. Suspended inorganic sediments in Big Hurricane Branch showed large increases, principally during road building

and logging, when use of roads was greatest (Fig. 4). The
highest suspended sediment concentration was recorded in June
1976, following the two large storms mentioned previously.
Suspended sediment concentrations then declined until they again
rose in the period of logging and site preparation (March-October
1977). Concentrations generally declined thereafter.

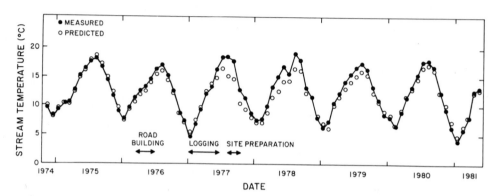

Fig. 3.   Measured mean monthly water temperatures in Big
           Hurricane Branch compared with temperatures predicted
           from regression on water temperatures in a stream
           draining an adjacent, undisturbed watershed, WS 10.

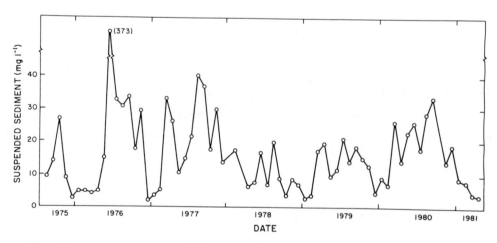

Fig. 4.   Mean monthly suspended sediment concentrations in Big
           Hurricane Branch.

Sediment which accumulated in the weir pond, primarily bed load, averaged 32 kg·d^{-1} prior to disturbance. During road construction, this fraction of the sediment load increased to 765 kg·d^{-1}. Much of this increase can be attributed to the effects of the two large storms which occurred as the roads were being built. These storms washed large amounts of material from roadbeds into the stream channel. In the subsequent period of road stabilization, sediment load decreased to 47 kg·d^{-1}. During logging this fraction of the sediment load increased to 80 kg·d^{-1}, and it further increased to 188 kg·d^{-1} during early 1978. There has been some subsequent decrease in sediment accumulation in the weir pond, though it still continues much above pre-disturbance levels. Much of the post-logging sediment has been fairly large particles (>0.05 mm), and we think that most of this material entered the stream during the 1976 storms and is continuing to be washed out of the streams, mainly during storms. Except for a brief period including the spring 1976 storms, sediment transport in undisturbed streams at Coweeta has remained low (c.f. Gurtz et al., 1980).

Leaf Fall and Blow-in

Leaf fall into Big Hurricane Branch was measured with six 0.4045-m^2 litter traps located over or adjacent to the main channel. Two traps were placed at each of three sites located on lower, middle, and upper reaches of the stream. Four 40-cm wide blow-in traps were also placed at each of these three sites, two traps on each bank. Material accumulated in the litter traps was collected approximately bi-weekly in the fall, separated to species if possible, oven dried, and weighed.

Prior to disturbance, leaf fall was estimated to be 259.2 g DW · m^{-2} · y^{-1}, a fairly typical value for eastern deciduous forests (Table 3). The species composition was dominated by oaks, hickories, yellow poplar, and rhododenron. One year after logging, leaf fall dropped to nearly zero, but two years later was up to about half the pre-logging input. Post-logging leaf fall was dominated by herbaceous material (e.g., Aster curtisii), woody shrubs such as blackberry (Rubus sp.) and green briar (Smilax sp.), and rapidly sprouting tree species (yellow poplar, black locust, sourwood, sassafras).

Blow-in, material that moves laterally from the stream bank into the stream, was measured as 174.8 g DW · m^{-1} · y^{-1} prior to logging. Based on an average channel width of 1.65 m (from measurements made on the entire WS 7 stream network), blow-in represented 45% of the total allochthonous input. From fall 1978

Table 3. Leaf fall to Big Hurricane Branch before and after
          clearcutting. The "other" category includes all leaf
          species comprising less than 2% of the total and all
          unrecognizable leaf fragments.

	Percent Composition			
	1974	1978	1979	1980
White Oak Quercus alba	14.2	–	–	–
Red Oak Quercus rubra	13.4	–	–	–
Rhododendron Rhododendron maximum	11.6	26.5	–	–
Hickories Carya spp.	11.4	–	–	–
Yellow Poplar Liriodendron tulipifera	9.1	10.8	–	5.0
Birches Betula spp.	5.4	15.7	11.8	–
Chestnut Oak Quercus prinus	4.8	–	–	–
Red Maple Acer rubrum	4.8	–	–	–
Basswood Tilia americana	4.2	–	–	–
Beech Fagus grandifolia	–	–	–	–
Dogwood Cornus florida	–	3.9	5.2	2.0
Sourwood Oxydendrum arboreum	–	–	4.9	4.2
Black Locust Robinia pseudoacacia	–	–	4.1	2.3
Sassafras Sassafras albidum	–	–	–	2.4
Other	21.1	43.1	74.0	84.1
Total annual leaf fall (g DW·m^{-2})	259.2	4.2	43.3	124.9

through summer 1979, blow-in was 38.6 g DW . m^{-1} . y^{-1} (92% of
total input), and the next year it was 41.0 g DW . m^{-1} . y^{-1} (53%
of total input). Initially, blow-in was not as greatly affected
by logging as leaf fall, since much of the blow-in may have been
from litter accumulated on the ground in previous years.
However, blow-in is apparently not recovering as rapidly as
direct leaf fall, perhaps due in part to the thick undergrowth.

Periphyton Primary Production

Measurements of periphyton primary production were made in
both Big Hurricane Branch and Hugh White Creek following logging
of WS 7. Approximately monthly measurements were made from
October 1977 through February 1979, and an additional seven
measurements were made between June 1980 and May 1981. All
primary production measurements were made using ^{14}C uptake on
natural substrates exposed in circulating water chambers similar
to those used by Rodgers et al. (1978) and Hornick et al. (1981).

Periphyton primary production in Hugh White Creek averaged
0.3 mg C . m^{-2} . hr^{-1} (Fig. 5). This rate of carbon fixation is
similar though somewhat smaller than rates reported for other
small, forested eastern streams (Minshall, 1967; Elwood and
Nelson, 1972; Hornick et al., 1981). Applying this rate to Big
Hurricane Branch prior to logging (assuming that rates were
similar in the two streams prior to disturbance and that
measurements taken from small areas of rocks can be expanded to
stream area, carbon is 45% of dry weight, production occurs 12 hr
per day, and ^{14}C fixation approximates net primary production)
periphyton primary production accounted for 2.9 g DW . m^{-2} . y^{-1},
approximately 0.6% of that provided by allochthonous inputs.

Following logging, primary production in Big Hurricane
Branch was significantly (p < 0.05) greater than in Hugh White
Creek (average 1977-1978 = 8.9 mg C . m^{-2} . hr^{-1}). Using the
assumptions above, autochthonous input in 1977-78 was 86.6 g DW .
m^{-2} . y^{-1}, slightly larger than total allochthonous inputs and
much larger than direct leaf fall during the same period.
However, two years later periphyton primary production had
decreased to 0.9 mg C . m^{-2} . hr^{-1} and accounted for only about
5% of that provided by allochthonous inputs. A similar increase
in periphyton primary production following logging was reported
by Gregory (1979).

The high level of periphyton primary production in 1977-78
may be attributed to opening of the canopy, increased dissolved

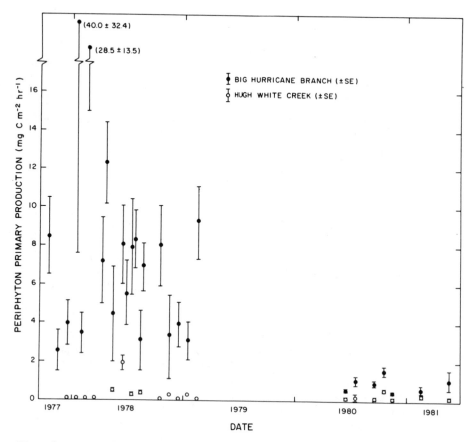

Fig. 5.   Periphyton primary production in Big Hurricane Branch
and Hugh White Creek.   Error bars are ± SE.

nutrient concentrations, and possibly a sparse grazer fauna (see
below).   Several factors may have been involved in the subsequent
decrease in primary production including increased shading,
increased number of grazers, and scouring by the large sediment
particles in transport beginning in early 1978.   Further details
on primary production of both algae and moss in Big Hurricane
Branch and Hugh White Creek were reported by Hains (1981).

Dissolved Organic Carbon

    Water samples for dissolved organic carbon (DOC) analysis
were collected biweekly from both Big Hurricane Branch and Hugh
White Creek from July 1979 through June 1980.   Estimates for DOC
budgets for the two streams were based on samples collected from
the main streams just above the weir ponds, precipitation
(throughfall), tributaries, and subsurface water seeps.   Eleven

storms were intensively sampled to develop concentration-discharge regressions. All samples were analyzed on a Dohrmann DC-54 Carbon Analyser, which used UV-catalyzed oxidation in the presence of persulfate.

DOC concentrations in the reference stream, Hugh White Creek, were highest during the growing season, averaging about 1 mg $C \cdot l^{-1}$ during spring and summer (Fig. 6). Concentrations in biweekly samples from the disturbed stream, Big Hurricane Branch, stayed at about 0.5 mg $C \cdot l^{-1}$ throughout the study. During baseflow conditions in the growing season, the reference stream showed a consistent increase in DOC concentration from the headwaters to the base of the watershed. This increase was not observed during dormancy, nor was it observed at any time in Big Hurricane Branch. DOC concentration increased to as high as 5 mg C l^{-1} during storms on both watersheds, and there was no difference in concentration-discharge regressions between the two streams or between rising and falling limbs of the hydrograph. This is in marked contrast to observations on fine particulate organic matter in these streams (Gurtz et al., 1980).

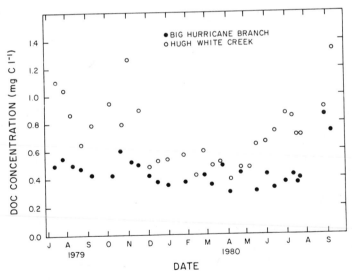

Fig. 6. Dissolved organic carbon concentrations in Big Hurricane Branch and Hugh White Creek.

Decreased DOC concentrations in Big Hurricane Branch cannot be attributed entirely to dilution because of higher discharge. During the period beginning two years after logging, annual DOC export was substantially less from the logged watershed (10.6 kg C·ha^{-1} vs. 14.8 kg C·ha^{-1} from WS 14). Three factors appear to be responsible for the lower export from Big Hurricane Branch: (1) The rate at which DOC was taken up by the biota was greater in Big Hurricane Branch. Experimentally added sucrose and leaf leachate was removed from the water column more rapidly than in the reference stream (J. Meyer, unpubl. data). This was due in part to warmer water temperatures and greater microbial biomass (as ATP, J. Meyer, unpubl. data). During storms, the uptake capacity of the stream was exceeded, DOC moved downstream faster than it could be taken up, and more similar concentrations were observed in the two streams. (2) Within stream production of DOC due to leaching of leaf and other litter was lower in Big Hurricane Branch due to lower standing stock of this material. The lack of a consistent downstream increase in DOC concentration in this stream supports this observation. (3) Less DOC was entering Big Hurricane Branch. Budget calculations showed DOC inputs in Big Hurricane Branch were about 70% of the inputs to the reference stream. The difference was primarily due to decreased inputs from tributaries and subsurface water.

During the year studied, both streams showed a net loss of DOC. This may be due to an underestimation of the within stream sources of DOC and the extraordinarily wet year; annual runoff was 52% greater than the 44 year average. Further details of DOC dynamics in Big Hurricane Branch and Hugh White Creek are being prepared for publication (Meyer and Tate, in press).

Benthic Particulate Organic Matter

Prior to logging WS 7, samples of benthic particulate organic matter (BPOM) were collected for two years from Big Hurricane Branch. Ten random samples (one sample every 100 m starting at a random point in the first 100 m) were taken with a square foot Surber sampler (0.57-mm mesh opening) each month. Collected material was separated into woody and non-woody (primarily leaves and moss), dried, and weighed. Monthly samples were taken from January 1977 through September 1978 using a somewhat different sampling scheme. Four samples (Surber, 0.30-mm mesh openings) were taken from each of four different habitat types: moss (moss covered boulders and outcrops), cobble riffle (riffles dominated by cobble-sized rocks, i.e., 64-256 mm), pebble riffle, and sand (predominantly sand and gravel substrate). To make these data comparable with the pre-logging samples, we used weighted means based on the percent of stream bottom area occupied by each habitat type.

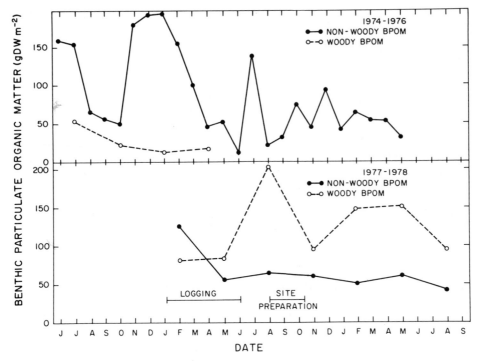

Fig. 7. Benthic particulate organic matter (BPOM) in Big
Hurricane Branch. Woody BPOM and 1977-1978 non-woody
BPOM are quarterly means of monthly samples.

In the two years prior to logging, non-woody BPOM in Big
Hurricane Branch averaged 86.2 g DW·m^{-2} and woody BPOM was 27.0 g
DW·m^{-2} (Fig. 7). In comparison, non-woody BPOM in the year
following logging (October 1977 through September 1978) was down
to 54.7 g DW m^{-2}, but woody material had increased to 120.6 g DW·
m^{-2}. Though slash was removed from some parts of the main stream
during site preparation, it is evident that large amounts of
woody material entered Big Hurricane Branch as a result of
logging. In the first fall following logging, the usual pulse of
leaf detritus was absent. However, there was a considerable
amount of leaf detritus which entered the stream primarily from
rhododendron slash over and adjacent to the stream.

Leaf Breakdown Rates

Breakdown rates in Big Hurricane Branch of three leaf
species, dogwood, white oak, and rhododendron, were measured
before (1974-1975), during (1976-1977), and after (1977-1978)

logging. Prior to logging, senescent leaves were collected just prior to abcision from trees on WS 7 or adjacent undisturbed watersheds. After logging, leaves were collected only from the adjacent watersheds so that changes in breakdown rates would not be caused by altered quality of successional vegetation. Two to four grams of air dried leaves were placed in nylon mesh bags (10 x 10 cm, 3-mm mesh openings). These bags were placed in the stream in autumn and retrieved periodically until most leaf material had disappeared. In the laboratory, leaves were rinsed to remove attached sediment, air dried, and weighed. Subsamples were ashed to determine ash-free dry weights (AFDW). Exponential breakdown rates were calculated as the slopes of semi-log regressions of AFDW remaining against time (Jenny et al., 1949; Olson, 1963; Petersen and Cummins, 1974). All regression slopes were significantly different from zero (p < 0.05). Breakdown rates were compared using analysis of covariance (Sokal and Rohlf, 1969).

In both the before-logging and during-logging studies, dogwood leaves broke down significantly (p < 0.05) faster, and rhododendron broke down significantly slower than the other two leaf species (Table 4). Although dogwood breakdown was still significantly faster after logging, rates for rhododendron and white oak were not significantly different. Breakdown rates of all three species were significantly slower during logging than before logging. Dogwood and white oak were affected similarly, declining to about 60% of their original rate, while rhododendron was more strongly affected, slowing to about 30% of its original rate. In contrast, white oak and rhododendron breakdown rates increased after logging as compared to before logging. Dogwood broke down at the same rate in both study periods, white oak increased 40%, and rhododendron was again most affected, increasing about 180% over its before-logging rate.

We analyzed in detail various factors which might have affected these results, including water temperature, dissolved nutrient concentrations, stream flow, sediment transport, and faunal abundance (Webster and Waide, in press). We have concluded that sediment and fauna are probably the two most important factors. In the study conducted during logging, burial of leaves in the sediment and decline of the dominant shredder decreased rates of leaf breakdown. The next year sedimentation was much less, but the shredder fauna remained low. However, because of the very low allochthonous leaf input that year (Table 3), our bags of leaves represented islands of a scarce food source. Dogwood leaf breakdown was not greatly affected since it is a rapidly conditioned and preferred shredder food (Wallace et al., 1970). However, rhododendron leaves, which are normally ignored by shredders at least during the first half year of conditioning, were eaten because of the scarcity of preferred

Table 4. Breakdown rates (d^{-1}) of three leaf species measured
          before, during, and after logging. Each rate is given
          with its 95% confidence interval. Numbers in
          parentheses are the number of samples and the
          coefficient of determination, r^2, of the semi-log
          regression.

	Before Logging 1974-1975	During Logging 1976-1977	After Logging 1977-1978
Dogwood	0.0219 ± 0.0025 (104, 0.74)	0.0134 ± 0.0023 (45, 0.77)	0.0219 ± 0.0024 (84, 0.81)
White Oak	0.0064 ± 0.0006 (143, 0.78)	0.0038 ± 0.0004 (44, 0.91)	0.0090 ± 0.0010 (84, 0.79)
Rhododendron	0.0037 ± 0.0007 (156, 0.37)	0.0011 ± 0.0003 (45, 0.65)	0.0105 ± 0.0020 (82, 0.59)

food. White oak leaves were affected at an intermediate level.
Results of this study are discussed more fully by Webster and
Waide (in press).

Particulate Organic Matter Transport

Transport of particulate organic matter (POM) was measured
by two techniques. Large particulate organic matter (LPOM) in
Big Hurricane Branch was measured using a 1-mm mesh drift net
placed in the stream a few meters above the weir pond where a
wooden flume directed all flow into the net. Samples were
collected at approximately weekly intervals over 4 or 24-hour
periods for 18 months prior to logging and 14 months following
logging. Collected material was dried, washed to remove
inorganic sediment, re-dried, and weighed.

Periodic samples of total POM were collected from Big
Hurricane Branch and Hugh White Creek from 1977 through 1981.
These samples were processed by filtering measured volumes of
water through a series of stainless steel screens (Gurtz et al.,
1980). Material collected on the screens and aliquots of
material passing the finest screen were collected on ashed and
preweighed glass fiber filters. The filters were dried (50°C),
weighed, ashed (500°C), and reweighed.

Monthly averages of LPOM concentrations in Big Hurricane
Branch during non-storm periods showed a substantial increase in

the year following logging (Fig. 8). We suspect that most of
this increase was due to logging debris that fell in the stream.
Total POM concentration showed a pattern very similar to LPOM
(Fig. 9). POM concentrations in Big Hurricane Branch have been
significantly (t-test, p < 0.05) greater than Hugh White Creek
since WS 7 was logged. This has resulted in substantially
greater export of POM from Big Hurricane Branch than Hugh White
Creek (Fig. 10). However, comparison of summer 1981 samples with
summer 1978 samples suggests that the elevated POM transport in
Big Hurricane Branch may be declining.

In addition to elevated transport during low flows, samples
taken during storms in 1977 and 1978 (Gurtz et al., 1980) and in
June 1981 (Fig. 11) indicate greatly increased storm transport.
The most recent storm samples were taken during a thunderstorm on
30 June - 1 July 1981. The storm lasted about two hours and was
very intense during the first hour. Flows returned to near

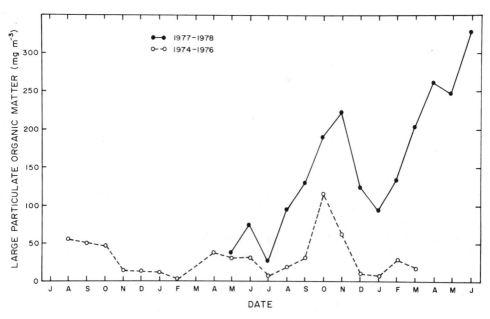

Fig. 8. Mean monthly estimates of large particulate organic
         matter (> 1 mm) concentrations in Big Hurricane Branch
         before (1974-1975) and after (1977-1978) logging. All
         samples included in this figure were taken during
         non-storm periods.

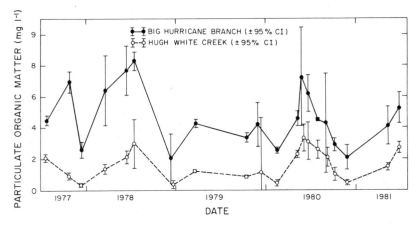

Fig. 9.   Total particulate organic matter (POM) concentrations in
Big Hurricane Branch and Hugh White Creek following
logging WS 7.   Only samples taken during non-storm
periods are included.   In all sampling periods, the POM
concentration in Big Hurricane Branch was significantly
($p < 0.05$, t-test) higher than the POM concentration in
Hugh White Creek.

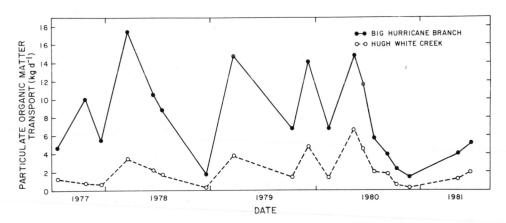

Fig. 10.   Particulate organic matter transport in Big Hurricane
Branch and Hugh White Creek following logging WS 7.
Only non-storm periods are included.

baseflow two hours after the rain ended. The concentration of
POM in Big Hurricane Branch peaked at the very beginning of the
storm (Fig. 11). Bilby and Likens (1979) also noticed this peak
in POM concentration which occurs prior to the discharge peak and
attributed it to entrainment of materials in the streambed. This
phenomenon was magnified in Big Hurricane Branch where, with the
absence of a tree canopy, direct impact of rain drops on the
stream bed helped initiate particle movement.

Based on the particle size fraction analysis, there was
little difference between the particle size distributions in Big
Hurricane Branch following clearcutting and Hugh White Creek.
The median particle size ranged from 0.041 to 0.093 mm and 0.037
to 0.101 mm in the two streams, respectively. However from
October 1977 through June 1978, the median POM particle size in
Big Hurricane Branch was consistently and significantly (t-test,
$p < 0.05$) larger than in Hugh White Creek. This was a period
when we also observed elevated transport of large (> 1 mm)
inorganic particles (Gurtz et al., 1980).

Benthic Macroinvertebrates

Collections of benthic macroinvertebrates were made in both
Big Hurricane Branch and Hugh White Creek for a 21-month period
beginning in January 1977 with the onset of logging. A Surber
sampler (mesh opening 0.3 mm) was used to collect four samples
monthly (three in Hugh White Creek) in each of four common
substrates: moss, cobble riffle, pebble riffle, and sand.
Samples were preserved in the field with approximately 10%
formalin solution, and invertebrates were picked from the samples
with the aid of a microscope at 7X magnification. Further
details of this study were given by Gurtz (1981); our concern
here is with faunal changes that occurred in response to logging.
Changes in the invertebrate fauna were compared between the
streams using two-way analysis of variance on log-transformed
data. The overall obaundance estimates in Figure 12 were calculated by weighting habitats according to their occurrence in
each stream. In the discussion below, functional feeding groups
for each taxon are those assigned by Merritt and Cummins (1978).

Among the statistically significant changes that occurred in
Big Hurricane Branch relative to Hugh White Creek following logging, the greatest change occurred in the collector-gatherer (and
possibly scraper) mayflies (Baetis spp. and Ephemerella spp.).
Initial abundances were similar in the two streams, but within
several months of logging, abundances in Big Hurricane Branch
were much higher than in the reference stream. These increases
occurred in all habitats for Baetis with the most pronounced
increases in the moss habitat. Species in the genus Ephemerella
increased in all habitats except sand.

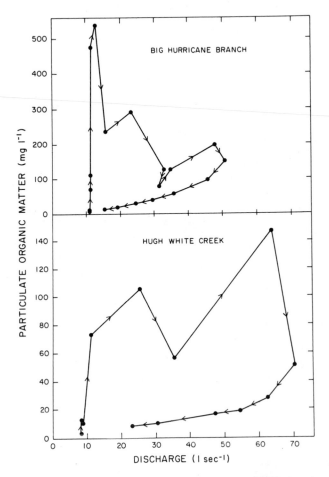

Fig. 11. Particulate organic matter concentrations in Big Hurricane Branch and Hugh White Creek during a storm on 30 June - 1 July 1981. Arrows indicate the time sequence of samples. The total sampling period was about four hours. Samples from Hugh White Creek were taken every 15 min. Samples from Big Hurricane Branch were taken even more frequently when the stream was rising. The time between the onset of the storm and peak discharge was 50 min. in Big Hurricane Branch and 65 min. in Hugh White Creek.

The dominant shredder in low-order Coweeta streams, Peltoperla (Tallaperla) maria declined in all habitats in Big Hurricane Branch relative to Hugh White Creek throughout the study period. Much of the decrease in Peltoperla was due to significant declines in the smallest size classes. Samples collected for an additional year after the study period reported here indicate that Peltoperla was nearly eliminated from Big Hurricane Branch within three years after logging.

Caddisflies represent a variety of functional groups and responded accordingly. Grazing caddisflies, including Glossosoma nigrior, Micrasema sp., and Neophylax spp., increased in moss samples and in pebble riffles as well as overall. Lype diversa, a caddisfly which constructs its retreat in grooves of decaying wood, was more abundant in Big Hurricane Branch than in Hugh White Creek; it increased over time in Big Hurricane Branch in each habitat in which it was commonly found, perhaps in response to decay of wood which entered the stream channel during logging. Hydatophylax argus and Pycnopsyche spp., relatively common shredding caddisflies in the study stream, decreased significantly in Big Hurricane Branch in moss and sand substrates as well as overall. No significant trends were found in abundance of the net-spinning hydropsychids Diplectrona modesta and Parapsyche cardis. Abundances of predaceous caddisflies, including Rhyacophila spp. and Polycentropus sp., did not change in response to logging.

All aquatic Coleoptera in the present study, including the psephenid Ectopria nervosa and the elmids Optioservus immunis, O. ovalis, and Oulimnius spp., can be considered scrapers. Elmids increased in the moss habitat as well as overall. This increase was much slower to develop than the increase in mayflies, in part because of the longer life cycle of the elmids.

Chironomidae was the most abundant taxon in both streams overall, in all habitats in Big Hurricane Branch, and in all habitats except moss in Hugh White Creek, where it was second to Parapsyche. Chironomidae increased in the moss habitat and decreased in sand, with little overall difference between the streams; increases in moss were largely due to Orthocladiinae, especially Eukiefferiella spp.

Summaries of insect abundance data by functional feeding group are presented in Figure 12. Chironomids were excluded from this analysis since they were identified below family level for only a subset of the samples and since this family represents a wide variety of functional groups.

A decline in shredder abundance as a response to the decreased BPOM food resource did not occur overall, despite the

decline in <u>Peltoperla</u> (Fig. 12). In fact, total shredder
abundance increased in the moss habitat, where FPOM accumulates
and LPOM is scarce. This suggests that taxa categorized as
shredders (e.g., <u>Amphinemura wui</u>) may in fact be feeding on FPOM.

The mayflies that increased were categorized as
collector-gatherers, although some may be scrapers. Thus it is
unclear whether the increase in collector-gatherers (Fig. 12) was
in response to elevated amounts of FPOM in Big Hurricane Branch
or to increased availability of algae. Largest increases
occurred in moss, which serves both as a trap for FPOM and as a
substrate for growth of attached algae.

Scraper taxa increased near the end of the study period
(Fig. 12), with a temporal lag similar to that observed with

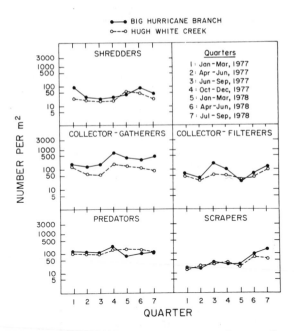

Fig. 12. Densities of benthic macroinvertebrate functional
    feeding groups in Big Hurricane Branch and Hugh White Creek.
    Data points are quarterly means of monthly abundance
    estimates. Big Hurricane Branch and Hugh White Creek
    densities were significantly different (p < 0.05) for
    shredders, quarter 1; collector-gatherers, quarters 4-7; and
    collector-filterers, quarter 3.

Elimidae alone. Significantly higher abundances in Big Hurricane
Branch compared to the reference stream occurred only in moss and
cobble substrate, while lower abundances were found in sand.
This latter finding suggests that siltation may have inhibited
establishment of periphytic algae in depositional substrates.
The increase in scrapers can probably be attributed to the
increase in periphyton production (Fig. 5). Also some insects
have been observed to feed on moss (e.g., Micrasema, Chapman and
Demory, 1963), and their increase may reflect an increase in moss
production.

    No significant trends were apparent among the filter feeders
(Fig. 12). Predators apparently responded to increases in
potential prey organisms. Though they changed little overall
(Fig. 12), they increased in moss and decreased in sand,
reflecting similar trends observed for total organisms.

    Thus, the stream invertebrate community responded to logging
in a direction corresponding to changes in food availability. A
differential response by the invertebrates depending on habitat
was also observed. For example, more taxa increased in moss
compared to other habitats. For many taxa the pattern was
highest increase in moss, progressively less increase in cobble
and pebble, and least increase or a decrease in sand. Of those
taxa which declined in any habitat, more did so in sand.
Explanation of these trends includes the probable increase in
food availability (FPOM and periphytic algae) in the moss
habitat, while this habitat was probably least affected by
sediment. Depositional areas, characterized by slower current
velocity and smaller substrate particle size, may have been most
adversely affected by sediments which entered the stream during
road building. This added sediment increased the physical
instability of these low-gradient reaches.

CONCLUSIONS

    In our study of the response of Big Hurricane Branch to
logging of WS 7, we found the following changes:

    (1) Abiotic changes included increased streamflow during the
growing season, higher summer water temperatures, increased
levels of some nutrients, and increased sediment load. There was
also an obvious increase in light to the stream bed. After four
years, most of these changes were showing significant return
toward pre-logging levels.

    (2) Allochthonous inputs to Big Hurricane Branch were
greatly reduced by logging. These inputs increased significantly
within three years after logging with the regrowth of terrestrial
vegetation.

(3) Periphyton primary production is generally a small portion of the energy base of Coweeta streams. However, following logging, periphyton primary production increased to levels greater than allochthonous inputs. Three years after logging, periphyton production was again fairly low, though still higher than reference levels.

(4) Two years after logging, DOC levels were significantly lower than reference levels due primarily to more rapid uptake, less instream production, and less watershed input.

(5) With reduced leaf fall and blow-in, BPOM decreased following logging. At the same time there was an increase in woody material which remained from the logging operation.

(6) Leaf breakdown rates were initially slowed and later accelerated. A variety of factors were probably involved, but sediment and fauna appear most important.

(7) Concentration and transport of POM were significantly increased by logging. Recent samples suggest that POM levels are returning toward reference levels.

(8) Stream invertebrates responded to logging in a direction corresponding to changes in food availability. While previously dominant shredder species declined, collectors and scrapers increased, as did taxa associated with wood. These changes were modified by substrate stability.

Interpreting these results in the context of stream ecosystem stability, we focus on attributes which reflect ecosystem function. One measure of stream stability is the response of the stream ecosystem to changes in the quantity and quality of energy inputs. This response should be reflected in the flora and fauna of the recovering stream, as well as in the quantity and quality of materials in transport (i.e., export). The concepts of resistance and resilience, while difficult to quantify, provide a useful framework for discussing the mechanisms involved in ecosystem stability.

The annual energy subsidy of allochthonous inputs is especially important to ecosystem function in headwater streams of forested watersheds. Its removal was directly related to the observed decline in storage of benthic particulate organic matter, a parameter associated with the resistance component of stability. The decrease in BPOM and the elevated levels of POM in transport point toward decreased efficiency of utilizing this energy resource. The change in energy base of the stream ecosystem suggests low resistance.

Resistance of a headwater stream in a forested watershed to disturbances associated with logging is low because of the near total dependence of the stream ecosystem on allochthonous inputs. Higher order streams have a greater diversity of energy sources (Vannote et al., 1980); hence, removal of streamside vegetation would have less impact in such a stream. In addition, sediment additions to Big Hurricane Branch had a greater effect (e.g., on macroinvertebrates) than might occur in a larger stream. The headwater study stream had low power and therefore, low ability to move large grain sediments. In fact, the greatest impact of sediment on the invertebrate fauna appeared to be in the slower current reaches of sandy substrate. In a larger stream, with greater power, a much greater fraction of the sediment would have been immediately washed away from the area of disturbance.

On the other hand, low stream power may also be viewed as a mechanism contributing to resistance. Debris dams trap and stabilize sediment (Bilby and Likens, 1980). Organic inputs from logging slash were also trapped in debris dams where they could be utilized rather than washed out of the system. Another factor which contributed to resistance in Big Hurricane Branch was the presence of moss-covered bedrock outcrops. These areas of stable substrate apparently provided refuges for many stream organisms. They served as traps for POM as well as substrates for growth of periphytic algae.

Resistance of the stream may not differ greatly from that of the surrounding forest. The energy base was changed in both ecosystems. While the stream shifted from allochthonous to a more autochthonous energy base, the terrestrial system energy base went from trees to primarily herbaceous plants (Boring et al., 1981). Further, consumers in both ecosystems underwent changes in response to the changed energy base and physical conditions (Seastedt and Crossley, 1981; Schowalter et al., 1981).

Resilience of the stream ecosystem cannot be adequately evaluated within the time frame of the present study. Return to pre-disturbance conditions requires recovery of the terrestrial ecosystem, i.e., canopy regrowth and renewal of allochthonous inputs. However, certain of the observed changes in Big Hurricane Branch can be attributed to the resilience component of stability.

While algal densities are typically low in undisturbed Coweeta streams, a rapid increase in algal growth in Big Hurricane Branch followed removal of the canopy. This increase in autochthonous production compensated in part for the reduction of allochthonous inputs. With this energy base shift, there was a corresponding shift in the invertebrate fauna. Reduced abundance of the dominant shredder, <u>Peltoperla</u>, was offset by numerical increases in taxa probably functioning as scrapers and more dependent on an autochthonous energy base. Invertebrates categorized as collector-gatherers increased in abundance, utilizing the FPOM resource or perhaps algae. There were also indications that some taxa which typically feed as shredders switched to a collector-gatherer feeding mode. The ability of the stream consumers as a group to quickly shift in response to a modified energy base clearly contributed to the resilience of the stream ecosystem.

It is clear from our study that the ability of the stream to recover from this particular disturbance was not entirely a function of the resilience mechanisms of the stream itself. Because of the dependence of the stream ecosystem on allochthonous inputs, complete recovery cannot occur until the quantity and quality of inputs have returned to predisturbance levels (Gurtz et al., 1980; Haefner and Wallace, 1981). We do not feel, however, that the resilience of the stream is exactly equal to that of the surrounding forest. Rather, the stream ecosystem has a much higher potential resilience that is not realized because of the continuing effects of the disturbance (Webster and Patten, 1979). Our study further reinforces the observation that the dynamics of a stream are very closely tied to characteristics of its watershed (e.g., Hynes, 1975), illustrating in another way the nature of some of these linkages.

In this study we attempted to evaluate some of the factors affecting the stability of a stream ecosystem in response to logging, both resistance to disturbance and resilience following disturbance. We emphasized functional characteristics of the ecosystem rather than the more traditional use of community structure as an indicator of stability. Logging caused a multifarious disturbance to Big Hurricane Branch, including physical alteration of habitat, changes in chemical and thermal characteristics of the water, and changes in food resources. While Big Hurricane Branch or another stream might respond in quite different ways to other types or combinations of disturbances, many of the factors investigated in this study would operate in a similar manner to ameliorate disturbance. Stauffer et al. (1978) attempted to evaluate the resistance and resilience (inertia and elasticity in their terminology) of a series of streams affecting by coal mining using a largely subjective rating system (Cairns and Dickson, 1979) and data on fish community structure. It would be extremely useful to make a similar but objective evaluation of the functional characteristics considered in this study for different types of streams in a wide geographical area. From such a classification it would be possible to initiate a meaningful management program. As suggested by Stauffer et al. (1978), certain streams, perhaps

those with high resistance but low resilience, would be given
maximum protection. Other streams, those with higher resilience,
could sustain short term disturbance, while others could be used
for long term disturbance. Such a classification and an
understanding of the differences in functional response induced
by disturbance should be an important long term objective of
lotic ecology.

ACKNOWLEDGMENTS

     This study was supported primarily by grants from the
National Science Foundation's Ecosystem Studies Program.

REFERENCES

Bakeless, J. E. 1961. The eyes of discovery. Dover, New York.
Ball, R. C., and Hooper, F. F. 1963. Translocation of
     phosphorus in a trout stream ecosystem. In: V. Schultz and
     A. W. Klement, Jr. (eds.), Radioecology. Reinhold, New
     York.
Bartram, W. 1791. The travels of William Bartram. Dover (1929
     edition), New York.
Bates, H. W. 1863. The naturalist in the River Amazons. Dover
     (1975 edition), New York.
Bilby, R. E., and Likens, G. E. 1979. Effect of hydrologic
     fluctuations on the transport of fine particulate organic
     carbon in a small stream. Limnol. Oceanog. 24:69-75.
Bilby, R. E., and Likens, G. E. 1980. Importance of organic
     debris dams in the structure and function of stream
     ecosystems. Ecology 61:1107-1113.
Boring, L. R., Monk, C. D., and Swank, W. T. 1981. Early
     regeneration of a clearcut southern Appalachian forest.
     Ecology 62:1244-1253.
Bormann, F. H., and Likens, G. E. 1979. Pattern and process in
     a forested ecosystem. Springer Verlag, New York.
Bormann, F. H., Likens, G. E., and Eaton, J. S. 1969. Biotic
     regulation of particulate and solution losses from a forest
     ecosystem. BioScience 19:600-611.
Botkin, D. B., and Sobel, M. J. 1975. Stability in time-varying
     ecosystems. Amer. Nat. 109:625-646.
Cairns, J., Jr. 1977a. Aquatic ecosystem assimilative capacity.
     Fisheries 2:5-7.
Cairns, J., Jr. 1977b. Quantification of biological integrity.
     In: P. K. Ballentine and L. J. Guarraia (eds.), The
     integrity of water. U.S. Govt. Printing Office, Washington,
     D.C.
Cairns, J., Jr., and Dickson, K. L. 1977. Recovery of streams
     from spills of hazardous materials. In: J. Cairns, Jr., K.
     L. Dickson, and E. E. Herricks (eds.), Recovery and
     restoration of damaged ecosystems. Univ. Press of Virginia,
     Charlottesville.

Campbell, I. C. 1981. A critique of assimilative capacity. J. Wat. Poll. Cont. Fed. 53:604-606.

Chapman, D. W., and Demory, R. 1963. Seasonal changes in the food ingested by aquatic larvae and nymphs in two Oregon streams. Ecology 44:140-146.

Child, G. I., and Shugart, H. H., Jr. 1972. Frequency response analysis of magnesium cycling in a tropical forest ecosystem In: B. C. Patten (ed.), Systems analysis and simulation in ecology, Vol. II. Academic Press, New York.

Cline, L. D., Short, R. A., Ward, J. V., and Carlson, C. A. 1979. The inertia and resilience of a mountain stream to construction impact. In: Proceedings of the mitigation symposium. U.S. Dept. of Agriculture, Ft. Collins.

Connell, J. H., and Slatyer, R. O. 1977. Mechanisms of succession in natural communities and their role in community stability and organization. Amer. Nat. 111:1119-1144.

Copeland, B. J. 1970. Estuarine classification and response to disturbances. Trans. Amer. Fish. Soc. 99:826-835.

Cummins, K. W. 1975. The ecology of running waters; theory and practice. In: Proceeding of the Sandusky River Basin symposium. Tiffin, Ohio.

DeAngelis, D. L. 1975. Stability and connectance in food web models. Ecology 56:238-243.

Douglass, J. E., and Swank, W. T. 1975. Effects of management practices on water quality and quantity; Coweeta Hydrologic Laboratory, North Carolina, USDA Forest Service General Technical Report NE-13.

Dunbar, M. J. 1960. The evolution of stability in marine environments. Natural selection at the level of the ecosystem. Amer. Nat. 94:129-136.

Elton, C. S. 1958. The ecology of invasions by animals and plants. Methuen, London.

Elwood, J. W., and Nelson, D. J. 1972. Periphyton production and grazing rates in a stream measured with a ^{32}P material balance method. Oikos 23:295-303.

Fisher, S. G. 1977. Organic matter processing by a stream-segment ecosystem: Fort River, Massachusetts, U.S.A. Int. Revue ges. Hydrobiol. 62:701-727.

Fisher, S. G., and Likens, G. E. 1973. Energy flow in Bear Brook, New Hampshire: An integrative approach to stream ecosystem metabolism. Ecol. Monogr. 43:421-439.

Gardner, M. R., and Ashby, W. R. 1970. Connectedness of large dynamic (cybernetic) systems: Critical values of stability. Nature 228:784.

Golley, F. B. 1974. Structural and functional properties as they influence ecosystem stability. In: Proceedings of the First International Congress of Ecology. The Hague, Netherlands.

Goodman, D. 1975. The theory of diversity-stability
    relationships in ecology. Quart. Rev. Biol. 50:237-266.
Gregory, S. V. 1979. Primary production in streams of the
    Cascade Mountains. PhD Thesis, Oregon State Univ.,
    Corvallis.
Gurtz, M. E. 1981. Ecology of stream invertebrates in a
    forested and a commercially clear-cut watershed. PhD
    Thesis, Univ. of Georgia, Athens.
Gurtz, M. E., Webster, J. R., and Wallace, J. B. 1980. Seston
    dynamics in southern Appalachian streams: Effects of
    clearcutting. Can. J. Fish. Aq. Sci. 37:624-631.
Haefner, J. D., and Wallace, J. B. 1981. Shifts in aquatic
    insect populations in a first-order southern Appalachian
    stream following a decade of old field succession, Can. J.
    Fish. Aq. Sci. 38:353-3.
Hains, J. J. 1981. The response of stream flora to watershed
    perturbations. M.S. Thesis, Clemson Univ., Clemson, South
    Carolina.
Hairston, N. G., Allan, J. D., Colwell, R. K., Futuyma, D. J.,
    Howell, J., Lubin, M. D., Mathias, J., and VanDermeer, J. H.
    1968. The relationship between species diversity and
    stability: An experimental approach with protozoa and
    bacteria. Ecology 49:1091-1101.
Hall, C. A. S. 1972. Migration and metabolism in a temperate
    stream ecosystem. Ecology 53:585-604.
Harrel, R. C., Davis, B. J., and Dorris, T. C. 1967. Stream
    order and species diversity of fishes in an intermittent
    Oklahoma stream. Amer. Midl. Nat. 78:428-437.
Harrell, R. C., and Dorris, T. C. 1968. Stream order,
    morphometry, physiochemical conditions, and community
    structure of benthic macroinvertebrates in an intermittent
    stream system. Amer. Midl. Nat. 80:220-251.
Holling, C. S. 1973. Resilience and stability of ecological
    systems. Ann. Rev. Ecol. Syst. 4:1-24.
Hornick, L. E., Webster, J. R., and Benfield, E. F. 1981.
    Periphyton production in an Appalachian Mountain trout
    stream. Amer. Midl. Nat. 106:22-36.
Hubbell, S. P. 1973. Populations and simple food webs as energy
    filters. II. Two-species systems. Amer. Nat. 107:122-151.
Hurd, L. E., Mellinger, M. W., Wolf, L. L., and McNaughton, S. J.
    1971. Stability and diversity in three trophic levels in
    terrestrial successional ecosystems. Science
    173:1134-1136.
Hurd, L. E., and Wolf, L. L. 1974. Stability in relation to
    nutrient enrichment in arthropod consumers of old-field
    successional systems. Ecol. Monogr. 44:465-482.
Hutchinson, G. E. 1963. The prospect before us. In: D. G. Frey
    (ed.), Limnology in North America. Univ. Wisconsin Press,
    Madison.

Hynes, H. B. N. 1975. The stream and its valley. Verh. Int.
    Ver. Limnol. 19:1-15.
Jenny, H., Gessel, S. P. and Bingham, F. T. 1949. Comparative
    study of decomposition rates of organic matter in temperate
    and tropical regions. Soil Sci. 68:419-432.
Johnson, E. A. 1952. Effect of farm woodland grazing on
    watershed values in the southeast. J. For. 50:109-113.
Kuehne, R. A. 1962. A classification of streams, illustrated by
    fish distribution in an eastern Kentucky creek. Ecology
    43:608-614.
Larson, P. F. 1974. Structural and functional responses of an
    oyster reef community to a natural and severe reduction in
    salinity. In: Proceeding of the First International
    Congress of Ecology. Hague, Netherlands.
Leopold, L. B., Wolman, M. G., and Miller, J. R. 1964. Fluvial
    processes in geomorphology. Freeman, San Franscisco.
Lewontin, R. C. 1969. The meaning of stability. Brookhaven
    Symp. Biol. 22:13-24.
MacArthur, R. 1955. Fluctuations of animal populations and a
    measure of community stability. Ecology 36:533-536.
Margalef, R. 1957. La teoria de la informacion en ecologia.
    Mem. Real. Acad. Ciencias y Artes de Barcelona 32:373-449.
    (English translation, W. Hall, 1957, Information theory in
    ecology. Gen. Syst. 3:36-71.)
Margalef, R. 1960. Ideas for a synthetic approach to the
    ecology of running waters. Int. Revue ges. Hydrobiol.
    45:133-153.
Margalef, R. 1963. On certain unifying principles in ecology.
    Amer. Nat. 97:357-374.
Margalef, R. 1968. Perspectives in ecological theory. Univ.
    Chicago Press, Chicago.
May, R. M. 1971a. Stability in multispecies community models.
    Math. Biosci. 12:59-79.
May, R. M. 1971b. Stability in model ecosystems. Proc. Ecol.
    Soc. Aust. 6:17-56.
May, R. M. 1972. Will a large, complex system be stable.
    Nature 238:413-414.
May, R. M. 1973. Stability and complexity in model ccosystems.
    Princeton Univ. Press, Princeton.
McIntosh, R. P. 1980. The relationship between succession and
    the recovery process in damaged ecosystems. In: J. Cairns, Jr.
    (ed.), The recovery process in damaged ecosystems. Ann
    Arbor Science, Ann Arbor.
McNaughton, S. J. 1977. Diversity and stability of ecological
    communities: a comment on the role of empiricism in
    ecology. Amer. Nat. 111:515-525.
Merritt, R. W. and Cummins, K. W. (eds.) 1978. An introduction
    to the aquatic insects of North America. Kendell-Hunt,
    Dubuque, Iowa.

Meyer, J. L. 1979. The role of sediments and bryophytes in phosphorus dynamics in a headwater stream ecosystem. Limnol. Oceanogr. 24:365-376.

Meyer, J. L., and Tate, C. M. The effects of watershed disturbance on dissolved organic carbon dynamics of a stream. Ecology, in press.

Minshall, G. W. 1967. Role of allochthonous detritus in a trophic structure of a woodland springbrook community. Ecology 48:139-149.

Minshall, G. W., R. C. Petersen, K. W. Cummins, T. L. Bott, J. R. Sedell, C. E. Cushing, and R. L. Vannote. 1983. Interbiome comparison of stream ecosysytem dynamics. Ecol. Monogr. 53:1-25.

Monk, C. D., Crossley, D. A., Jr., Swank, W. T., Todd, R. L., Waide, J. B., and Webster, J. R. 1977. An overview of nutrient cycling research at Coweeta Hydrolic Laboratory. In: D. L. Correll (ed.), Watershed research in Eastern North America. Smithsonian, Washington, D.C.

Motten, A. F., and Hall, C. A. S. 1972. Edaphic factors override a possible gradient of ecological maturity indices in a small stream. Limnol. Oceanogr. 17:922-926.

Mulholland, P. J. 1981. Organic carbon flow in a swamp-stream ecosystem. Ecol. Monogr. 51:307-322.

Murdoch, W. W., Evans, F. C., and Peterson, C. H. 1972. Diversity and pattern in plants and insects. Ecology 53:819-829.

Naiman, R. J., and Sedell, J. R. 1979. Benthic organic matter as a function of stream order in Oregon. Arch. Hydrobiol. 87:404-422.

Nelson, D. J., and Scott, D. C. 1962. Role of detritus in the productivity of a rock outcrop community in a Piedmont stream. Limnol. Oceanogr. 7:396-413.

Newbold, J. D., Elwood, J. W., O'Neill, R. V., and Van Winkle, W. 1981. Measuring nutrient spiralling in streams. Can. J. Fish. Aq. Sci. 38:860-863.

Odum, E. P. 1953. Fundamentals of ecology. Saunders. Philadelphia.

Odum, E. P. 1969. The strategy of ecosystem development. Science 164:262-270.

Olson, J. S. 1963. Energy storage and the balance of producers and decomposers in ecological systems. Ecology 44:322-332.

O'Neill, R. V. 1976. Ecosystem persistence and heterotrophic regulation. Ecology 57:1244-1253.

O'Neill, R. V., Elwood, J. W., and Hildebrand, S. G. 1979. Theoretical implications of spatial heterogeneity in stream ecosystems. In: G. S. Innis and R. V. O'Neill (eds.), Systems analysis of ecosystems. Int. Cooperative Pub. House, Fairland, Maryland.

O'Neill, R. V., Harris, W. F., Ausmus, B. S., and Reichle, D. E. 1975. A theoretical basis for ecosystem analysis with particular reference to element cycling. In: F. G. Howell, J. B. Gentry, and M. H. Smith (eds.) Mineral cycling in Southeastern ecosystems. ERDA Symposium Series, Washington, D.C.

O'Neill, R. V. and Reichle, D. E. 1980. Dimensions of ecosystems theory. In: R. H. Waring (ed.), Forests: fresh perspectives from ecosystem analysis. Oregon St. Univ. Press, Corvallis.

Orians, G. H. 1974. Diversity, stability and maturity in natural ecosystems. In: W. H. van Dobben and R. H. Lowe-McConnell (eds.), Unifying concepts in ecology. The Hague, Netherlands.

Patrick, R. 1970. Benthic stream communities. Amer. Sci. 58:546-549.

Patten, B. C., and Witkamp, M. 1967. Systems analysis of ^{134}Cs kinetics in terrestrial microcosms. Ecology 48:813-824.

Peterman, R. M. 1980. Influence of ecosystem structure and perturbation history on recovery processes. In: J. Cairns, Jr. (ed.), The recovery process in damaged ecosystems. Ann Arbor Science, Ann Arbor.

Petersen, R. C., and Cummins, K. W. 1974. Leaf processing in a woodland Stream. Freshwat. Biol. 4:343-368.

Pimentel, D. 1961. Species diversity and insect population outbreaks. Ann. Ent. Soc. Amer. 54:76-86.

Platts, W. S. 1979. Relationships among stream order, fish populations, and aquatic geomorphology in an Idaho river drainage. Fisheries 4:5-9.

Preston, F. W. 1969. Diversity and stability in the biological world. Brookhaven Symp. Biol. 22:1-12.

Roberts, A. 1974a. The stability of a feasible random ecosystem. Nature 251:607-608.

Rodgers, J., Dickson, K. L., and Cairns, J., Jr. 1978. A chamber for in situ evaluations of periphyton productivity in lotic systems. Arch. Hydrobiol. 84:389-398.

Schowalter, T. D., J. W. Webb, and Crossley, D. A., Jr. 1981. Community structure and nutrient cycles of canopy arthropods in clearcut and uncut forest ecosystems. Ecology 64:1010-1019.

Seastedt, T. R., and Crossley, D. A., Jr. 1981. Microarthropod response following cable logging and clearcutting in the southern Appalachians. Ecology 62:126-135.

Sedell, J. R., Naiman, R. J., Cummins, K. W., Minshall, G. W., and Vannote, R. L. 1978. Transport of particulate organic material in streams as a function of physical processes. Verh. Int. Ver. Limnol. 20:1366-1375.

Sheldon, A. 1968. Species diversity and longitudinal succession in stream fishes. Ecology 49:333-351.

Smith, F. E. 1972. Spatial heterogeneity, stability, and
    diversity in ecosystems. Trans. Conn. Acad. Arts Sci.
    44:309-335.

Sokal, R. R., and Rohlf, F. J. 1969. Biometry. Freeman, San
    Francisco.

Stauffer, J. R., Hocutt, C. M., Hendricks, M. L., and Markham, S.
    L. 1978. Inertia and elasticity as a stream classification
    system: Youghiogheny River case history evaluation. In: D.
    E. Samuel, J. R. Stauffer, and C. M. Hocutt (eds.), Surface
    Mining and Fish/Wildlife Needs in Eastern United States.
    U.S. Fish and Wildlife Service, Washington, D.C.

Swank, W. T., and Douglass, J. E. 1977. Nutrient budgets for
    undisturbed and manipulated hardwood forest ecosystems in
    the mountains of North Carolina. In: D. L. Correll (ed.),
    Watershed research in Eastern North America. Smithsonian,
    Washington, D.C.

Swank, W. T., and Waide, J. B. 1980. Interpretation of nutrient
    cycling research in a management context: Evaluating
    potential effects of alternative management strategies on
    site productivity. In: R. H. Waring (ed.), Forests: fresh
    perspectives from ecosystem analysis. Oregon St. Univ.
    Press, Corvallis.

Swift, L. W., Jr., and Messer, J. B. 1971. Forest cuttings raise
    temperatures of small streams in the southern Appalachians.
    J. Soil Wat. Conservation 26:111-116.

Swift, L. W., Jr., Swank, W. T., Mankin, J. B., Luxmore, R. J.,
    and Goldstein, R. A. 1975. Simulation of
    evapotransporation and drainage from natural and clearcut
    deciduous forest and young pine plantation. Wat. Res. Res.
    11:667-673.

Tramer, E. J., and Rogers, P. M. 1973. Diversity and
    longitudinal zonation in fish populations to two streams
    entering a metropolitan area. Amer. Midl. Nat. 90:366-375.

Van Voris, P., O'Neill, R. V., Emanuel, W. R., and Shugart, H.
    H., Jr. 1980. Functional complexity and ecosystem
    stability. Ecology 61:1352-1360.

Vannote, R. L., Minshall, G. W., Cummins, K. W., Sedell, J. R.,
    and Cushing, C. E. 1980. The river continuum concept.
    Can. J. Fish. Aq. Sci. 37:130-137.

Velz, C. J. 1976. Stream analysis - forecasting waste
    assimilative capacity. In: H. W. Gehm and J. I. Bregman
    (eds.), Water resources and pollution control. Van Nostrand
    Reinhold, New York.

Waide, J. B., Krebs, J. E., Clarkson, S. P., and Setzler, E. M.
    1974. A linear systems analysis of the calcium cycle in a
    forested watershed ecosystem. In: R. Rosen and F. M. Snell
    (eds.), Progress in theoretical biology, Vol. III. Academic
    Press, New York.

Waide, J. B., and Webster, J. R. 1975. Engineering systems
    analysis: applicability to ecosystems. In: B. C. Patten
    (ed.), Systems analysis and simulation in ecology, Vol. IV.
    Academic Press, New York.
Wallace, J. B., Webster, J. R., and Woodall, W. R., Jr. 1977.
    The role of filter feeders in flowing waters. Arch.
    Hydrobiol. 79:506-532.
Wallace, J. B., Woodall, W. R., Jr., and Sherberger, F. F. 1970.
    Breakdown of leaves by feeding of Peltoperla maria nymphs.
    Ann. Ent. Soc. Amer. 63:562-567.
Watson, V., and Loucks, O. L. 1979. An analysis of turnover
    times in a lake ecosystem and some implications for systems
    properties. In: E. Halfon (ed.), Theoretical systems
    ecology. Academic Press, New York.
Watt, K. E. F. 1964. Comments on fluctuations of animal
    populations and measures of community stability. Can.
    Entomol. 96:1434-1442.
Watt, K. E. F. 1968. A computer approach to analysis of data on
    weather, population fluctuations, and disease. In: W. P.
    Lowry (ed.), Biometerology. Oregon St. Univ. Press,
    Corvallis.
Webber, M. I. 1974. Food web linkage complexity and stability
    in a model ecosystem. In: M. B. Usher and M. H. Williamson
    (eds.), Ecological stability. Chapman and Hall, London.
Webster, J. R. 1975. Analysis of potassium and calcium dynamics
    in stream ecosystems on three southern Appalachian
    watersheds of contrasting vegetation. PhD Thesis, Univ. of
    Georgia, Athens.
Webster, J. R., and Patten, B. C. 1979. Effects of watershed
    perturbation on stream potassium and calcium dynamics.
    Ecol. Monogr. 49:51-72.
Webster, J. R., and Waide J. B. In press. Effects of forest
    clearcutting on leaf breakdown in a southern Appalachian
    stream. Freshwat. Biol.
Webster, J. R., Waide, J. B., and Patten, B. C. 1975. Nutrient
    cycling and stability of ecosystems. In: F. G. Howell, J.
    B. Gentry, and M. H. Smith (eds.), Mineral cycling in
    Southeastern ecosystems. ERDA Symposium Series, Washington,
    D.C.
Whiteside, B. G., and McNatt, R. M. 1972. Fish species in
    relation to stream order and physio-chemical conditions in
    the Plum Creek drainage basin. Amer. Midl. Nat. 88:90-101.
Wilhm, J. 1972. Graphical and mathematical analysis of biotic
    communities in polluted streams. Ann. Rev. Entomol.
    17:223-252.
Williams, J. G. 1954. A study of the effect of grazing upon
    changes in vegetation on a watershed in the southern
    Appalachian mountains. M.S. Thesis, Michigan St. Univ.,
    East Lansing.

INDEX